P9-BUI-669

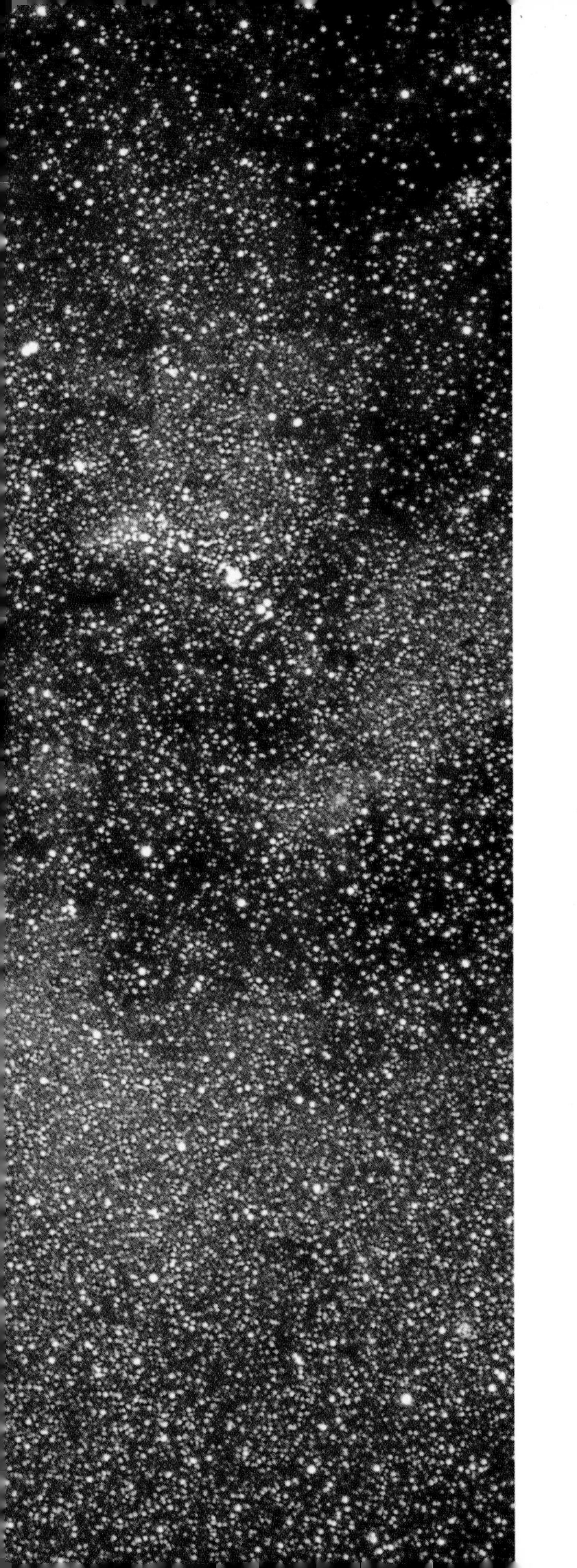

STARS & PLANETS

A VIEWER'S GUIDE

GUNTER D. ROTH

Sterling Publishing Co., Inc.
New York

Photo on page 1:
Las Campanas, the mountain observatory of the Carnegie Institute in the Andes of Chile at a height of 2280 meters (7524 feet), is an ideal location for researching the southern sky.

Photo on pages 2–3:
A view into the Milky Way in the direction of the constellation Sagittarius. Through a pair of binoculars or a small telescope, we see magnificent star clusters and nebulae.

Photo on the right:
The Helix Nebula, a planetary nebula in the constellation Aquarius. Planetary nebulae mark the end of a star's red giant stage. There are 1500 known planetary nebulae in our Milky Way.

Photo on pages 6–7:
Bright clouds of gas at the edge of the sun called prominences rise up to 50,000 or more kilometers (31,000+ miles) above the photosphere (surface of the sun).

Library of Congress Cataloging-in-Publication Data

Roth, Günter Dietmar.
[Sterne und Planeten. English]
Stars & planets : a viewer's guide /
by Günter D. Roth.
p. cm.
Includes index.
ISBN 0-8069-9906-3
1. Astronomy—Popular works.
2. Astronomy—Observers' manuals. 3. Stars—Observers' manuals. 4. Planets—Observers' manuals. I. Title.
QB44.2.R6813 1998
520—dc21 97-48369
CIP

Photo credits and permissions are on page 176.

10 9 8 7 6 5 4 3 2

Published by Sterling Publishing Company, Inc.
387 Park Avenue South, New York, N.Y. 10016
Originally published in Germany by BLV Verlagsgesellschaft mbH under the title *Sterne und Planeten: erkennen und beobachten*
© 1996 by BLV Verlagsgesellschaft mbH
English translation © 1998 by Sterling Publishing Co., Inc.
Distributed in Canada by Sterling Publishing
c/o Canadian Manda Group, One Atlantic Avenue, Suite 105
Toronto, Ontario, Canada M6K 3E7
Distributed in Great Britain and Europe by Cassell PLC
Wellington House, 125 Strand, London WC2R 0BB, England
Distributed in Australia by Capricorn Link (Australia) Pty Ltd.
P.O. Box 6651, Baulkham Hills, Business Centre, NSW 2153, Australia
Printed in China
All rights reserved

Sterling ISBN 0-8069-9906-3

Preface

The interest in astronomy is stimulated by the desire to learn more about the extraterrestrial worlds and, at the same time, to better understand our human existence. The starry sky conveys the message of our prehistory. The light of the stars traveled to us over millions and billions of years. With our observations, we can take a cosmic look back and study the beginning of the universe.

We can have astronomical experiences from simple observations with our naked eyes or with the help of small binoculars. This book gives you suggestions on how to do this as well as information on the orientation of the sky.

Each spread contains a section of the sky and two specially-selected individual objects for viewing with binoculars and telescopes. Twenty-seven of the sky sections show the entire starry sky as it is seen with the naked eye from the northern and southern hemisphere.

With precise information, anyone can recognize the world of the stars. By simply looking with your naked eyes, or by using binoculars and a camera, you can find different objects in the sky (up to the 5th magnitude) and can record and follow their various appearances. The observations range from sunspots and moon craters to the pulsating life of changing stars and the bizarre forms of cosmic nebulae.

Modern astronomical researchers nowadays use high-tech instruments with digital detectors and computers for their observations. In recent years, the use of radiation-detecting satellites and space probes, in addition to astronomical observation from Earth, have become increasingly important. In order to deepen your own experiences of stars, this book offers, in the second section, the current status of knowledge about all objects of the sky. In a separate section, the modern astronomical measuring methods are presented and research goals are described. Also included are an overview of special sky-events in the upcoming years (e.g., solar and lunar eclipses) and a glossary that explains important technical terms.

This "sky guide" is the result of the encouragement of Dr. Georg Walterspiel. In order to make this book an easily-remembered manual for anyone to use in looking out for the stars and exploring the cosmic landscape, clear illustrations are included with explanatory text. The graphic designer, Barbara von Damnitz, and the internationally-known and respected astrophotographer, Dr. Hans Vehrenberg, took a large part in creating the look of this book. I thank the both of them as well as the other photographers for their work and ideas.

I would also like to thank M. Rosa and R.M. West (European South Observatory, Garching), L.D. Schmadel (Astronomical Mathematics Institute, Heidelberg), H.J. Staude (editor of Stars and Space, Heidelberg), J. Trümper (Max Planck Institute for Extraterrestrial Physics, Garching) as well as K. Löchel, Jena (eclipses) and J. Meeus Erps-Kwerps (ephemerides).

—Gunter D. Roth

Contents

Experiencing the Starry Sky

Every person has, at some point in his or her life, wondered with the secretively glimmering stars in the night sky. When there are no streetlights or clouds (outside in the country, on the mountains, or at the beach), they still are there—the dark nights and the sky that is covered with stars. Then thoughts awake about human existence and outer space. The desire springs up to learn more about the universe and the extraterrestrial landscape. During the day, the sun reminds us of the extraterrestrial, while the white light of the full moon cannot be completely thrust aside by any kind of light of the big city. Your interest is awakened and the path for your first observations has begun.

People who were living on earth as hunters and collectors centuries ago followed the events in the sky and adapted to its natural laws in their everyday lives. These laws govern the change of day and night, the daily sunrise and sunset, the sun's varying positions in the sky in the course of a year, and the phases of the increasing and decreasing moon. These events had two effects on the people. First, they were a sign of a large world order. Second, they suggested ways of marking time: the day, month, and year.

As civilization developed, people became more settled, governments became more structured, and trade relations between cultures were more intertwined. As a result, a systematic calendar and the necessity for time order became very important. In later centuries, navigational knowledge was a precondition for the discovery of the world. Thereby, the precise reading of stars was indispensable.

Why Is It So Dark at Night?

The night sky allows us to see the stars with our naked eyes. But why is it so dark at night? In modern times, astronomers have searched for an answer to this question again and again. They were motivated by the idea that this simple observation seemed to be connected to the construction of the universe. One of the first astronomers who asked this question was Johannes Kepler in 1610. Then came the comet discoverer Edmond Halley in 1720 and, about 100 years later, Wilhem Olbers in 1826.

Olbers' paradox stated that: If the universe was evenly filled with stars, infinite in space and time, and unchanging, then the sky at night would shine brightly. It would be a brightness, which would correspond approximately to the medium surface-brightness of the stars, such as our sun.

We see the opposite. In the 19th century, the famous short-story writer Edgar Allan Poe tried to come up with an explanation that came close to the solution. In the universe, one observes stars only up to a certain distance. After that, we look into darkness in which stars do not yet exist. Therefore, it is dark at night. Thereby, Poe assumed that stars exist only from a certain point in time on. The astrophysicists of the 20th century confirmed this assumption. The universe, which we observe at the time being, has a finite age and it is constantly changing because it expands. This is why the night is dark.

The Formation of the Constellations

At first glance, it is not easy for the beginning astronomer to find the constellations in the night sky. The number of the bright and less bright stars, which you see with your naked eyes, is too confusing. The longer you stare into the darkness until your eyes adapt and become more sensitive to the weaker points of light, the more difficult it will seem to place the stars into clearly defined constellations. But constellations, in addition to the sun and moon, have been a part of civilization since the time when people became occupied with the starry sky, and have served as a means of establishing orientation. The rise or set of a certain constellation was used as time marker, as a sign for the beginning of a new season, or as a guide for events of everyday life. For instance, the appearance of the bright star Sirius in the constellation Canis Major (Great Dog) at dawn was a signal to the ancient Egyptians of the beginning of the Nile floodings, which were and still are so incredibly important for their agricul-

A view of the impressive constellation Orion with the famous nebula. What is unique about Orion is the unmistakable star line consisting of three stars of the magnitude 2 near the celestial equator, which marks the belt of Orion.

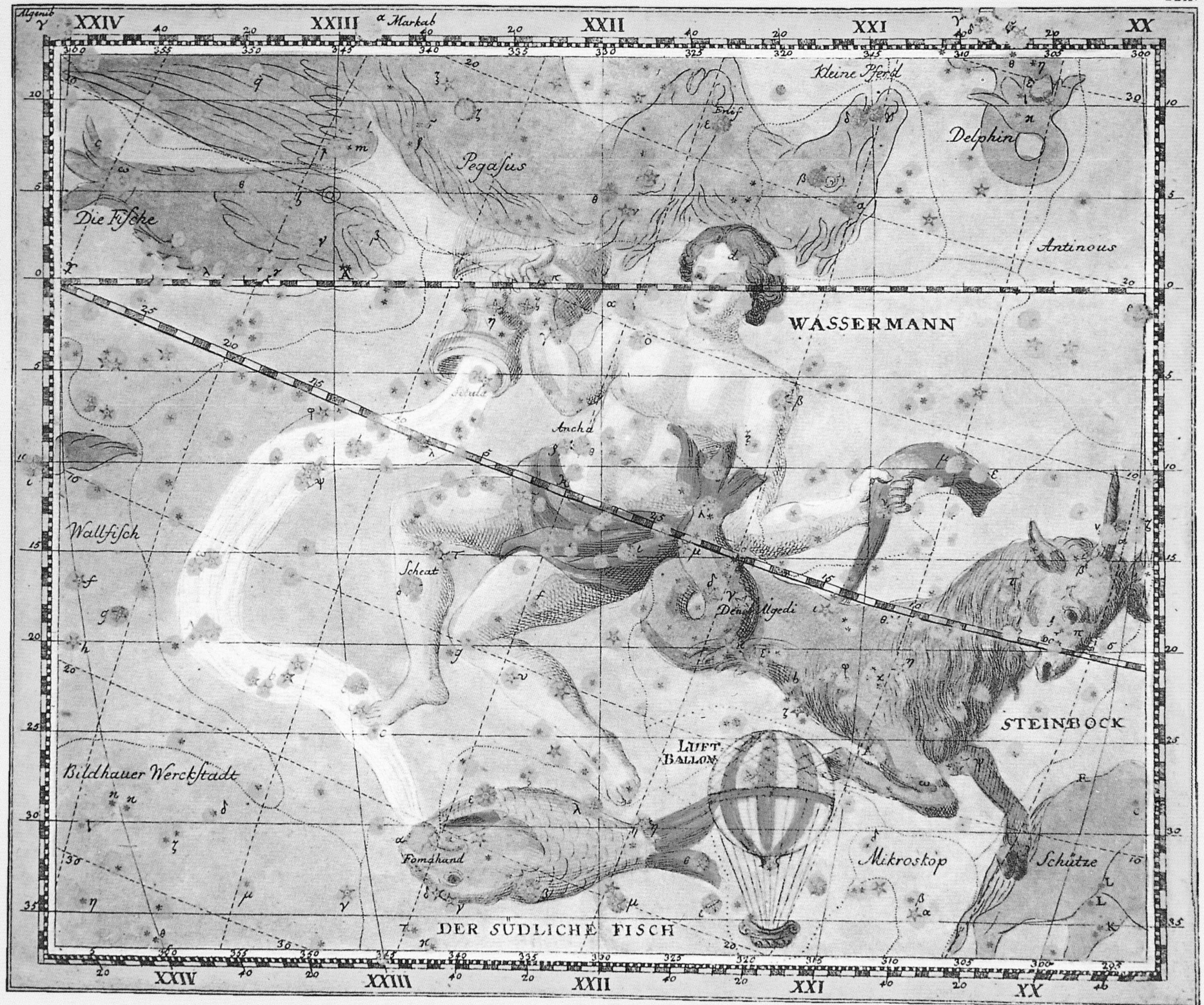

The constellation Aquarius from the atlas "Presentation of the Stars" (published in 1872) by the German astronomer Johann Elert Bode. The entire sky, including the southern hemisphere, is depicted on thirty maps. Also included is an index of five thousand fixed stars with descriptions of their coordinates and brightnesses.

ture. (Why many constellations are only visible at certain times of the year will be explained in the next chapter.)

The strong interrelation throughout history between humans and the cosmos is also expressed in the religious beliefs, which have paired stars and constellations with gods. For instance, the ancient Egyptian goddess Sopdet (Greek "Sothis") is the personification of the star Sirius, because the ancient Egyptians saw her as the goddess of fertility and indispensable wetness (the Nile floodings). In ancient Persia, the star god Tishtrya (responsible for moisture and rain) was thought to be related to the star Sirius.

The signs of the zodiac belong to the oldest constellations. The sign of the zodiac Aquarius is supposed to have been known for 20,000 years. Its appearance in the morning sky announced the rainy season to the ancient Assyrians. The twelve constellations of the zodiac were gradually formed according to their figure and name. According to the number mysticism of the Sumerians (the original inhabitants of Assyria who placed a lot of importance on the number 7), the zodiac could have consisted of seven constellations. But later, the sacred number 12 became the central point, which expanded the zodiacal signs to twelve. The zodiacal signs do not consist of the brightest stars in the sky. Instead, special attention is paid to them because of the apparent yearly course of the sun on its path through the sky, which is caused by the movement of the earth around the sun that leads through these constellations. The imaginary line that traces the apparent path of the sun through the sky is called the ecliptic.

Solar and lunar eclipses can only take place when the moon is a new moon or full moon near the ecliptic. These relationships were already known to the people in antiquity, including the movement of the five bright planets in the area of the ecliptic: Mercury, Venus, Mars, Jupiter, and Saturn. These signs of the zodiac had the important function of marking positions. With them, a beginning was made to divide the sky into constellations.

Rotation Around the Sun

The great observer of antiquity, the Greek astronomer Hipparchus, discovered in 150 B.C. an interesting change of position of the stars for the observer on earth. The stars apparently wandered forward parallel to the ecliptic opposite the first point of Aries. About 2000 years ago, the intersection of ecliptic and celestial equator (compare p. 20) was in the sign of the zodiac of Aries. This intersection is also called the vernal equinox, because when the sun crosses there, the hours of daylight and darkness are equal. Because of the westerly precession of the equinoxes, the vernal equinox now lies in the sign of the zodiac of Pisces. It remains on an average 2150 years in each sign of the zodiac. The next one will be in the constellation Aquarius, and in 25,800 years the vernal equinox will move again from Aries into Pisces.

Although Hipparchus discovered this phenomenon, it was the great English physicist Sir Isaac Newton who came up with the explanation in the 17th century. On its course around the sun, the earth's axis completes a rotation every 25,800 years. This phenomena, called precession, is described in detail in the glossary on page 172.

Two more star movements, which are triggered by the movement of the earth, belong to the apparent star: the daily turn of the earth around its axis and the course of the earth around the sun. (More about that on pages 18 and 20.)

Constellations are not for eternity, even though stars rise and set in the eyes of the observer in seemingly fixed constellations in the night sky. But constellations are products of human imagination. It is often not easy to recognize, in the grouping of stars, a certain god, mythical figure, or animal. Some are more discernible than others, such as the constellations Aquila and Cygnus (which belong to the bright stars of the Summer Triangle), or the constellation Scorpius (which forms in southern latitudes the impressive figure of a scorpion in the sky).

Nevertheless, all constellations are the work of humans, and what appears to be a constellation consists, in reality, of stars that go along their path in the universe mostly far apart from each other and have nothing physically to do with each other. Star clusters are exceptions. An example is globular clusters that are formed towards the beginning of the evolution of the Milky Way galaxy. (More about that is explained on p. 142.) Because most stars did not develop together, the look of most of the constellations will change over the course of time.

Leading Constellations

At some point, probably every person has heard or read something about the constellation Ursa Major (Big Dipper or Great Bear). Many people have seen this unmistakable constellation in the spring high up in the sky. It never completely disappears from the sky in our latitudes. On every clear night, it is ready to point the observer towards the north star and, thus, to the North Pole of the sky (see p. 21). Ursa Major is known as an asterism—a distinct, easily recognized group of stars within a constellation. Such constellations make it easy for the observer to find his or her way among the stars. Because of its yearlong visibility in the Northern Hemisphere, Ursa Major is also called circumpolar.

Each season has its leading constellations that are easily remembered and formed by bright to very bright stars. The most important leading constellations are presented on the following pages with their zodiacal signs. These constellations are described in detail together with the others of the northern and southern sky on the maps from pages 34 to 87. There, you will find the encouragement for the observer who wants to go out to discover with binoculars or with a small telescope. An overview table of all eighty-eight constellations is on page 14.

Leading constellations can also be signposts on earth. Ursa Major is a guide towards the north, while the constellations Orion and Virgo serve as guides for the east-west direction.

Leading Constellations

Constellations help you find your way in the sky by orienting you towards the two sky poles and the sky equator.

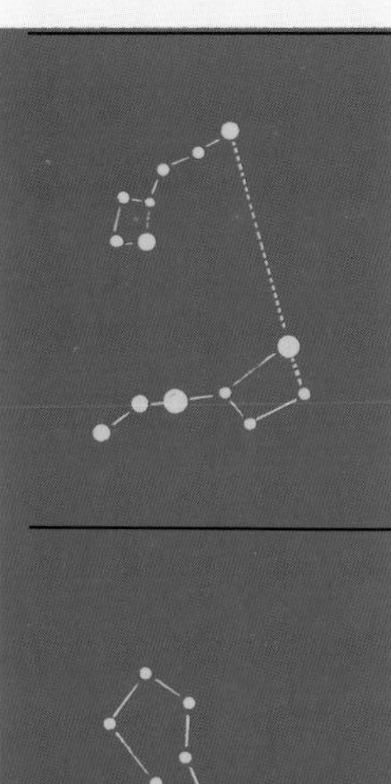

Ursa Major and Ursa Minor (Great and Little Bear): Ursa Major is the most striking constellation of the northern sky. The bottom two stars, Dubhe and Merak, of Ursa Major point towards the north star in Ursa Minor (see also pp. 40 and 42). The "M" of Cassiopeia (see p. 34) can also be used to find the north star. The constellation is circumpolar in the northern latitudes, and it never sets.

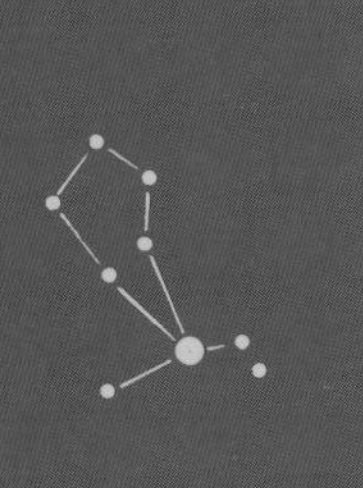

Boötes (The Herdsman): Boötes can be found by following the curving handle of the Big Dipper. Its brightest star, Arcturus, in the center of the arc, is a strikingly yellowish-red color (see also p. 62). It points down towards Spica, the main star in constellation Virgo. In May and June, Arcturus stands in the central northern latitudes high above the southern horizon.

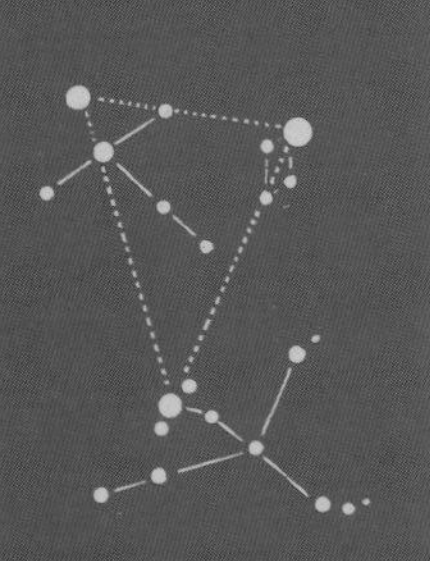

Summer Triangle: It is composed of the stars Vega (constellation Lyra), Deneb (constellation Cygnus the Swan), and Altair (constellation Aquila the Eagle). It is an almost isosceles triangle in the northern summer sky (see also p. 48). The straight line, prolonged towards the east, Vega-Deneb points towards the constellations Pegasus and Andromeda.

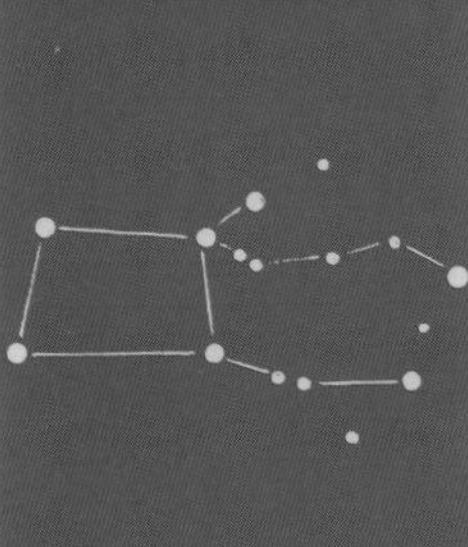

The Great Square of Pegasus: The large constellation Pegasus helps to orient the northern fall sky. The northeast corner of the Pegasus square joins with the constellation Andromeda and its famous spiral galaxy (see also p. 68). The western side of the square points towards the direction β-α Pegasi to the south onto the bright star Fomalhaut in Piscis Austrinus (Southern Fish).

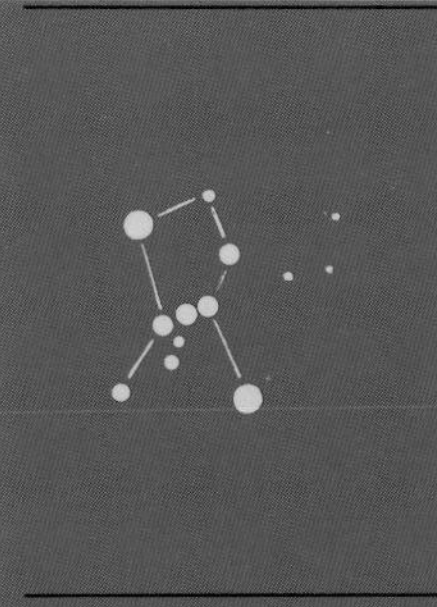

Orion: It has three bright "belt stars" near the celestial equator. Below the belt stars is the Orion Nebula. To the southeast of Orion is Canis Major (Greater Dog) with Sirius (see also p. 56). Orion is approximately in the center of the bow, which connects the bright stars Capella in the constellation Auriga (see p. 38) in the north and Canopus in the constellation Carina in the south.

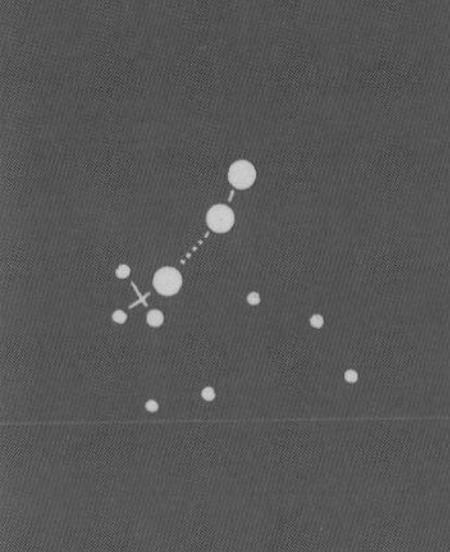

Crux and Centaurus (The Southern Cross and Centaur): Crux is the smallest constellation in the entire sky. Together with the two main stars of Centaurus (Alpha is a magnificent double star), it helps to orient the southern sky (see also p. 78). Almost the same distance to the pole, opposite Crux, is the very bright fixed star Achernar (see p. 72) in the constellation Eridanus.

Constellations of the Zodiacal Signs

The constellations of the zodiacal signs have the same names as the signs of the zodiac. But they must not be confused, since the constellations do not correspond anymore to the signs. That means, for instance, when in astrology, one speaks about the fact that a wandering star (the moon or a planet) is in the sign of Taurus, this star is in fact within the constellation Aries.

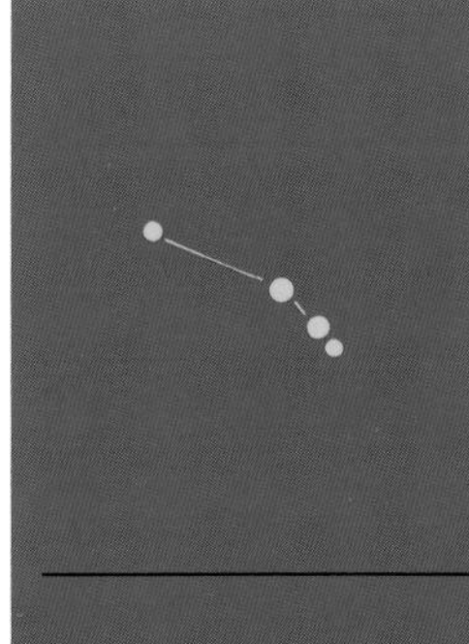

Aries: The first point of Aries designated the position of the sun on March 21 about 2000 years ago when, at that time, the sun's position in Aries coincided with the vernal equinox—the day in which the hours of darkness equal the hours of daylight (see also p. 52). Its length in the ecliptic is 26°–50°. The beginning point for the sign of the zodiac Aries in the ecliptic is 0°.

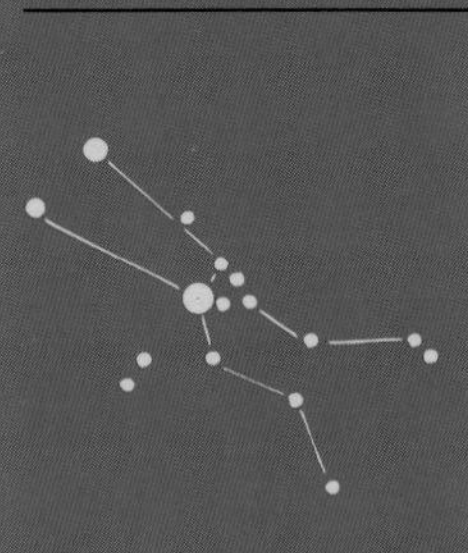

Taurus: The main star is the reddish Aldebaran. In the constellation, there are the well-known "seven sister" stars, the Pleiades, which are a young, open cluster and the subject of many fairytales (see also p. 54). Its length in the ecliptic is 50°–89°. The beginning point for the sign of the zodiac Taurus in the ecliptic is 30°.

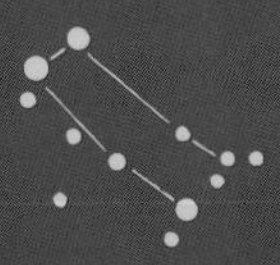

Gemini: This sign of the zodiac has the prominent stars Castor and Pollux. In the constellation Gemini, the sun reaches on June 21 its northernmost position (see also p. 56). Its length in the ecliptic is 89°–119°. The starting point for the sign of the zodiac Gemini in the ecliptic is 60°.

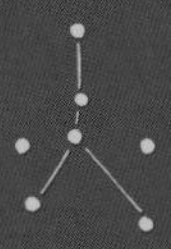

Cancer: This is an inconspicuous constellation between Gemini and Leo with the open cluster of the Praesepe (the Beehive), in the center on the connecting line between the stars Pollux and Regulus (see also p. 58). Its length in the ecliptic is 119°–139°. The starting point for the sign of the zodiac Cancer in the ecliptic is 90°.

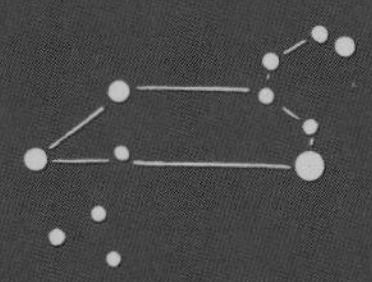

Leo: This constellation is with the 1st magnitude star Regulus, which in the legend is referred to as the heart of the lion. Regulus is the brightest star in immediate proximity to the ecliptic. It is covered by the moon and planets (see also p. 58). Its length in the ecliptic is 130°–174°. The starting point for the sign of the zodiac Leo in the ecliptic is 120°.

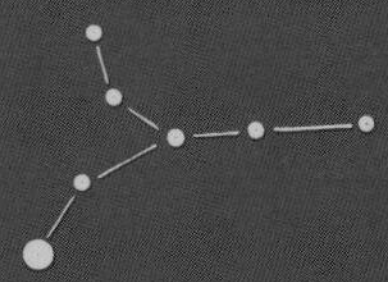

Virgo: This constellation has the main star Spica, whose name means "ear of wheat." This is the constellation in which the sun is standing at the point of the autumnal equinox (when the length of the day and night are equal) at the beginning of fall (September 23; see also p. 60). Its length in the ecliptic is 174°–214°. The starting point for the sign of the zodiac Virgo in the ecliptic is 150°.

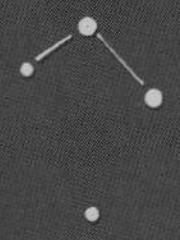

Libra: The Babylonians did not yet count this constellation independently and allocated the stars to the constellation Scorpio. It was the Romans who regarded the stars as a separate constellation. In Roman times, the sun entered Libra on the autumnal equinox. Libra is a balance or scale. Its length in the ecliptic is 214°–239°. The starting point for the sign of the zodiac Libra in the ecliptic is 180°.

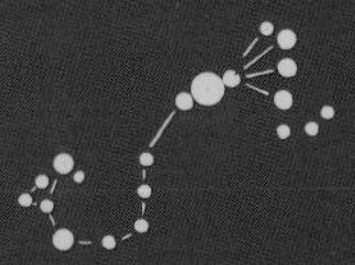

Scorpius: This constellation clearly shows the shape of a scorpion. It is the most magnificent constellation in the sky near the equator. The red main star Antares (meaning "rival of Mars" in Latin; see also p. 65) is striking. Its length in the ecliptic is 239°–245°. The starting point for the sign of the zodiac Scorpio in the ecliptic is 210°.

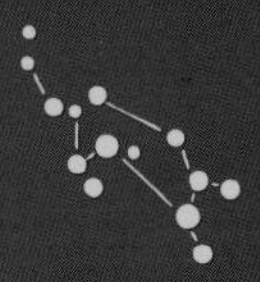

Sagittarius: In this constellation, the sun reaches its southernmost position in the sign of the zodiac on December 22. The band of the Milky Way is the brightest in Sagittarius (see also p. 66). Its length in the ecliptic is 265°–301°. Between 245° and 265° is the constellation Ophiuchus (The Serpent Bearer). The starting point for the sign of the zodiac Sagittarius in the ecliptic is 240°.

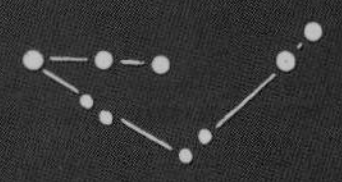

Capricornus: This constellation was once called "Sea Goat," which indicates an animal that was half goat and half fish. How it changed to Capricornus is somewhat mysterious (see also p. 66). Its length in the ecliptic is 301°–329°. The starting point for the sign of the zodiac Capricorn in the ecliptic is 270°.

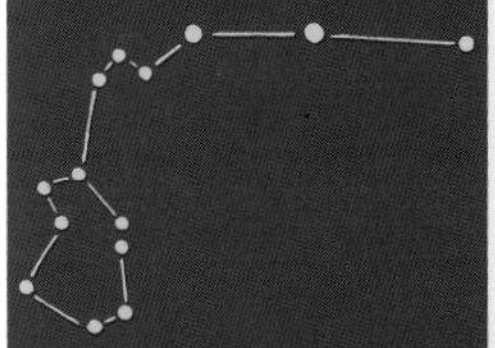

Aquarius: This is supposedly one of the oldest constellations. The name of this constellation is estimated to be 20,000 years old. The name of most of the other constellations of the signs of the zodiac are estimated to be about 5000 years old (see also p. 68). Its length in the ecliptic is 329°–351°. The starting point for the sign of the zodiac Aquarius in the ecliptic is 300°.

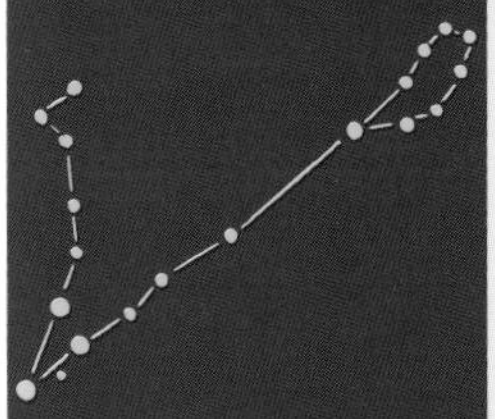

Pisces: This is the last constellation of the signs of the zodiac. Today, the sun stands at the point in time of spring equinox (March 21) in this constellation (First point Aries; see also p. 52). Its length in the ecliptic 351°–26°. The starting point for the sign of the zodiac Scorpio in the ecliptic is 330°.

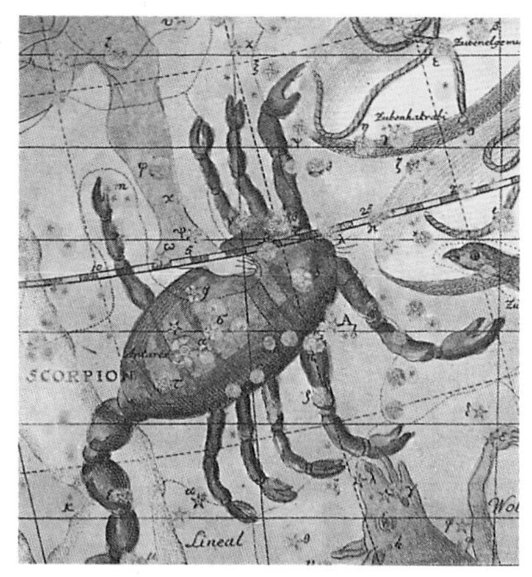

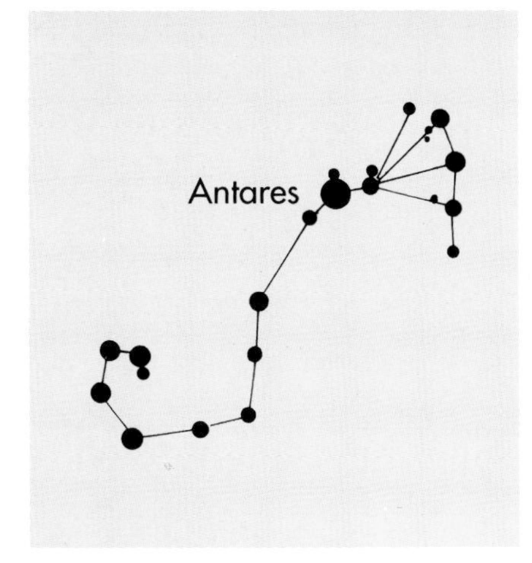

Three depictions of Scorpius:
Top, 18th-century historical illustration;
Center, briefly-exposed photo;
Bottom, drawing of the constellation.

The 88 constellations of the northern and southern sky

The form and number of the constellations have not always been the same. The constellations were the result of a long, historical development by the Greek astronomers. (Almagest was an astronomical work compiled by Ptolemy of Alexandria around A.D. 140.) Since the beginning of modern times, especially during the era of the discoveries of the 18th century, new constellations were created.

Scientifically, 88 constellations are recognized today, which were staked out in 1928 by the astronomer E. Delporte on instruction of the International Astronomical Union. In scientific publications, the names of the constellations are usually abbreviated with three letters. The overview table contains, in alphabetical order, the Latin names of the constellations with their three- and four-letter abbreviations, their translation, and their surface area (in square degrees) according to the international constellation boundaries.

Latin name	Three-letter abbrev.	Four-letter abbrev.	Translation	Surface area (in square degrees)
Andromeda	And	Andr	Andromeda	722
Antlia	Ant	Antl	Air Pump	239
Apus	Aps	Apus	Bird of Paradise	206
Aquarius	Aqr	Aqar	Aquarius	980
Aquila	Aql	Aqil	Eagle	652
Ara	Ara	Arae	Altar	238
Aries	Ari	Arie	The Ram	441
Auriga	Aur	Auri	The Charioteer	657
Boötes	Boo	Boot	The Herdsman	905
Caelum	Cae	Cael	The Chisel	125
Camelopardalis	Cam	Caml	Giraffe	757
Cancer	Cnc	Canc	The Crab	506
Canes venatici	Cvn	Cven	Hunting Dogs	467
Canis majoris	Cma	Cmaj	Great Dog	380
Canis minoris	Cmi	Cmin	Little Dog	183
Capricornus	Cap	Capr	Capricorn or Goat	414
Carina	Car	Cari	Keel	494
Cassiopeia	Cas	Cass	The Queen	598
Centaurus	Cen	Cent	Centaur	1060
Cepheus	Cep	Ceph	The King	588
Cetus	Cet	Ceti	Whale	1231
Chamaeleon	Cha	Cham	Chameleon	132
Circinus	Cir	Circ	Drawing Compass	93
Columba	Col	Colm	Dove	270
Coma Berenices	Com	Coma	Berenice's Hair	386
Corona australis	CrA	CorA	Southern Crown	128
Corona borealis	CrB	CorB	Northern Crown	179
Corvus	Crv	Corv	The Raven	184
Crater	Crt	Crat	The Cup	282
Crux	Cru	Cruc	The Southern Cross	68
Cygnus	Cyg	Cygn	Swan	804
Delphinus	Del	Delf	Dolphin	189
Dorado	Dor	Dora	The Goldfish	179
Draco	Dra	Drac	Dragon	1083
Equuleus	Equ	Equl	The Little Horse	72
Eridanus	Eri	Erid	River	1138
Fornax	For	Forn	The Furnace	398

Latin name	Three-letter abbrev.	Four-letter abbrev.	Translation	Surface area (in square degrees)
Gemini	Gem	Gemi	Twins	514
Grus	Gru	Grus	Crane	366
Hercules	Her	Herc	Hercules	1225
Horologium	Hor	Horo	Clock	249
Hydra	Hya	Hyda	Water Snake	1303
Hydrus	Hyi	Hydi	Lesser Water Snake	243
Indus	Ind	Indi	Indian	294
Lacerta	Lac	Lacr	Lizard	201
Leo major	Leo	Leon	Lion	947
Leo minor	Lmi	Lmin	Little Lion	232
Lepus	Lep	Leps	Hare	290
Libra	Lib	Libe	Scales	538
Lupus	Lup	Lupi	Wolf	334
Lynx	Lyn	Lync	Lynx	545
Lyra	Lyr	Lyra	Lyre	286
Mensa	Men	Mens	Table Mountain	153
Microscopium	Mic	Micr	Microscope	210
Monoceros	Mon	Mono	Unicorn	482
Musca	Mus	Musc	Fly	138
Norma	Nor	Norm	Level	165
Octans	Oct	Octn	Octant	291
Ophiuchus	Oph	Ophi	Serpent Bearer	948
Orion	Ori	Orio	Hunter	594
Pavo	Pav	Pavo	Peacock	378
Pegasus	Peg	Pegs	Winged Horse	1136
Perseus	Per	Pers	Hero	615
Phoenix	Phe	Phoe	Phoenix	469
Pictor	Pic	Pict	Painter	247
Pisces	Psc	Pisc	Fish	889
Piscis austrinus	PsA	PscA	Southern Fish	245
Puppis	Pup	Pupp	The Stern	673
Pyxis	Pyx	Pyxi	Compass	221
Reticulum	Ret	Reti	Net	114
Sagitta	Sge	Sgte	Arrow	80
Sagittarius	Sgr	Sgtr	The Archer	867
Scorpius	Sco	Scor	Scorpion	497
Sculptor	Scl	Scul	Sculptor	475
Scutum	Sct	Scut	Shield (of Sobieski)	109
Serpens	Ser	Serp	Snake	637
Sextans	Sex	Sext	The Sextant	314
Taurus	Tau	Taur	Bull	797
Telescopium	Tel	Tele	Telescope	252
Triangulum australe	TrA	TrAu	Southern Triangle	110
Triangulum (boreale)	Tri	Tria	(Northern) Triangle	132
Tucana	Tuc	Tucn	Toucan	295
Ursa major	UMa	UMaj	Great Bear	1280
Ursa minor	UMi	UMin	Little Bear	256
Vela	Vel	Velr	Ship's Sail	500
Virgo	Vir	Virg	The Maiden	1294
Volans	Vol	Voln	Flying Fish	141
Vulpecula	Vul	Vulp	Fox	278

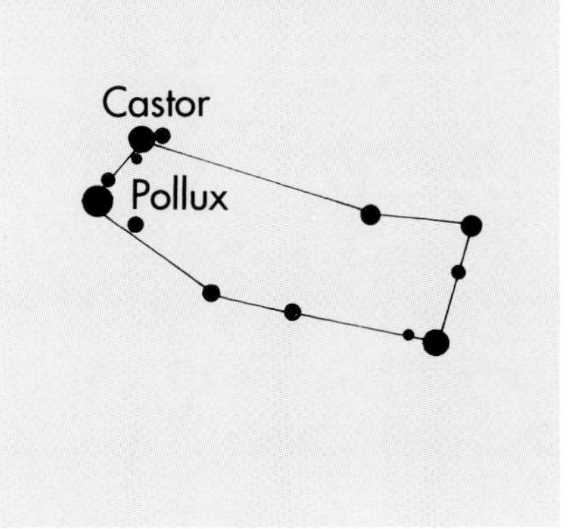

Three depictions of Gemini:
Top, 18th-century historical illustration;
Center, briefly-exposed photo;
Bottom, a drawing of the constellation.

The Rotation of the Earth and Yearly Course of the Sun

There are things in life that we take for granted. For instance, every morning we wake up and realize that it is daytime again, especially early in the summer and late in the winter. Sometimes the sun is shining, sometimes it is cloudy, or it rains or snows. In any case, the night is again followed by the day. This simple example shows how the cosmic influences shape our lives on a daily basis.

The sun rises in the east in the morning and the day begins. In the west, the sun sets in the evening and the night begins. Why? Because our earth rotates. Of course, the spot on the horizon where the sun rises and sets is not the same place every day. It moves within certain boundaries and is dependent on the geographical latitude.

The Seasons

The sun rises exactly in the east only

- on the day of the vernal equinox (March 21 in the northern hemisphere; September 23 in the southern hemisphere)

and

- on the day of the autumnal equinox (September 23 in the northern hemisphere; March 21 in the southern hemisphere).

On these days, the sun then sets exactly in the west. Otherwise, there are deviations to the north and south of where the sun sets and rises. The largest northern deviation takes place

- on the day of the summer solstice in the northern hemisphere (June 21), which is the day of the winter solstice in the southern hemisphere.

The largest southern deviation takes place

- on the day of the winter solstice in the northern hemisphere (December 22), which is the day of the summer solstice in the southern hemisphere.

The course of the seasons is caused by the rotation of the earth around the sun and by the inclination of its axis towards the plane of the orbit of the earth (the ecliptic). The earth's axis of rotation is tilted at an angle of 23.4° towards the plane of the orbit of the earth (compare picture on p. 19).

The Length of the Day

Depending on the geographical latitude, the duration of day and night is different:

- At the equator (0° geographical latitude), the length of the longest and of the shortest day of the year is 12 hours each—i.e., the days are always of the same length.
- At 45° geographical latitude, the longest day has 15 hours and 26 minutes, the shortest day 8 hours and 34 minutes—i.e., a difference of almost 7 hours.
- At the poles (90° geographical latitude), a polar day lasts 186 days (North Pole) and 179 days (South Pole); accordingly, the polar night lasts 179 days (North Pole) and 186 days (South Pole).

Day and night are the results of the daily rotation of the earth (1 day = 24 hours). The yearly movement of the earth around the sun, on the other hand, results in the seasons—spring, summer, fall, and winter. Because of the tilt of the earth's axis of rotation, the northern hemisphere and the southern hemisphere have opposite seasons. When it is winter in the northern hemisphere, summer occurs in the southern hemisphere.

Dusk

Everyone knows from experience that it does not become day and night all of a sudden. Between them are dusk and dawn. The atmosphere of the earth is responsible for that as well. We also owe to the atmosphere of the earth the diffusion of the sunlight in the sky during the day and, thus the blue sky (with clouds, the gray sky).

Since Nicolaus Copernicus (1473–1543), the heliocentrical view of the world with the sun in the center has replaced the geocentrical one with the earth in the center. The planets in the picture are (from left to right) Neptune, Jupiter, Saturn, and Uranus.

And when the sun is below the horizon, sunlight still reaches the upper atmosphere up to a depth of 18°. The air of the upper atmosphere diffuses the sunlight and produces dawn and dusk.

Each second there is

- bright daylight on about half of the earth's surface,
- dusk or dawn on 15% of the earth's surface,
- night on 35% of the earth's surface.

The short length of dusk and dawn in the tropics is due to the course of the sun, which stands especially high on the horizon in that area. The "white nights" in the moderate latitudes in the height of summer are caused by the low position of the sun under the horizon. Practically, it is dusk or dawn throughout the entire night for a couple of weeks.

A dark night is the precondition for the visibility of the stars. Moonlight and dusk disturb the view of the starry sky. Even though brighter stars are also visible under these circumstances (with some skill, you can even find the planet Venus or the extraordinarily bright fixed star Sirius and some other bright stars in the sky during the day), a dark, clear night is necessary for seeing the entire starry sky.

Daily Rotation of the Sky

The beginning of any kind of astrology came about from people's awareness, and eventually worship, of the precision of the stars and sky. After looking at the starry sky for a long period of time, one gets the impression that a hollow globe, pasted with stars, constantly rotates around a fixed axis, which goes through the stationary earth. Yes, one appears to be simply the center of the rotating celestial sphere: This daily rotation of the sky is the most impressive and fastest of all movements in the sky for the observer on earth. All suns, moons, planets, and stars are subject to it.

It is known that the earth rotates around its axis from west to east (counterclockwise as seen from the a point above the North Pole). This causes objects in the sky to appear to travel from east to west.

The point directly (90°) above the observer is the observer's zenith. If you look from the zenith down so that you are at the corner of a right angle, you will see the horizon (strictly speaking, the mathematical horizon), which has the height of 0°. In practice, it is the landscape horizon that lies higher than the mathematical horizon.

The line that runs from the exact north point on your horizon up through the zenith and down to the exact south point on your horizon forms the large circle of the celestial meridian. The stars seemingly move in our latitudes between the south point and celestial North Pole through the meridian where they reach their greatest altitude (upper culmination). Between the celestial North Pole and north point, stars cross the meridian at their lowest altitude (lower culmination).

The camera is oriented towards the southern sky pole and exposed for 2 hours. H. Vehrenberg took this picture on one of his South African expeditions with a small astro-camera with a 71 mm opening and a 250 mm focal distance on Ilford HPS plate.

Photo of the Earth's Rotation

With the help of a camera, the sky's rotation can be beautifully documented. Load your camera with highly sensitive black-and-white film. Open the stop completely, set to infinity, exposure to B (wire trigger with clamp screw), and expose the film for a couple of hours. Place the camera on a tripod and face it towards the celestial north or celestial south pole. During the time of exposure, the sky seemingly continues to rotate (in reality, the earth rotates), and on the film, the stars are depicted in more or less long light trails that are bent towards the pole (compare the photo). The picture also shows that the stars draw parallel circles during the daily apparent rotation of the celestial vault. A moonless, clear night is recommended for this photo experiment of the starry sky.

Celestial north pole and celestial south pole are the bearings in which the axis of the seemingly rotating celestial sphere runs. You can see from your location, at any time, onto only one celestial pole—either the northern or southern. That half of the earth, from which the celestial north pole is visible, is called the northern hemisphere; the other half with the view onto the celestial south pole is called the southern hemisphere. The equator separates in the geographical coordinate-system the two earth-hemispheres. Accordingly, there is also the celestial equator, which is in all its points exactly 90° away from the two celestial poles.

These star trails on both sides of the celestial equator taken with longer time of exposure. Only stars on the celestial equator form a straight-lined track. The star trails, which are to the north and south of the celestial equator, clearly show a curve towards the celestial North Pole and/or towards the celestial South Pole.

Equatorial Coordinates

It doesn't take much to imagine a simple earth-sky celestial system. The foundation is the ecliptic plane of the earth, which is at the same time also the ecliptic plane of the imagined celestial globe. The rotating axis stands vertical to the earth equator and celestial equator. A star fits into this system as follows:

1. Specification of the angle from the celestial equator. Stars, which have the same distance from the equator, lie on the same celestial latitude called declination.
2. Specification from which place above or below the equator a star lies—the declination of a star. Stars, which go through the same point on the celestial equator, have the same right ascension—the celestial equivalent of longitude on earth.

The earth's axis of rotation is inclined 23.4° towards the plane of the ecliptic. The angles of incidence of the sun's rays are different, according to the spot of the earth on its orbit. That explains the seasons. These are on the northern hemisphere opposite the southern hemisphere. The yearly deviations at the beginning of the seasons of about 1 day are due to the leap year, which returns on our calendar every 4 years.

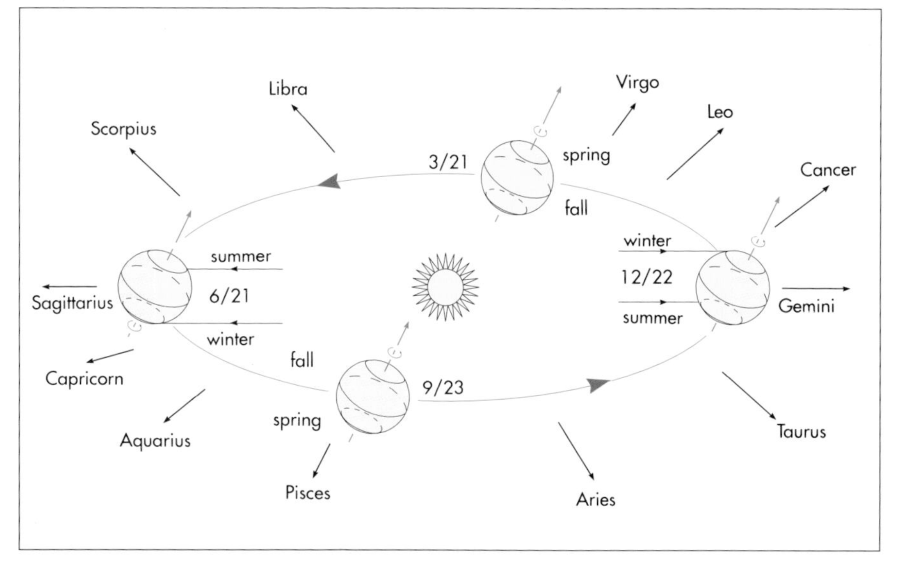

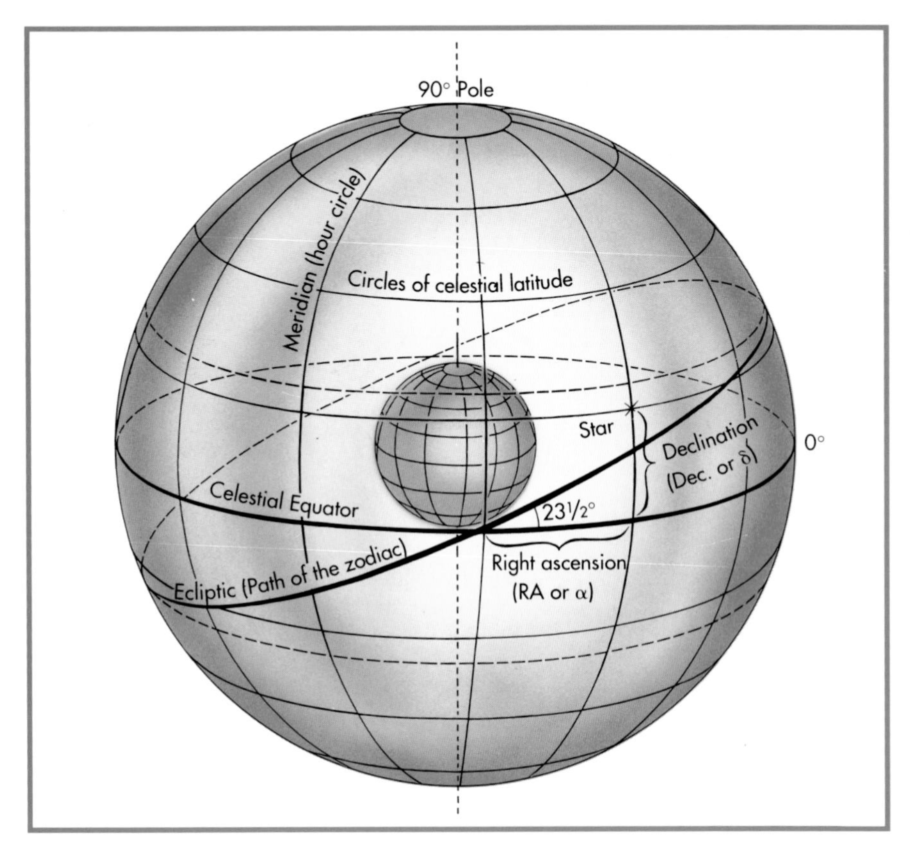

A simple relation system for the coordinates of objects in the sky and on the earth: Equatorial coordinate system.

Is this somewhat too complicated? The above picture will hopefully make it more clear and plain. The circles of celestial latitude are of different sizes. They become smaller and smaller near the pole. They become largest near the equator. On the other hand, the meridians are always of the same length. They approach each other on their path to the poles. The two descriptions, mentioned before, for the determining the location of a star, led to fixed coordinates for the earth:

Earth North Pole	– Celestial North Pole
Earth South Pole	– Celestial South Pole
Geographical latitude	– Declination
Geographical longitude	– Right ascension
Point of intersection	– Point of intersection
Line of Earth orbit with earth zero meridian (Greenwich)	– Ecliptic (zodiac) with celestial equator (zero point)

The zero point of the geographical length is Greenwich (London). Was this an arbitrary choice? Well, the British were, when the matter became of immediate interest (1675), on their way to becoming the leading navigator nation on earth. Therefore, nobody disputed this decision. It was even officially acknowledged in 1884 by all states.

The zero point at the celestial sphere globe for the right ascension is the point at which the sun crosses the celestial equator—March 21. Astronomers measure right ascensions eastward from the zero point of the celestial sphere in degrees, from 0° up to 360° (from 0 hours to 24 hours).

We already talked briefly about the course of the sun on page 17. It is again the earth, which, on its rotation around the sun (1 rotation = 1 year) causes the yearly course of the sun in the sky. It is good to remember these connections. The yearly course of the sun takes place on a large circle, which is inclined 23.4° towards the celestial equator. This large circle is better known by the name zodiac or ecliptic ("line of the darknesses"), because solar and lunar eclipses are only possible when the moon's orbit crosses the ecliptic.

The stars of the twelve constellations of the signs of the zodiac group themselves around the orbit of the earth in outer space. Seen from the earth, the sun wanders, during the yearly rotation of the earth around the sun, once through all the constellations of the signs of the zodiac and causes, thereby, their varying visibility in the course of the year. The stars, which are standing directly behind or adjacent to the sun, are invisible for the observer from the earth. Therefore, when the sun is standing in the constellation of Aries, this constellation is at that moment not visible from the earth (see picture on p. 19). You can best "calculate" the constellation, in front of which the sun is standing, in the morning dawn or evening dusk, since then you can recognize the adjacent constellations. Every 6 months, these constellations then have their best visibility in the night sky.

We also find the moon and the planets of the sky within the range of the constellations of the signs of the zodiac, because they extend only about 8° north and south of the ecliptic.

The Yearly Rotation of the Sky

The starry sky does not only change in the course of one night (daily rotations of the sky as result of the earth's rotation). Its visibility also changes gradually in the course of a year. This is because of the summer or winter constellations. The sun wanders in the course of the year through the zodiac constellations of Aries, Taurus, Gemini, Cancer, Leo, Virgo, Libra, Scorpius, Sagittarius, Capricorn, Aquarius and Pisces.

Day after day, the earth goes a bit farther on her course around the sun. And day after day, the view of the sky shifts a little bit. If someone could observe the stars at the same spot and in clear weather every night at 24 hours (midnight), he or she would then notice a small change each time—i.e.,

- on the northern hemisphere counterclockwise;

NORTH

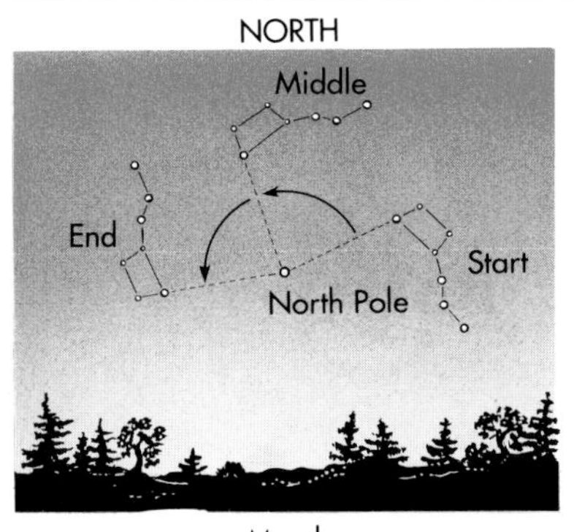

March

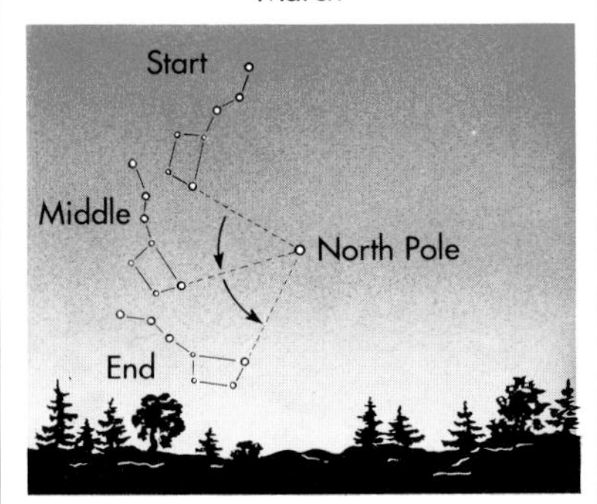

June

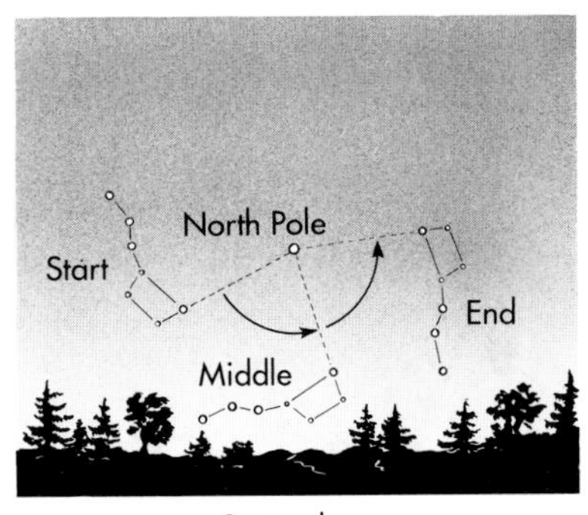

September

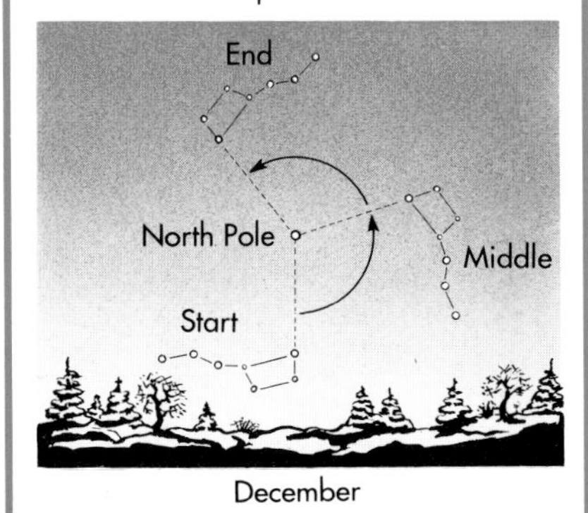

December

- on the southern hemisphere clockwise.

This means that the direction of rotation is different on the northern hemisphere than on the southern hemisphere. And it is also different during the daily rotation of the sky (see pictures on the left and right):

- on the northern hemisphere, the daily rotation of the sky is counterclockwise;
- on the southern hemisphere, it is clockwise.

To prove this, take a pocket watch and hold it against the sky in such a way that the face of the clock points towards you and the backside faces the sky pole that is to be observed from this location. It is best to fixate a constellation above a roof, a chimney, or a tree, because that way the shifting after half an hour or after an hour can be better determined.

Of course, proving the rotation direction for the yearly rotation of the sky cannot be done quite as fast. Again, a fixed point on the earth is a useful marker. Observe the same constellation at the same time and within 1–2 weeks, and you can see a shift.

This simple project can help anyone learn about the cosmical movements. Spaceship earth is constantly on the road. The daily rotation of the earth around its axis and its yearly path around the sun give us two time markers: the day and year.

Left:
The yearly movement shown in the constellation of Ursa major (Great Bear), also called Big Dipper, at the beginning of the four seasons (March = spring, June = summer, September = fall, December = winter). The illustration is for 50° northern latitude. The terms "beginning," "middle," and "end" correspond to the course of the constellation's appearances in the night sky. The varying length of the rotation movement is due to the varying length of the night (and thus to the visibility of the constellations).

Right:
The yearly movement, shown at the example of the constellation Southern Cross and the stars α and β of the constellation Centaurus (Centaur) at the beginning of the four seasons (March = fall, June = winter, September = spring, December = summer). The illustration is for about 40° southern latitude. The terms "beginning," "middle," and "end" correspond to the course of the constellation's appearances in the night sky.

SOUTH

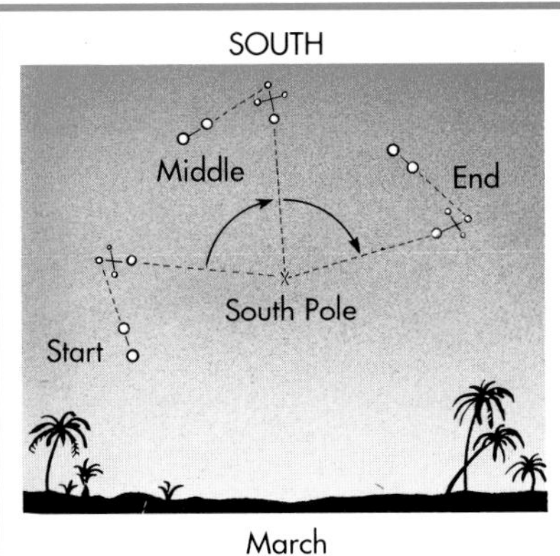

March

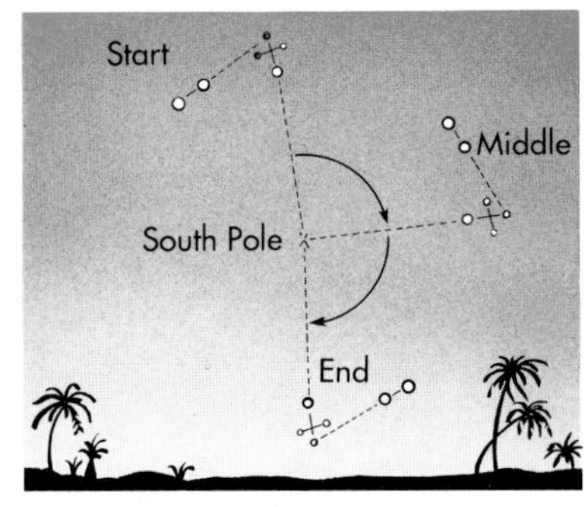

June

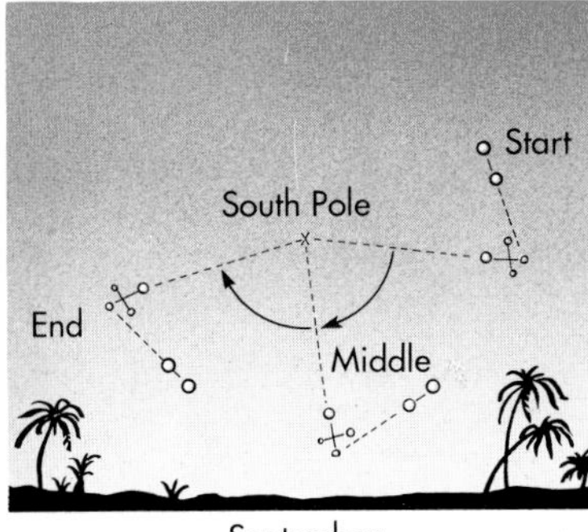

September

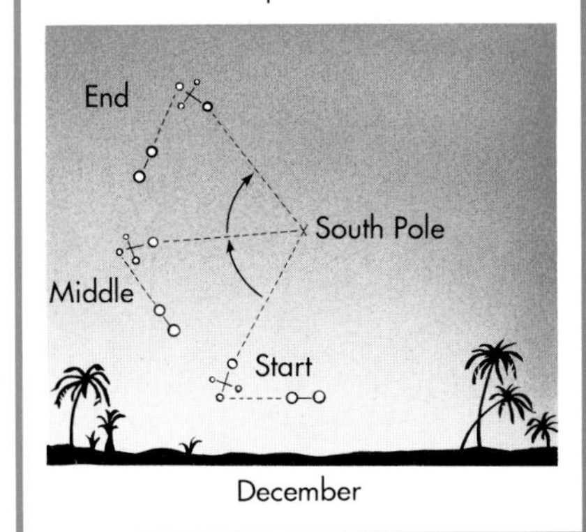

December

The Sun as Time Marker

There are different celestial ways of measuring of time. You can following the daily movement of the sky or of a star, such as the sun. But the sun is not a very precise time marker, because it moves with varying speed in the ecliptic, which is inclined towards the equator. There are two causes for the inequality:

- The orbit of the earth around the sun is an ellipse and not a circle. Therefore, the sun cannot travel the same section of orbit in the same time period.
- The projection of the orbit on the celestial equator, on which time is measured, produces uneven sections. This is because the earth makes one 360° revolution about the sun in 365 days (1 year) while the sun appears to move about 1° in our sky each day. This means that after the earth has made one complete 360° rotation, it must turn an extra 1° to bring the sun back to the meridian, which takes about 4 minutes.

The sundial is the oldest clock. It developed out of the gnomon, which is a long pointer that casts a shadow. The gnomon was mentioned for the first time in a Chinese document around 1100 B.C. The gnomon casts a shadow of measurable length and direction on a graduated horizontal base. The ratio of gnomon height to shadow length gives the tangent of the altitude angle. From the position of the shadow of the pointer onto which the sun shines, the real sun time (real local time) can be determined,

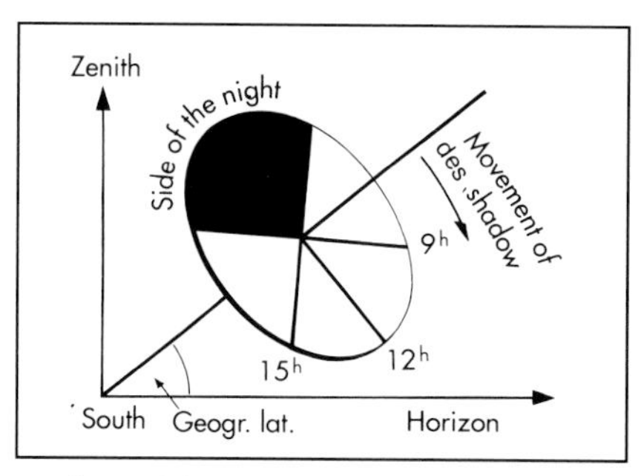

Simple sundial. The pointer (gnomon) is directed towards the zenith. The shadow is read on a disk that is uniformly divided and lying on the celestial equator.

A calibrated plate (upon which the gnomon casts its shadow) makes up the second part of the sundial. As a general principle, it can be attached to any kind of surface. But in the case of the sundial in the illustration, the projection plane for the shadow is parallel to the equator—i.e., vertical to the gnomon. In accordance with the uniform rotation of the earth around its axis, the shadow rotates each hour 15° around the pointer. In order to use it during the winter, the gnomon has to go through to the lower side of the shadow plane. With horizontal sundials, the shadow plane lies in the plane of the horizon. Here, the angles between the hour lines, over which the shadow runs, vary in size. That is also the case with the vertical sundials, whose calibrated plate stands vertically to the meridian plane (see photo on this page).

There is extensive trade literature for constructing sundials. *Astronomy: A Handbook* also contains a chapter with construction suggestions.

The time between two consecutive transits of the sun through the spring point (see p. 20) corresponds to one year. By observing the longest and shortest shadows that the gnomon is throwing at noon, you can determine approximately the times of the solstices (see p. 17). The shadow that a gnomon of a certain length draws noon after noon onto the ground is never of the same length; it depends on the seeming movement of the sun on the ecliptic. By proving the solstices, the gnomon indirectly also draws our attention to the inclined plane of the ecliptic (see p. 20).

For more about the different times (star, solar), see *Astronomy: A Handbook.*

Sundial on a steep Baroque gable (in Rothenburg, Germany) above the city clock (1683), a calendar, and eagle.

Observing the Sky

Observations from Different Geographical Latitudes

We have already discussed the change of the starry sky in the course of one night (daily rotation of the earth) and in the course of one year (rotation of the earth around the sun).

The horizon limits the view of the starry sky and confines it for each location of observation to one-half of the apparent celestial sphere. Thus,

- The same starry sky becomes visible in the course of one earth-rotation for all locations of observation on the same parallel of latitude on the earth. "In the course of one earth-rotation" means while viewing the same section of the sky at the same local time. The horizon changes with the geographical latitude—the visible half of the apparent celestial sphere shifts towards the south and/or north. Wherever we are observing on the earth, one half of the celestial equator is above the horizon while the other one is below the horizon.

New Technology Telescope (NTT) of the European Southern Observatory (ESO) in Chile. The telescope, with 3.58 m mirror diameter, is located in a novel type of protection building, which deviates from the usual dome shape. The building follows the movements of the telescope during the observation.

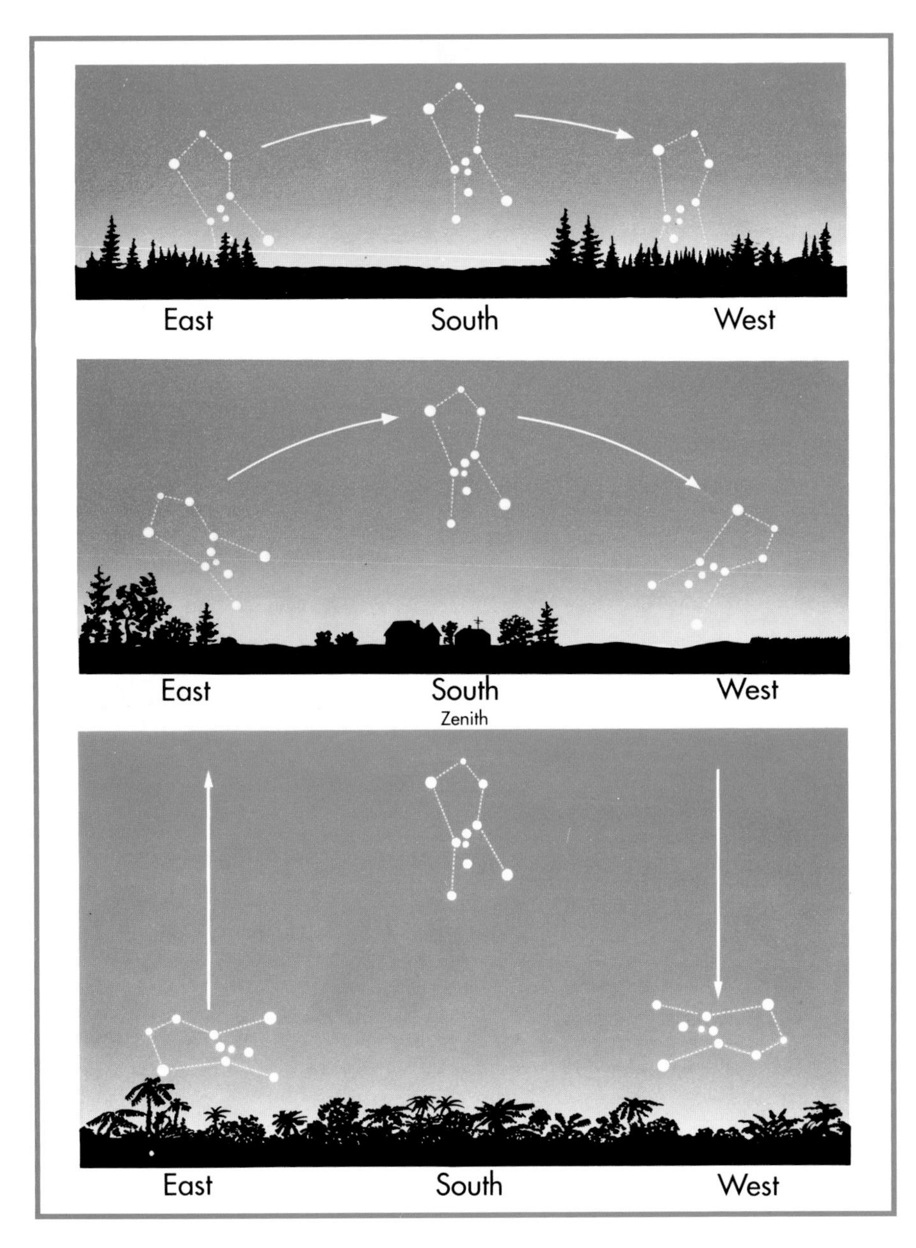

Duration of visibility of the stars

Geographical Latitude	Overall visibility of starry sky	Constantly visible	At times visible	Constantly invisible
North Pole	50 %	50 %	0	50 %
45° to the north	85 %	15 %	70 %	15 %
Equator	100 %	0	100 %	0
45° to the south	85 %	15 %	70 %	15 %
South Pole	50 %	50 %	0	50 %

The orbits of constellations in different geographical latitudes

The daily orbits of the stars vary in different geographical latitudes. We will use as an example the constellation Orion, which can be seen on the northern as well as southern hemispheres of the earth. Rise (eastern sky), culmination, set (western sky) in 66° northern latitude (top) in 45° northern latitude (center), in 0° latitude (at the equator; bottom). For the observer at the equator of the earth, the stars form half-circles around the north point of the horizon. The celestial pole lies in the horizon line. The constellations rise and set vertically.

Therefore, a star on the equator is visible for the duration of half an earth-rotation and not visible for the same amount of time. There are, at each place of observation, many constantly visible and invisible stars. That leads to the following classifications:

1. One zone with the celestial pole in the center that can be constantly observed. This is referred to as the circumpolar area, with circumpolar stars.
2. One zone with the celestial equator in the center that is visible at times. The stars rise and set in it.
3. One zone with the opposite celestial pole in the center that is constantly invisible.

It is important to note that the duration of the visibility of the stars varies with the geographical latitude as well as the size of the circumpolar area and the width of the partially-visible equatorial zone. (The three aforementioned zones are distributed across the earth as indicated in the table on this page).

All stars become visible to the observer when they stand above the horizon. That is the condition for the "geographical visibility." "Seasonal visibility" plays a major role, and it depends on the apparent course of the sun on the ecliptic. Helmut Werner's *From the Polar Star to the Southern Cross* (only available in used-book stores) is an excellent book that describes the appearances of the celestial sphere in all details. These two aspects are detailed in "The orbits of constellations in different geographical latitudes" on these two pages and the maps of the stars in the front and back jackets.

Rise, culmination, and set of Orion in 12° southern latitude (top), in 23° southern latitude (center), in 56° southern latitude (bottom). In the southern hemisphere, north and south are switched! A very simple relation determines where you can still see stars at the northern sky: 90° – star declination (north) = geographical latitude (south), where the star appears at the moment above the horizon. For example, the star Alpha Gemini (Castor in the constellation Gemini) has the declination of +32° in the northern hemisphere. Calculate: 90° – 32° = 58° southern latitude. This latitude is the most southern one in which it is still possible to see the star above the horizon (see below and page 32).

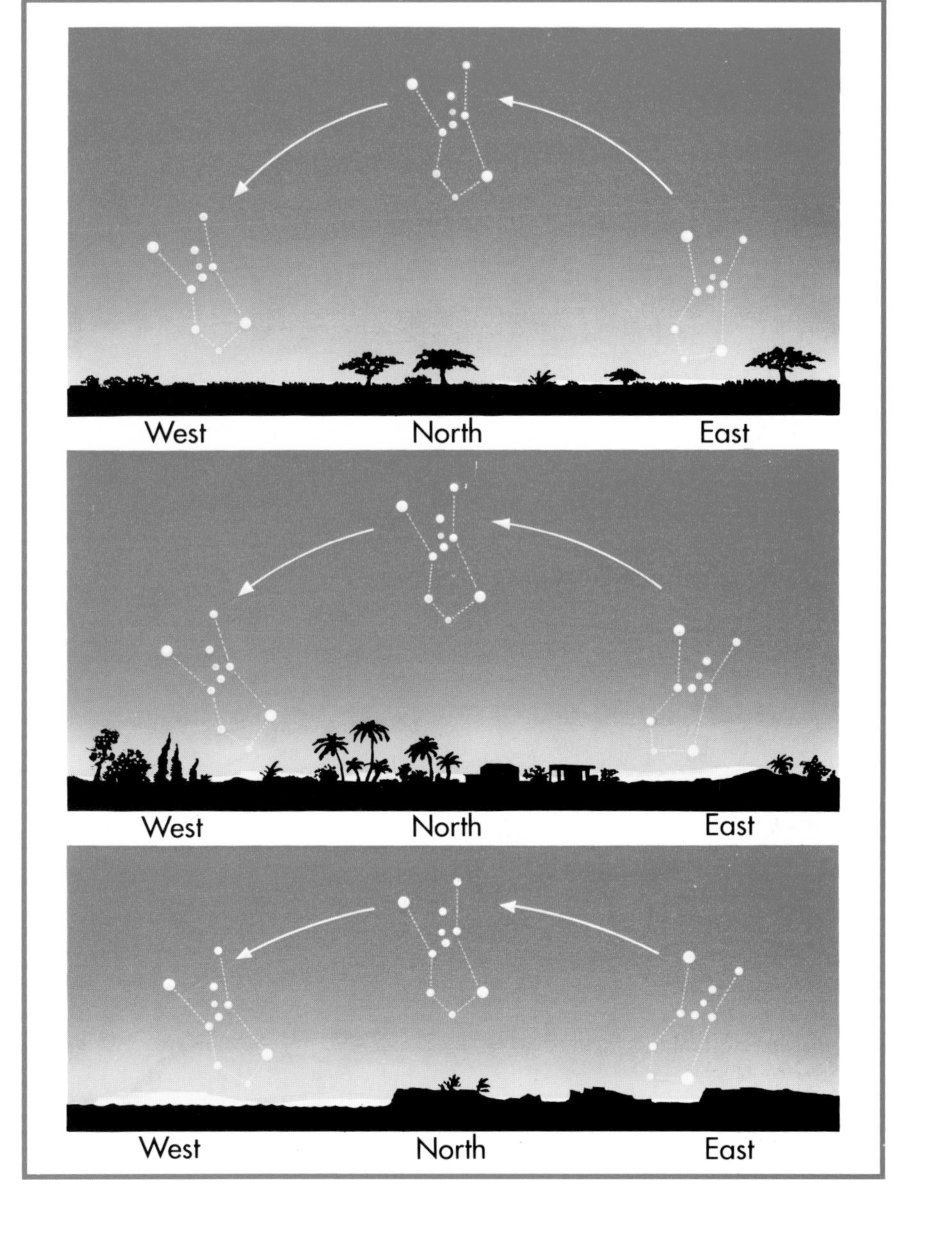

Height Measurements with Simple Tools

The measuring of the altitude (the angle above or below the plane of the observer's horizon) of stars and planets was already an important task for the astronomers in antiquity. A very simple as well as instructive instrument to do this is the protractor. Anyone can make it out of cardboard, wood, plastic, aluminum or brass.

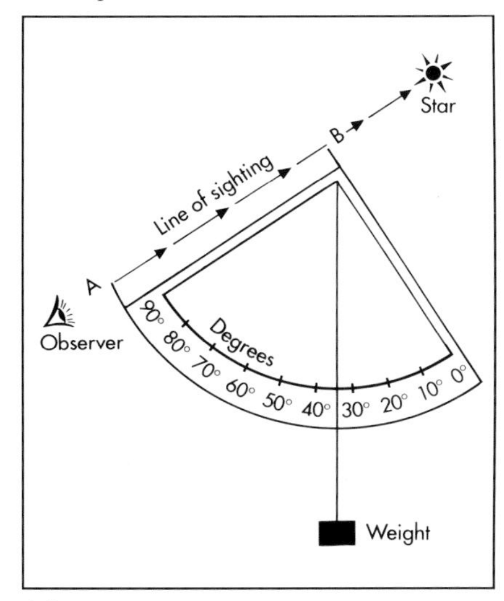

A Protractor

To use a protractor (see above picture), take aim at a star along the edge A–B. The weight indicates the degree of the altitude of the star. The more precisely the aiming edge of the protractor stands perpendicular to the zero line of the protractor, the more precise the measurements will be. What can you measure with a protractor? You can determine your geographical latitude by looking for a bright fixed star, for instance, Sirius in the constellation Canis Major (Great Dog), which stands out well in the evening sky from November until March. At the beginning of March, Sirius reaches his highest position above the horizon between 6:30 P.M. and 7:30 P.M. In intervals of 5 minutes, you can measure the altitudes of the star above the horizon with a protractor (the highest position above the horizon for you on the northern hemisphere looking southward, or for you on the southern hemisphere looking northward). The culmination point (the highest point above the horizon) is on February 1, 9 P.M., and the measured height is 23° at a certain place on the northern hemisphere. Add to this value, the declination (see p. 20) of the star. It is rounded off to –17°. The calculation is: 23 + 17 = 40. That corresponds to the altitude of the celestial equator above the horizon in the meridian (see p. 20). The illustration below makes it clear. The simple relation "90° – altitude of the celestial equator = geographical latitude" leads to the result of +50°. Remember that for the northern hemisphere:

- The declination with the sign plus is deducted from the altitude measurement.
- It is exactly the opposite on the southern hemisphere.
- The declination with the minus sign is added to the height measurement.

The declination and right ascension of bright stars is found in the text that accompanies the star maps on pages 34–87.

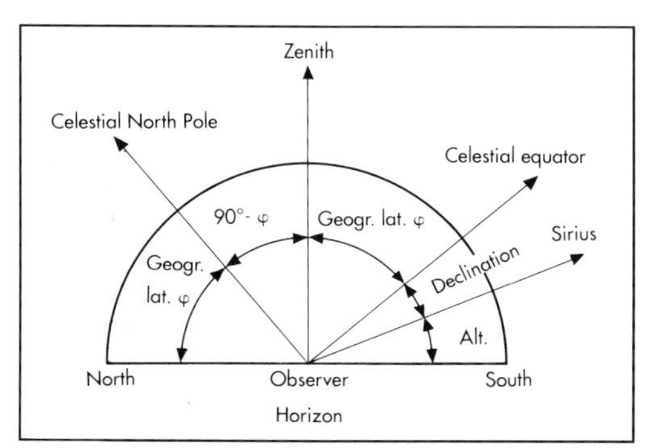

Altitudes for determining the geographical latitude with the help of a protractor.

Electronic Aids

Since 1993, determining the geographical latitude and longitude of a position is possible for everyone with the help of satellites following the "Global Positioning System" (GPS). What you need is a GPS receiver that picks up the location signals of the twenty-three Navstar satellites. This receiver costs only a little more than a cellular telephone. The position determination is up to 1–5 m precise. Microelectronics and computer technology make it possible to see the sky on the screen.

Astronomy via computers does not restrict itself to the calculation of positions of sun, moon, and planets. Today, there are already a number of astro programs that

The constellation Ursa Major (Great Bear) as an example for apparent angle distances of the stars in the sky. See the table on the right with the star names and distances (see also p. 42).

The starry sky on the screen. Using computers, anyone can today make his or her own stars with the help of software, like TheSky (see also p. 159).

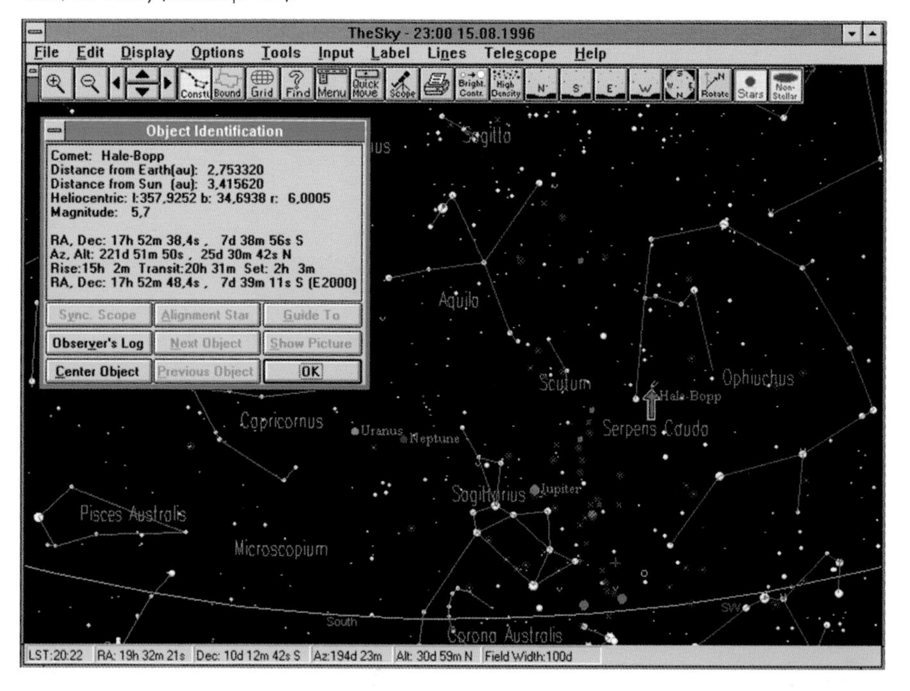

turn the computer into a home planetarium and let the starry sky appear on the monitor for a desired observation position at a certain day and hour. This picture can also be printed out as a ready-made map of the stars.

Angle Distances of the Stars

It is helpful to know the apparent angle-distances of stars in order to determine the visual field-diameter of a pair of binoculars or a telescope. The constellation Ursa Major (Great Bear), which (in latitudes north of 41°) is visible throughout the entire year, can be used as an example (compare the top photo on p. 26):

Stars		Distance
Benetnasch	– Mizar	4°53'
Dubhe	– Merak	5°38'
Mizar	– Megrez	9°78'
Mizar	– Phecda	13°15'
Merak	– Alioth	15°53'
Mizar	– Dubhe	19°35'

The Rays from Outer Space

Chemists and physicists both observe and experiment with nature through the laboratory or by studying test results from all fields of chemistry and physics—e.g., from the fields of optics, acoustics, or mechanics. Astronomers, in the literal sense of the word, have little in their hands, apart from moonstones, which probes brought to the earth, or meteorites, which landed on the earth's surface. By far, the predominant amount of information for astronomical research comes from the electromagnetic radiation that is picked up by different radiation receivers.

The Milky Way shot from a miniature camera. Any astronomy buff can do this!

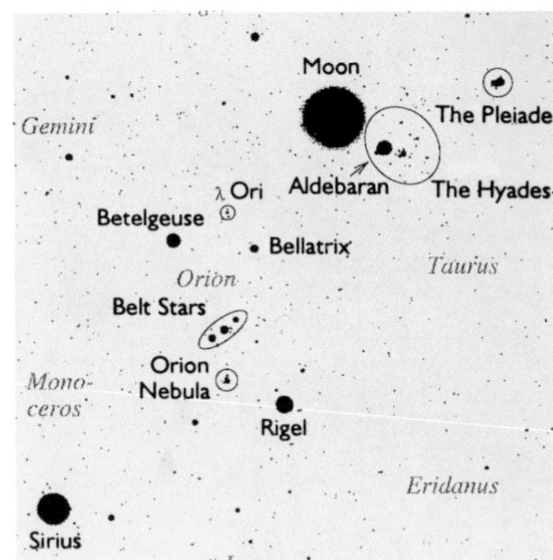

Various views of the same sky section, which were caught with different methods.

Left: The constellation Orion with moon (half moon) as seen with the naked eye.

Above: Map with moon and the most important stars of the Orion region.

Right: The constellation with moon (large bright circle top right) based on a photo with visible light.

Far right: The constellation with moon (small yellowish moon-crescent right top) using a device that picks up X-ray radiation.

Analysis of the rays from outer space gives us all the information about the characteristics of the celestial bodies. The spectrum of the electromagnetic radiation is wide. There is the visible light, which ranges from violet and blue and green to yellow and red. Infrared, which is invisible to the human eye, follows the visible red. There is the field of the warmth radiation and of the radiowaves. The shortest wave and richest in energy rays is the cosmic radiation and X rays. They are being swallowed up almost completely by the earth's atmosphere (see illustration on p. 133).

All rays from outer space are similar in that they are electromagnetic waves, which travel at the speed of light. Light travels at a speed of 186,000 miles per second. On the other hand, the types of rays are very different in reference to their wavelength, intensity, and composition.

Every Radiation Detector Has "Its Sky"

Anyone who steps out of a bright room into the dark night will at first be surprised by the small number of stars in the night sky. But as the eyes become accustomed to the darkness, you will gradually find the number of stars increasing.

This adjustment of the eyes to the darkness is called adaption. It takes half an hour or longer for the eyes to completely adapt themselves to the dark. The kind of astronomical observations that are done with your naked eye, a pair of binoculars, or a telescope are called visual observations. Every undesired light (haze) obstructs the adaption to the night and impairs the quality of the observation. Background illuminations of the night sky and especially moonlight diminish the boundary size of the stars that are visible with the naked eye. In a dark, clear night, the experienced observer perceives stars up to the 6th magnitude—a total of about 6000 stars.

The oldest receiver of the stars' rays, after the eyes, is photographic emulsion. In science, it has been mostly replaced by electronic radiation detectors. It still plays a role for the preparation of shots of large star-fields, and especially for the amateur photographer. Tips for celestial photography are on page 159.

Most of the shots of celestial objects in this book are photographed. Anyone who observes the sky and who uses a photo as comparison or for orientation will quickly see differences:

- the photos show in the comparison more stars,
- the contrast of photos is much more

intense than the observer can see,

- the photos show colors, which the observer does not see.

Photos that are taken with telescopes of different focal lengths and with shorter or longer exposure time convey different impressions. The photos of the Pleiades ("Seven Sisters") on pages 55 and 143 are examples of that. The photo on page 55 shows a larger field. The telescope used to take the shot had a shorter focal length than the one on page 143.

The photographic astro-shots contributed considerably to visualizing telescopes of different focal lengths. Their additive effect onto small radiation-impressions provides a large detail- and contrast-yield. There are publications devoted to the field of astrophotography as well as electronic radiation detectors, photo multipliers (electronic devices used to convert low-intensity light signals into electrical signals, which are amplified), and especially Charge-Coupled Devices (CCD—light-sensitive electronic detectors).

The three examples of the constellation Orion with the moon illustrate different types of radiation and how differently they are imaged with the help of different radiation detectors:

1. Illustration based on observation with the naked eye,
2. Photographic shot,
3. X-ray image taken by ROSAT (Röntgenstrahlen Satellit, a major German X-ray satellite).

Looking up at the night sky provides the non-professional astronomer with his or her most immediate and firsthand experience with the stars. But in order to study their orientation and develop a deeper understanding, you need photographs and X-ray images. Therefore, it is helpful to know the differences.

When observing with a pair of binoculars or a telescope, it is important to adjust the aperture of the instrument—and, to a certain degree, vary the enlargement—in order to study the boundary size of stars that are close enough. For that, the following overview:

Binoculars	7 × 50 bis 10m
	10 × 50 bis 10.2m
63 mm-refractor	V 34× bis 11.2m
	V 53× bis 11.5m
100 mm-refractor	V 40× bis 12.3m
	V 100× bis 12.8m
180 mm-reflector	V 45× bis 13.4m
(mirror)	V 112× bis 13.9m

As a comparison, a reflector-telescope with 1.5 m opening reaches stars near the 17th magnitude when its enlarged about 200 times. In astro-photography, the situation is completely different. In a few minutes, the 180 mm reflector of a CCD camera reaches the limiting value of the large 1.5 m telescope!

Galaxies are the building blocks of the universe. One of these large star systems is the Andromeda Nebula. It can be seen in the constellation Andromeda with the naked eye as a pale spot. The spiral structure of the galaxy and the resolution into individual stars can be seen through photos exposed for a long time.

The Constellations and Their Most Interesting Objects

Looking at the starry sky on a moonless, clear night can be both a pretty and a confusing experience. The stars seem to be legion, and it seems impossible to find an orientation in this large number. Or is it? Take some time to let your eyes rest and concentrate on the bright, striking stars. Yes, and there are the leading constellations that were described on pages 12–13.

A good orientation point begins at the celestial poles and the leading constellations that surround them. In the northern sky it is Ursa Major (the Great Bear or the Big Dipper). In the southern sky, it is the Southern Cross with the two main stars of the constellation Centaurus (Crux, Centaurus; see illustrations on p. 21).

The constellations Orion, Virgo, and Aquila are further leads in the sky. They are similar in that they lie on the celestial equator (or ecliptic). The strikingly square constellation Pegasus is a little to the north of the celestial equator. The table below gives an overview of the visibility of these constellations. On average, a constellation rises every month 2 hours earlier. Keep this rule in mind when you are using the star maps on p. 34ff.

Distances in Space

Anyone who looks at the night sky with binoculars or a telescope should give some thought to the ratios of size in space, especially to the distances. Based on our daily experiences, we tend to "comprehend" distances. Old farmers speak about the path-hour that separates two villages. An odometer in a car helps the driver determine how many miles he or she has driven that day. And we literally see these miles, because towns and landscapes are connected with them. It is the personal experience of the distance. Anyone who reads that the earth's circumference at the equator is about 40,000 km (24,800 miles) can still bring this in relation to number of miles he or she has driven in a year. But the distances grow fast when the human being leaves the earth. On July 20, 1969, the lunar spacecraft, Apollo II "The Eagle" landed on the moon. For the first time, human beings reached another celestial body. This space neighbor is about 385,000 km (238,700 miles) away from the earth.

Below are the distances of other large celestial bodies of the solar system from the earth (in millions of kilometers):

Sun	149.6
Moon	.384405
Mercury	75 to 225
Venus	35 to 255
Mars	60 to 400
Jupiter	600 to 970
Saturn	1200 to 1650
Uranus	2600 to 3150
Neptune	4350 to 4700
Pluto	4300 to 7500

The distances of the planets from the earth are not equal, because the respective position of the planets and of the earth during the rotation around the sun is different (see illustrations on p.106). The following lists how many miles the planets cover on their orbit around the sun (in millions of kilometers):

Visibility of Important Constellations in the Course of the Year

Constellation	Rises in the east	Midnight in the south (north. hemis.) in the north (south. hemis.)	Sets in the west
Orion	September	Mid-December	April
Virgo	November	Mid-April	July
Eagle	March	Mid-July	November
Pegasus	June	Mid-September	January

Mercury	360	Saturn	9000
Venus	680	Uranus	18,000
Earth	940	Neptune	28,000
Mars	1400	Pluto	37,000
Jupiter	4900		

This means, year after year, our earth runs the distance of 940 million km (582.2 million miles) around the sun. Thereby, it develops the medium orbit-speed of 29.8 km (18.4 miles) per second. On earth that equals 107,280 km/hr (66,513 mph).

The star system closest to our solar system is Alpha Centauri (see star map on p. 78), which is actually a system of three stars. It is "only" 4.3 light years away from the sun. A light year represents 9.4605 billion km (5.88 trillion miles).

Maybe the following proportionate model will be a help for a better understanding. In it, every million of kilometers is reduced here to a centimeter. Then:
The sun–earth distance is 1.5 m.
The sun–Jupiter distance is 7.78 m.
The sun–Pluto distance is 59.10 m.
The sun–Alpha Centauri distance is 410,000 m.

We could easily store this model of the solar system in the Empire State Building in New York. But in order to add Alpha Centauri, we would need to make the distance longer—from New York to Washington, D.C. Keep this model in mind the next time you observe the starry sky.

Star's Brightness

Stars have different brightnesses in the sky because they do not all give out the same amount of light and they lie at different distances from the earth. Astronomers refer to a star's brightness as its magnitude. The brightest stars are a 1st magnitude, the faintest (that the naked eye can see) are a 6th magnitude. Since a few stars are brighter than 1st magnitude, they have been accommodated by having magnitudes of 0 and then negative numbers. For instance, Sirius, the brightest star, has a magnitude of –1.4. (Also see p. 139 for further discussion about magnitude.)

The Star Maps

Pages 34–86 show the entire starry sky that is visible on earth with the stars up to magnitude 5 (from the North Star to the Southern Cross). Each spread focuses on a certain section of the sky.

Left pages: They show the section of the star map in relation to the entire starry sky. They also have descriptions of the most important stars and constellations. The section in the star map corresponds to the sky field of 40° × 60°. The pictured stars are all visible with the naked eye.

In order to make it easier to find the described celestial objects, they are circled in the large overview map. The map also contains a number of other celestial objects, which are marked with symbols. Each of the symbols is explained on the bottom left next to the map.

Right pages: They show photographs of two interesting celestial objects (double stars, clusters, nebulae, etc.) for observing with binoculars or a small telescope. The photos correspond to about 4° visual field. The visual field of a strong performance—e.g., Fujinon FMT-SX 16 × 70 or Zeiss 15 × 60—covers about 4° (more about suitable instruments on p. 155ff).

Visibility: The leading constellations (described on p. 12) and their visibility as well as the star maps for the seasonal visibility of the starry sky on the northern and/or southern hemisphere (see maps on the front and back jackets) will help you determine the visibility of the respective sections. You can use a simple formula to calculate the declination of those northernmost or southernmost stars, which are harder to see at the observation location.

Example for page 34: The observation location in South America has $\varphi = -40°$. Then: $90° - 40° = 50°$. All stars of the northern sky are visible up to the declination +50° as well as the stars β and γ Andromedae.

Example for page 60: The observation location in Europe has $\varphi = +55°$. Then: $90° - 55° = 35°$. All stars of the southern sky are visible up to the declination –35° as well as the stars γ Corvi and α Virginis.

The star maps begin
- Dec. +90° up to +30°, RA from 0 to 360° (map 1 to 9);
- Dec. +30° up to –30°, RA from 0 to 360° (map 10 to 18);
- Dec. –30° up to –90°, RA from 0 to 360° (map 19 to 27).

The star maps overlap a little in order to make the transition easier. The "bright stars" mentioned at the bottom right-hand page include their positions, which will help you compare them with the other stars in the other maps.

Names: The brighter stars usually have a Greek letter while the constellations have a Latin name. The name of the star in parentheses usually goes back to a proper name of Arabic origin. In the star maps, the Latin names of the constellations are given with their translation in parentheses. On the right pages, the information next to the photo gives the catalog number of the celestial object (e.g., M42; see also p. 144–145) and its proper/special name (e.g., Great Orion Nebula), if it has one. In addition, it mentions the type of object (e.g., cluster, nebula, etc.) and the following astronomical data:
- Right ascension (RA)—measured in hours (h) and minutes (m); see page 20;
- Declination (Dec.)—measured in arc-degree (°) and arc-minutes (′); see page 20. RA and Dec. refer to the equatorial coordinates in the celestial sphere.
- Apparent brightness—a relative value for the magnitude of a star (abbreviated m); see page 139.
- In the case of variable stars (as opposed to fixed stars, which, because they are farther away, do not appear to move)—the period measured in days (d).
- In the case of double stars and other interesting objects—the distance in arc-seconds (″).

Each spread describe the objects (left page) and give tips for observing them with a telescope or binoculars (right page). A number of technical terms concerning astronomical objects, technical procedures, and instruments, as well as foreign words and abbreviations, are briefly explained in the glossary on page 169ff.

A field shot of the two open clusters Hyades (bottom left) and Pleiades (top right), whose brightest stars can be well seen with the naked eye. The bright, reddish star on the photo is Aldebaran, the main (α) star of the constellation Taurus.

1 Andromeda and Cassiopeia

In 1923, Edwin Powell Hubble determined for the first time the distance of the Andromeda Nebula, which proved that this spiral nebula was really an independent galaxy.

Visibility At midnight over the southern horizon (mean local time): October. The stars in the section of the map stand each month about 2 hours earlier over the southern horizon: November 22^h, December 20^h.

Andromeda (Andromeda) As a continuation of the Pegasus square (see p. 68), this constellation, with its four brighter stars that form a slightly bent chain, is a striking object in the northern sky. The constellation is only partially circumpolar, i.e., visible throughout the entire year.

Cassiopeia (Cassiopeia) A circumpolar constellation (for latitudes only north of 50°), whose W-shaped formation of the brighter stars stands out in the sky. It is between the constellations Andromeda and Cepheus. This constellation was compared to ancient Greek keys, because the keys often had a W-shape.

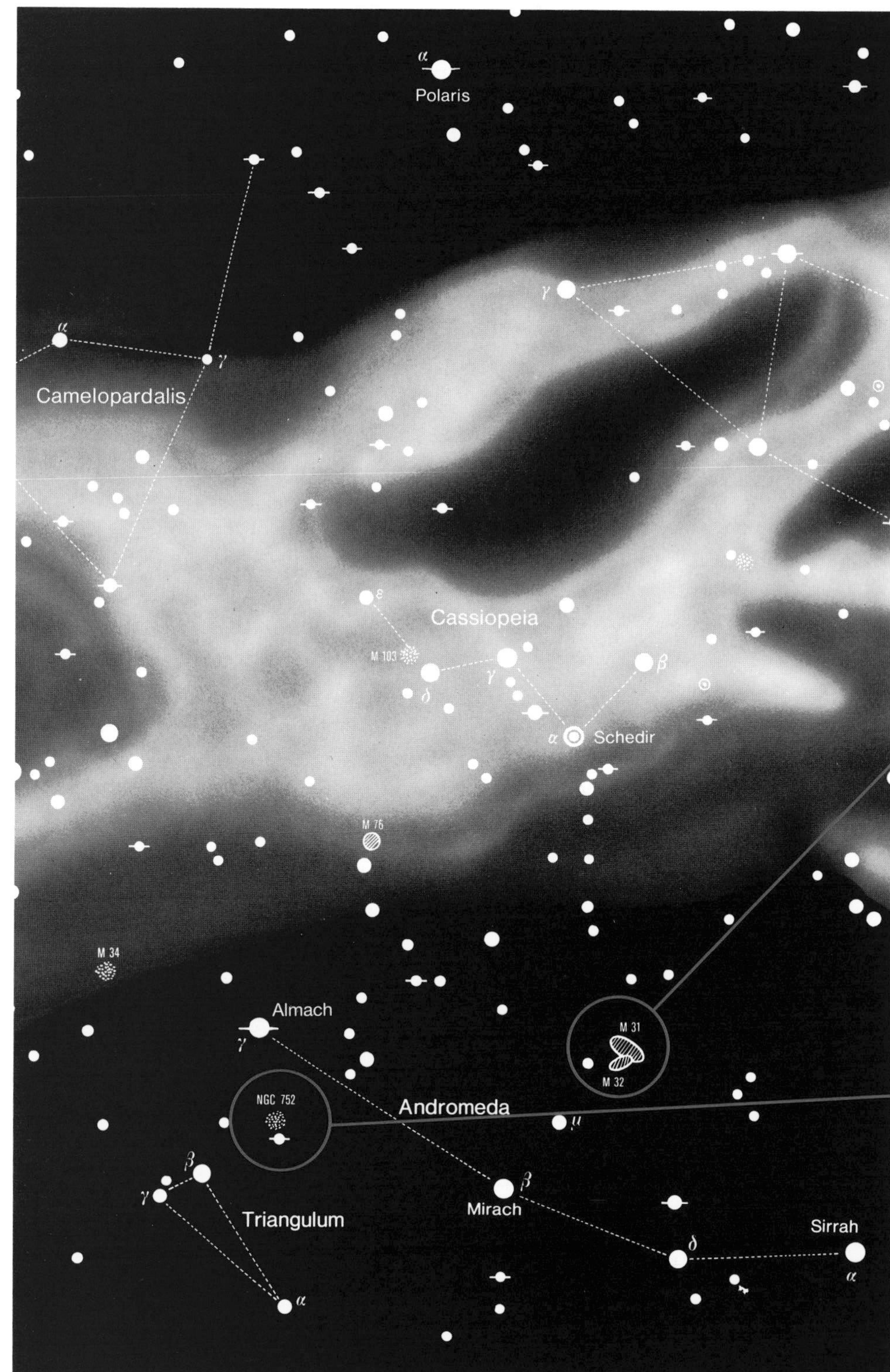

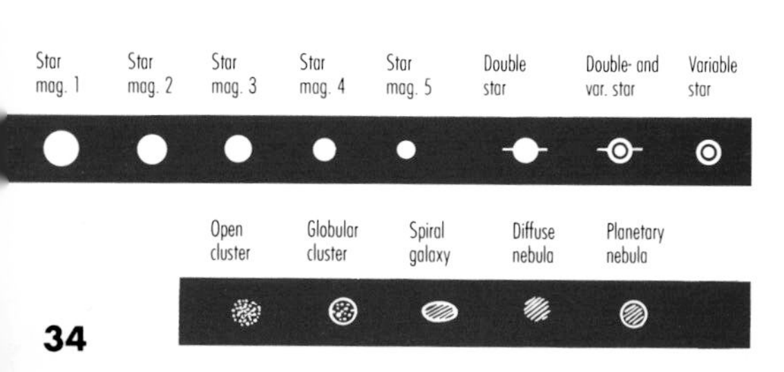

Andromeda Galaxy
= M31 = NGC 224.
Spiral galaxy;
RA 0^h42^m7;
Dec. $+41°16'$; 4^m.

The Andromeda Galaxy is one of the closest galaxies. It is faintly visible to the naked eye. The characteristic form is clearly recognizable, as shown in the photo.

NGC 752
Open cluster;
RA 1^h57^m8;
Dec. $+37°41'$; 5.7^m.

The open cluster NGC 752 is seen in the picture directly to the right of the center of the picture (arrow). The loose assembly of stars is unmistakable. The object is still within the boundaries of the constellation Andromeda.

Observation Tips for the Andromeda Galaxy

The most that you can see with an amateur telescope (3–4 inch aperture) is the bright core. It appears stronger, but you still do not recognize anything of the actual form. Only astro-photography can capture the spiral arms and individual stars (see photo on p. 30). At a distance of 1.5 light-years away from earth, the Andromeda Galaxy is the biggest neighboring galaxy. Based on its dimensions and its structure, it has a lot in common with our Milky Way Galaxy, which makes it a favorite object to research. With the naked eye, the galaxy can be seen as a longish bright spot in the dark night. Search for it starting at the stars β and μ Andromedae. The connecting line of the two stars leads straight to the Andromeda Galaxy.

Observation Tips for NGC 752

Look for it beginning at the bright star γ Andromedae. From there, go south in direction of the star β Trianguli. Using wide-angle binoculars (with a visual field of about 7°), the open cluster appears together with the star β Trianguli. The two brighter stars to the left at the edge of the picture are the stars 59 and 58 Andromedae, which lie on the line γ Andromedae after β Trianguli and can be used as the base for the star cluster.

NGC 752 is on the outskirts of the Milky Way. Although it is a striking object to observe through a pairs of binoculars or a small telescope, you cannot see it with the naked eye, even under the best conditions. The star cluster has the seeming brightness 5.7m and an apparent diameter of 45 arc-minutes (full moon 30).

The Bright Stars

β Andromedae (Mirach); RA 1^h09^m7; Dec. $+35°37'$; 2.4^m.

γ Andromedae (Almak); RA 2^h03^m9; Dec. $+42°20'$; 2.2^m.

α Cassiopeiae (Schedir); RA 0^h40^m5; Dec. $+56°32'$; 2.5^m.

β Cassiopeiae (Caph); RA 0^h09^m2; Dec. $+59°09'$; 2.4^m.

2 Camelopardalis and Perseus

The constellation Perseus sits next to his wife, Andromeda, and her mother, Cassiopeia.

Visibility Around midnight over the southern horizon (mean local time): November. The stars in the map section stand every month about 2 hours earlier above the southern horizon: December 22^h, January 20^h.

Camelopardus (Giraffe) A constellation that is circumpolar in medium latitudes—thus, visible throughout the entire year. It lies inconspicuously between the constellations Cassiopeia and Auriga.

Perseus (Perseus) A constellation that is partially circumpolar in medium northern latitudes. It is easily found between the constellations Andromeda (with the famous spiral galaxy) and Auriga. The name Perseus goes back to the myths of the ancient Greeks.

Star mag. 1	Star mag. 2	Star mag. 3	Star mag. 4	Star mag. 5	Double star	Double- and var. star	Variable star

Open cluster	Globular cluster	Spiral galaxy	Diffuse nebula	Planetary nebula

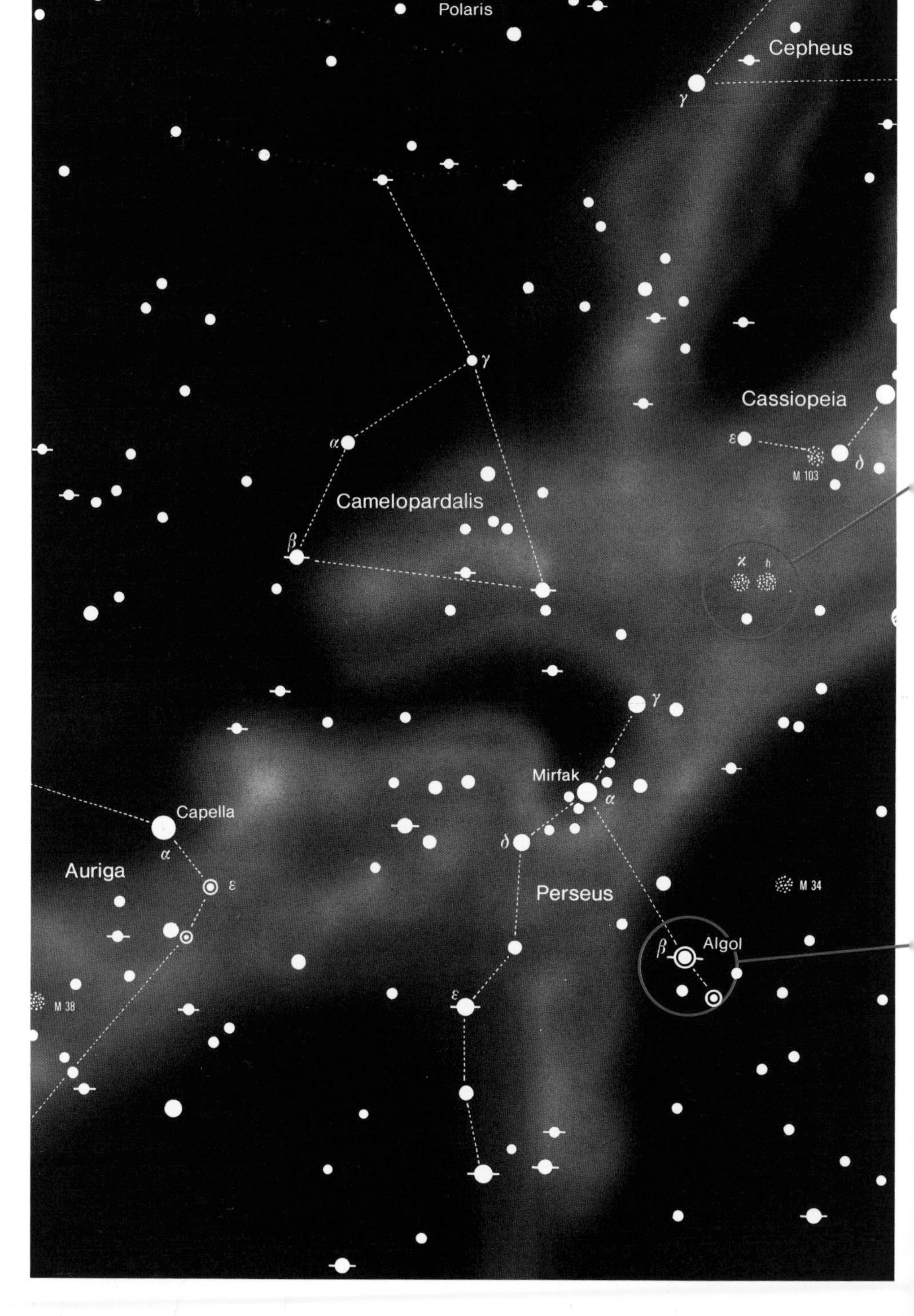

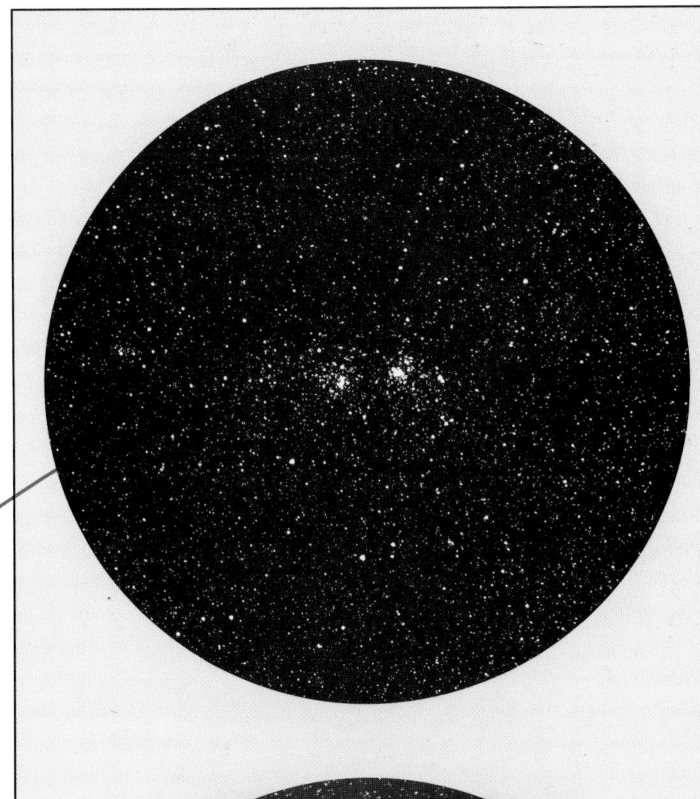

NGC 869 and 884
= h and χ Persei
Two open clusters (the double cluster);
h Persei; RA 2^h19^m0;
Dec. +57°09′;
4.3^m;
χ Persei: RA 2^h22^m4;
Dec. +57°07′;
4.4^m.

The two clusters, standing close to each other, appear to the naked eye as a milky, fog-like spot. They are in the center between the bright stars γ Persei and δ Cassiopeiae.

Algol
= β Persei.
Eclipsing variable star (two stars in mutual orbit);
RA 3^h08^m2;
Dec. +40°57′;
Max. 2.12^m, Min. 3.40^m, period 2.86^d.
Regularly in an interval of 2.86 days, the apparent brightness of the stars varies between 2.1 and 3.3 mag.

Observation Tips for NGC 869 and 884

Those two open clusters (double cluster) unfold their real splendor only through binoculars or an astronomical telescope whose enlargement is not too high (20–40 times). The color differences between the two clusters can be seen under good conditions. There are over 600 stars in both clusters together that are up to the 15th magnitude.

This double cluster is indicated by the experts as being 3 million years old. Astronomically, that means that it is very young! In this section, the left cluster is the object χ Persei and the right one is the object h Persei.

Observation Tips for Algol

With some practice, you can easily see the variations in brightness even without a telescope. Algol is, so to speak, the prototype of an entire class of stars—double stars or eclipsing binary stars. Today, we known that the variations in brightness are due to the movement mechanism of two stars that circle around each other. When the two stars are standing next to each other, the overall brightness is higher than when one star covers the other one. The naked eye does not see anything of the double star, but it registers the change in brightness. Since Algol is such a prominent representative of its type, the time indications of its light changes are published in yearbooks.

Algol is the bright star in the center in the section. It forms, together with the stars α Persei and γ Andromedae, a triangle that is approximately at a right angle.

The Bright Stars

α Persei (Algenib); RA 3^h24^m3; Dec. +49°52′; 1.8^m.

β Persei (Algol); RA 3^h08^m2; Dec. +40°57′; $2.1–3.4^m$ (double star–variable).

3 Auriga

Auriga is often depicted on stars maps as a figure carrying a goat on his back. The ancient Greeks associated the bright star Capella with Amalthes—the goat that nursed the infant Zeus—which was an independent constellation.

Visibility Around midnight just over the southern horizon (mean local time): December. The stars in the section of the map stand each month about 2 hours earlier above the southern horizon: January 22^{h}, February 20^{h}.

Auriga (The Charioteer) A prominent, partially circumpolar constellation that can be found to the north between the constellations Taurus and Gemini. The hexagonal shape, which the bright stars give to this constellation, helps us find it. The southernmost of these six stars belongs to the constellation Taurus (β Tauri). Almost directly on the connecting line between the star α Ursae majoris and α Tauri (the reddish bright Aldebran), lies the main star Capella (the "little she-goat"). Capella is the sixth brightest fixed star (after Sirius, Canopus, α Centauri A, Vega, and Rigel) that we can observe. In mid-northern latitudes, the star is circumpolar.

Star mag. 1 | Star mag. 2 | Star mag. 3 | Star mag. 4 | Star mag. 5 | Double star | Double- and var. star | Variable star

Open cluster | Globular cluster | Spiral galaxy | Diffuse nebula | Planetary nebula

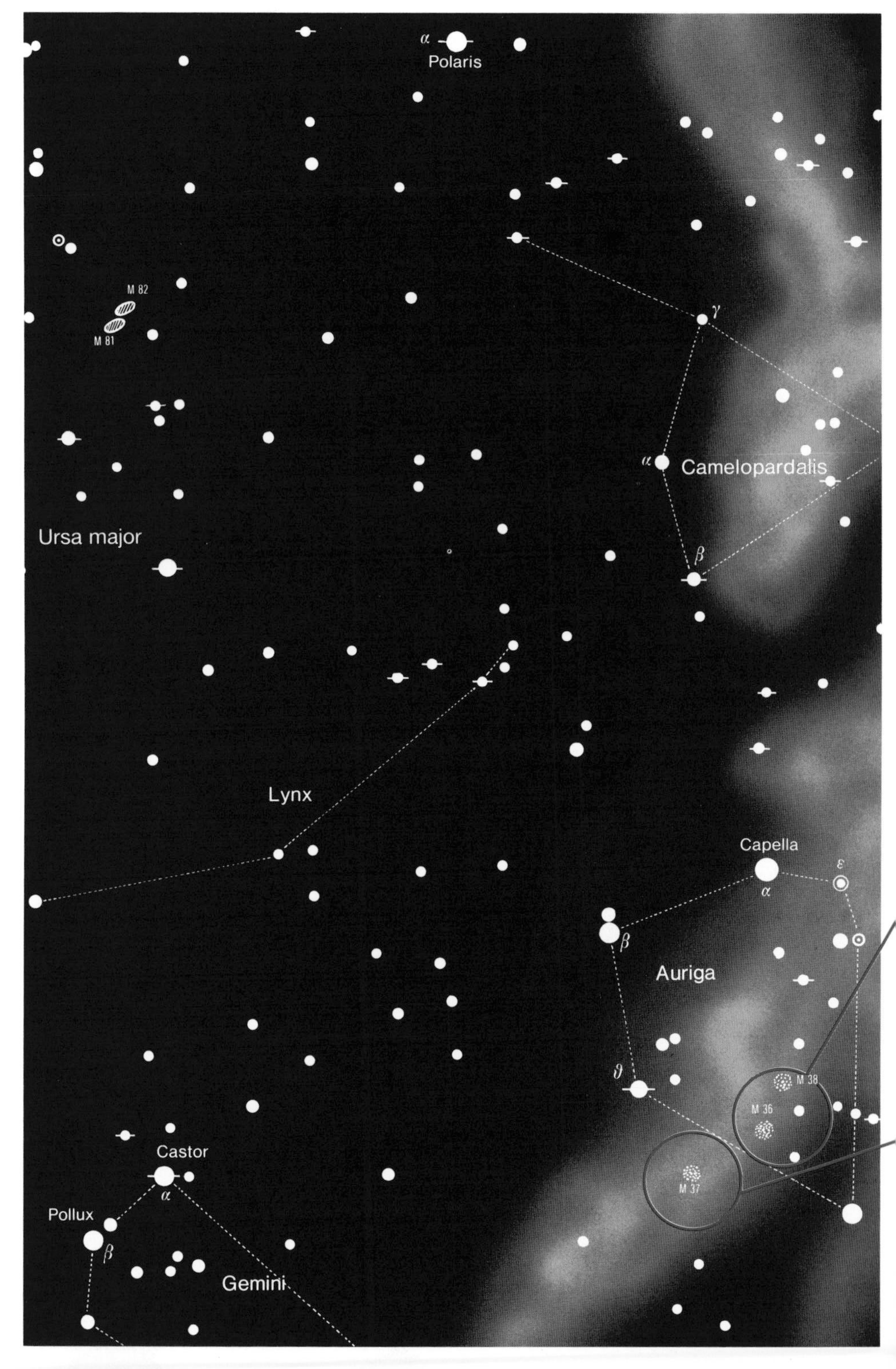

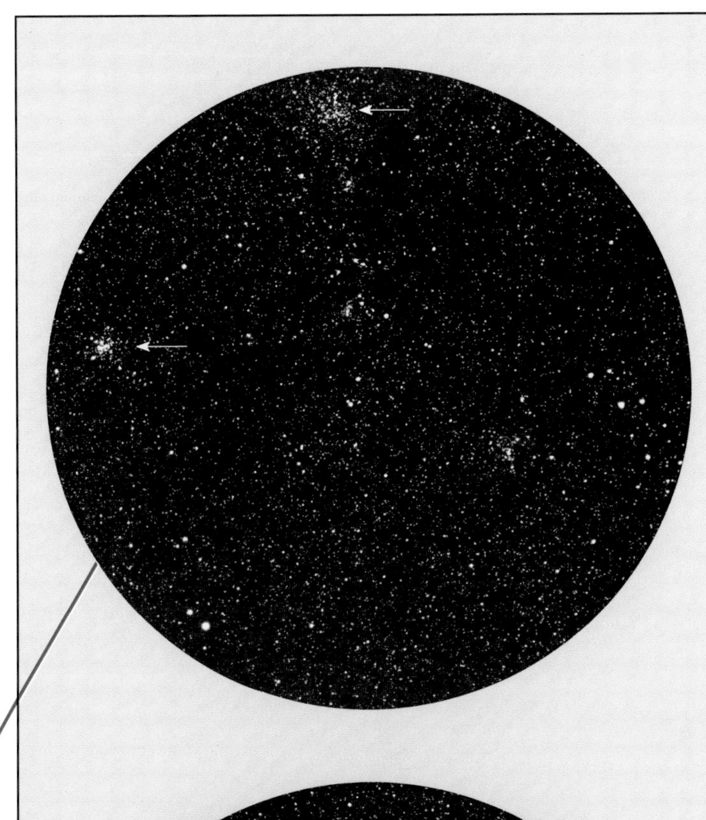

M36 and M38
M36 = NGC 1960
Open cluster;
RA 5^h36^m1;
Dec. +34°08′;
6.0^m.
M38 = NGC 1912
Open cluster;
RA 5^h28^m7;
Dec. +35°50′;
6.4^m.
M36 is on the left side of the picture and M38 is all the way at the top edge (see arrows). Below M38 is another open cluster, NGC 1907.

M37
= NGC 2099.
Open cluster;
RA 5^h52^m4;
Dec. +32°33′;
5.6^m.
The area of the Milky Way in the southern part of the constellation Auriga is rich in open clusters. M37 deserves special attention, because this open cluster has about 150 stars over an area about a third of a degree across.

Observation Tips for M36 and M38

The star φ Aurigae with a distinct grouping of stars is also shown in the center right of this section. The field, as a whole, lies along the Milky Way. The connecting line between the stars ϑ Aurigae and υ Aurigae helps you look for it: In the middle, between the two stars, is M36. Using wide-angle binoculars (see p. 160), locate M36 in the center of your visual field. Then you will also find M38 and M37 in the same visual field, since the distance between M38 and M37 is only about 7°. M37 is presented in detail below. For M38, one notices the special richness of stars.

Observation Tips for M37

Use a 3-inch telescope (with a 30–40 times enlargement) for the resolution into individual stars. H. Vehrenberg points out in his *Messier Book:* "Near the center is a striking orange-colored star." There are about 150 individual stars. They are about 200 millions years old and 1450 parsec away.

In order to find M37, locate M36 first, as described above. It is about 4° away from M37. That means you can get both objects at the same time into your visual field with most binoculars.

The Bright Stars

α Aurigae (Capella);
RA 5^h16^m7;
Dec. +46°00′;
0^m.

β Aurigae;
(Menkalinan);
RA 5^h59^m5;
Dec. +44°57′;
Max. 1.9^m,
Min. 2.0^m;
Period 3.96^d.

4 Ursa Major, Leo Minor, and Lynx

In this section of the sky, we find fewer stars than usual. The reasons are the construction of the Milky Way and the location of our solar system in it.

Visibility Around midnight over the southern horizon (mean local time): February. The stars in this sky section stand each month about 2 hours earlier above the southern horizon: March 22^h, April 20^h.

Ursa Major (The Great Bear) See page 42 for detailed description. According to some Greek legends, Callisto, a woman desired by Zeus, was turned into a bear by Hera, Zeus' wife, because she was jealous of Callisto. Zeus placed the bear in the sky.

Leo Minor (The Little Lion) An inconspicuous constellation between the constellations Ursa Major and Leo.

Lynx (The Lynx) The area is sparsely populated by stars. The constellation Lynx is circumpolar. You can find its stars southwest of Ursa Major and adjacent to Leo Minor. Near the star ∝ Lyncis is the double star 38 Lyncis. The two stars are only 2.8 arc-seconds away from each other, making it difficult to separate the pair.

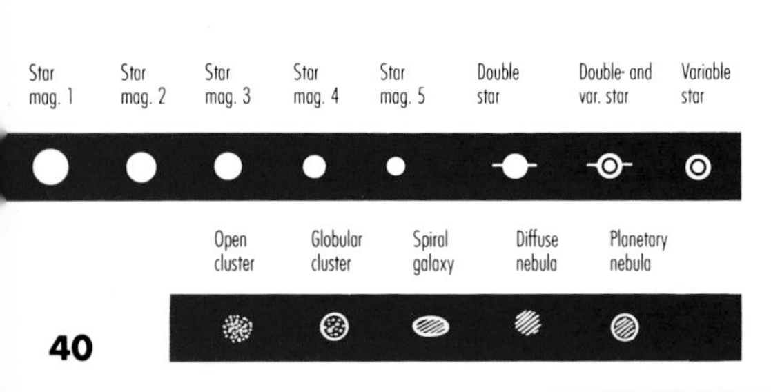

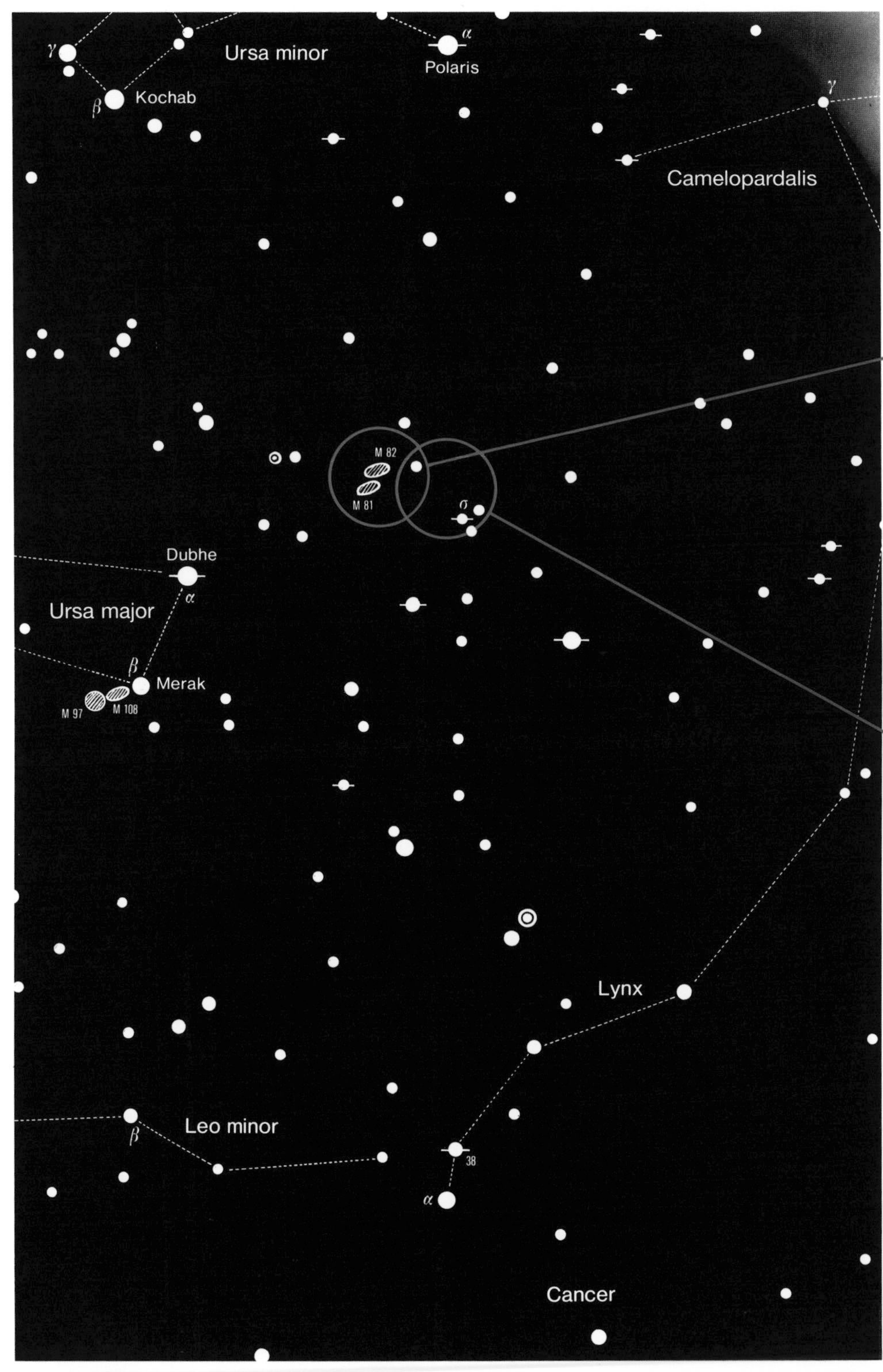

M81 and M82

M81 = NGC 3031.
Spiral galaxy;
RA 9^h55^m6;
Dec. +69°04′;
8^m.
M 82 = NGC 3034.
Irregular galaxy;
RA 9^h55^m8;
Dec. +69°42′;
9^m.
In our section, M82 is the upper and M81 (notice the spiral structure) is the lower object (see arrows).

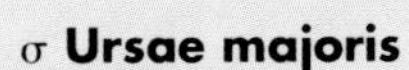

σ Ursae majoris

Double star;
RA 9^h10^m4;
Dec. +67°08′.
The bright star on the bottom right in this section is a double star (see arrow). But at a distance of 3.9 arc-seconds, it is not possible to see the separation in the photo. The brighter star has 4.85^m, the weaker one has 8.16^m.

Observation Tips for M81 and M82

The two galaxies are about equally bright. But in order to recognize them, it is especially important to view them on a dark, moonless night. Even then, with a pair of binoculars or a small telescope, one cannot count on seeing more than two washed-out foggy spots. Photos taken with a large telescope show that M81 has a stunning resemblance to the Andromeda Galaxy (it too is a spiral galaxy). M82 has become known because it is a radio source—it seems to be pulled by its longer neighbor.

Observation Tips for σ Ursae Majoris

Use a 4-inch telescope to see the individual stars of this double star. The larger the brightness difference between the two stars of the double star, the more difficult its becomes to resolve them: The brighter star outshines the accompanying star. Thus, it is possible that a telescope will not resolve the individual stars of this double star even if, by calculation, there is enough distance. Of course, sky conditions also play a role in its visibility.

The Bright Stars

β Ursae majoris (Merak);
RA 11^h01^m8;
Dec. +56°23′.

α Ursae majoris (Dubhe);
RA 11^h03^m7;
Dec. +61°45′.

5 Ursa Major and Canes Venatici

The constellation Ursa Major played a role in many cultures. At the time of the ancient Romans, the seven main stars were called the "Seven Thresh Oxen," which constantly move around the celestial pole.

Visibility At midnight over the southern horizon (mean local time): March. The stars in the map section stand each month about 2 hours earlier above the south-horizon: April 22h, May 20h.

Ursa Major (Great Bear) The constellation is also known as the Big Dipper. Seven stars make up this constellation: four bright stars form the "dipper" and three form the "handle." Weaker stars make up the shape of the bear. The connecting line between the stars Merak and Dubhe points towards the North Pole and North Star.

Canes Venatici (The Hunting Dogs) The brightest stars in this inconspicuous constellation are between the 3rd and 5th magnitudes. This constellation is known especially for the accumulation of spiral galaxies.

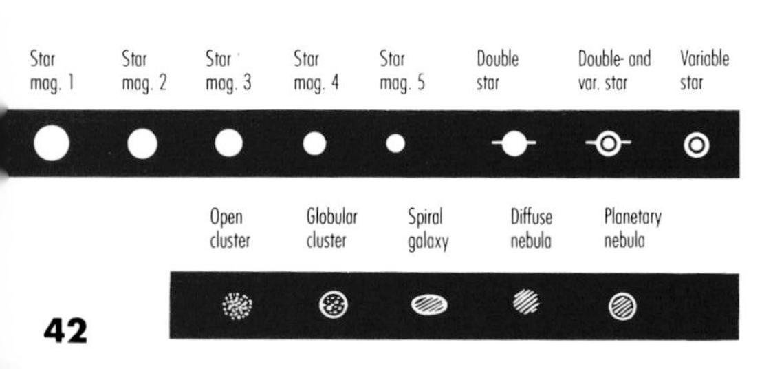

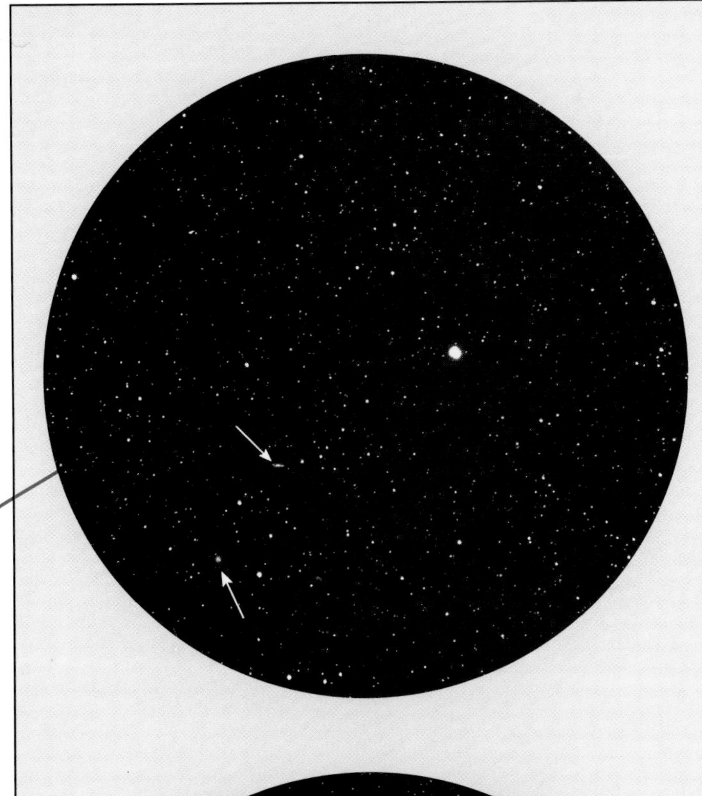

M97 and M108
M97 = NGC 3587
Planetary nebula;
RA 11^h14^m8;
Dec. +55°01′;
12.08^m.
M 108 = NGC 3556.
Spiral galaxy;
RA 11^h11^m5;
Dec. +55°40′;
10.1^m.
You may see these celestial bodies with a 4–5-inch telescope or a pair of 15 × 60 binoculars. M97 is indicated by the bottom arrow; M108 is indicated by the top arrow.

α Canum venaticorum
Double star;
RA 12^h56^m2;
Dec. +38°19′;
2.9^m and/or 5.5^m; 19.4″.
This brightest star in the constellation Canes Venatici (Hunting Dogs) is popularly called Cor Caroli (Charles's heart). Its designation number is 12 (see arrow).

Observation Tips for M97 and M108

A good starting place is the bright star Merak of the constellation Ursa Major (the bright star in the center of this section). If you view this star at the edge of the visual field in your binoculars, it is possible that you may find M97 and M108 at the same time (see the photo with the arrows). You are more likely to see M97, which may not be more than a pale washed-out spot, than M108. The visibility depends a lot on the condition of the atmosphere.

The planetary nebula M97 is also called "Owl Nebula" (more about planetary nebulae on pages 49 to 147). The apparent magnitude of M97 is 12.0^m, while M108 is 10.0^m.

Observation Tips for α Canum Venaticorum

It is easy to find this double star with a 2-inch astronomical telescope. The distance between the pair is almost 20 arc-seconds. The difference between the two stars is 2.9^m to 5.5^m.

The brighter of the two stars is at the same time also a variable star, whose seeming brightness varies about a tenth of a magnitude. The length of the period of the magnitude change (e.g., the time from one maximum to the next) is 5.47 days. This star belongs to the class of pulsating variable stars, which are stars that expand and contract.

The Bright Stars

γ Ursae majoris (Phecda); RA 11^h53^m8; Dec. +53°42′; 2.4^m.

α Ursae majoris (Alioth); RA 12^h54^m0; Dec. +55°58′; 1.8^m.

η Ursae majoris (Benatnasch); RA 13^h47^m5; Dec. +49°19′; 1.9^m.

6 Boötes and Ursa Minor

Constellations often served as navigational guides. The Phoenician navigators preferred the constellation Ursa Minor while the Greeks preferred Ursa Major.

Visibility At midnight over the southern horizon (mean local time): May. The stars in the map section stand each month about 2 hours earlier above the southern horizon: June 22^{h}, July 20^{h}.

Boötes (The Herdsman) In Greek mythology, Boötes represents Arcas, the son of Zeus and Callisto. On this map, we only see the northern part of the constellation (more about Boötes on map 15, page 62).

Ursa Minor (The Little Bear) The circumpolar constellation has the star Polaris, α Ursae minoris. It is only somewhat less than 1° away from the northern celestial pole. The distance to the North Pole does not remain the same; at this time, the star is approaching the pole (it will be closest to the pole in about the year 2105). Polaris is a cepheid variable star (a supergiant star that varies in brightness from as little as a few tenths of a magnitude to as much as 2nd magnitude).

Star mag. 1	Star mag. 2	Star mag. 3	Star mag. 4	Star mag. 5	Double star	Double- and var. star	Variable star

Open cluster	Globular cluster	Spiral galaxy	Diffuse nebula	Planetary nebula

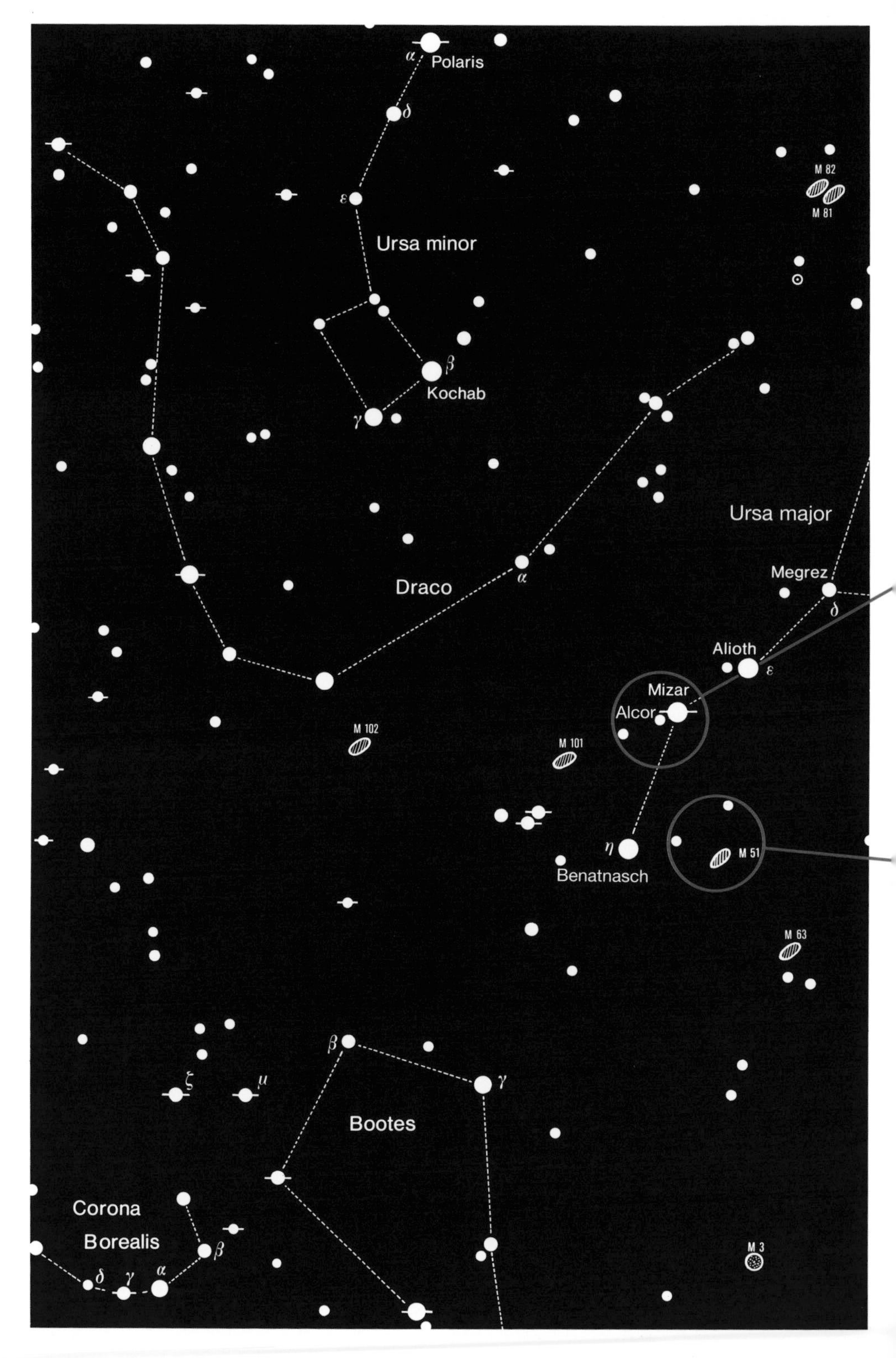

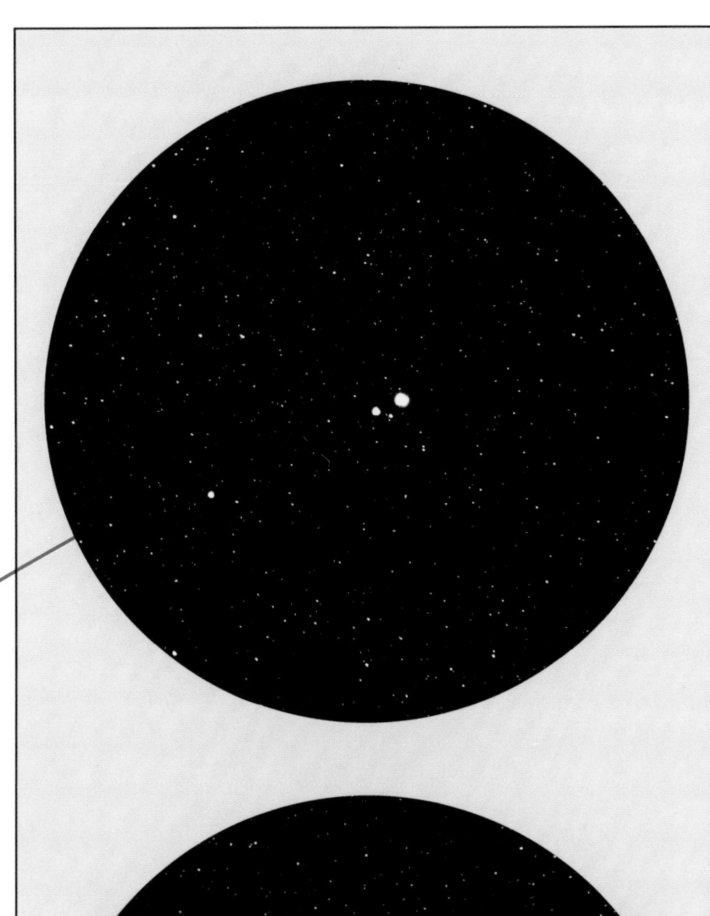

Mizar-Alcor
ζ Ursae majoris (Mizar) and 80 Ursae majoris (Alcor).
Optical double star system;
RA 13^h23^m6;
Dec. +54°55′;
2.3^m and/or 4.0^m;
708.7″.
This object could have been described along with the constellation Ursa Major. The sections, nevertheless, make the shift necessary.

M51
= NGC 5194.
Spiral galaxy;
RA 13^h29^m9;
Dec. + 47°12′;
8^m.
Strictly speaking, this constellation belongs to Canes Venatici (see p. 42). It is one of the most distinct spiral galaxies of the northern sky.

Observation Tips for Mizar-Alcor

Mizar is part of the handle of the Big Dipper. At a distance of 11 arc-minutes stands Alcor, the little rider, which is a star of the 4th magnitude. Anyone who has very good eyes can just manage to separate the two stars without the help of a telescope!

The brighter star, Mizar, is the first double star to be discovered telescopically. It is 2.3^m to 4^m bright. To separate this double star, you need powerful binoculars (22-fold enlargement) or, better, a 2-inch astronomical telescope. The distance is 14 arc-seconds.

Observation Tips for M51

This bright object in the 8th magnitude was discovered by the Frenchman Charles Messier, whose name is connected with the catalog of over a hundred star clusters, nebulae, and galaxies.

Through a small telescope, M51 looks almost like a washed-out double star, because of two brighter cores of the spiral galaxy. Very close to it (to the north) is another galaxy (NGC 5195), about 10.5^m bright. It is a so-called "background galaxy," because it lies behind M51.

H. Vehrenberg recommends the following rule for finding M51: "Divide the connecting line from η Ursae to α Canum Venaticorum into three parts and look for M51 at the ends of the first third." By the way, once you have made the line, look at the beginning of the third third for M63—a galaxy with a relative brightness of 9.5^m.

The Bright Stars

γ Boötis (Seginus);
RA 14^h32^m5;
Dec. +38°13′;
3.2^m.

α Ursae minoris (Polaris);
RA 2^h31^m8;
Dec. +89°16′;
2.0^m (variable).

β Ursae minoris (Kochab);
RA 14^h50^m7;
Dec. +74°09′;
2.0^m.

7 Draco and Hercules

The dragon personifies anarchy and chaos in the world of the legends of antiquity. The star α Draconis Thuban was the pole star 5000 years ago.

Visibility At midnight over the southern horizon (mean local time): June. The stars in the map section stand each month about 2 hours earlier above the southern horizon: July 22^h, August 20^h.

Draco (The Dragon) The circumpolar constellation stretches to the north of the constellation Hercules between Ursa Major and Ursa Minor. The constellation, which never sets in northern latitudes, does not stand out very much. The stars beta and gamma are seen as the wide open jaw of the dragon.

Hercules (Hercules) Only the northern half of this partially circumpolar constellation is shown here on the map. In it, we find two known globular clusters: M13 and M92. For the southern half of this constellation, see map 16 on page 64. The star η Draconis is a double star that could be seen with a 4-inch telescope. Even though the distance between the two stars is about 5", this star is difficult because of its brightness variation (2^m and $7–8.7^m$).

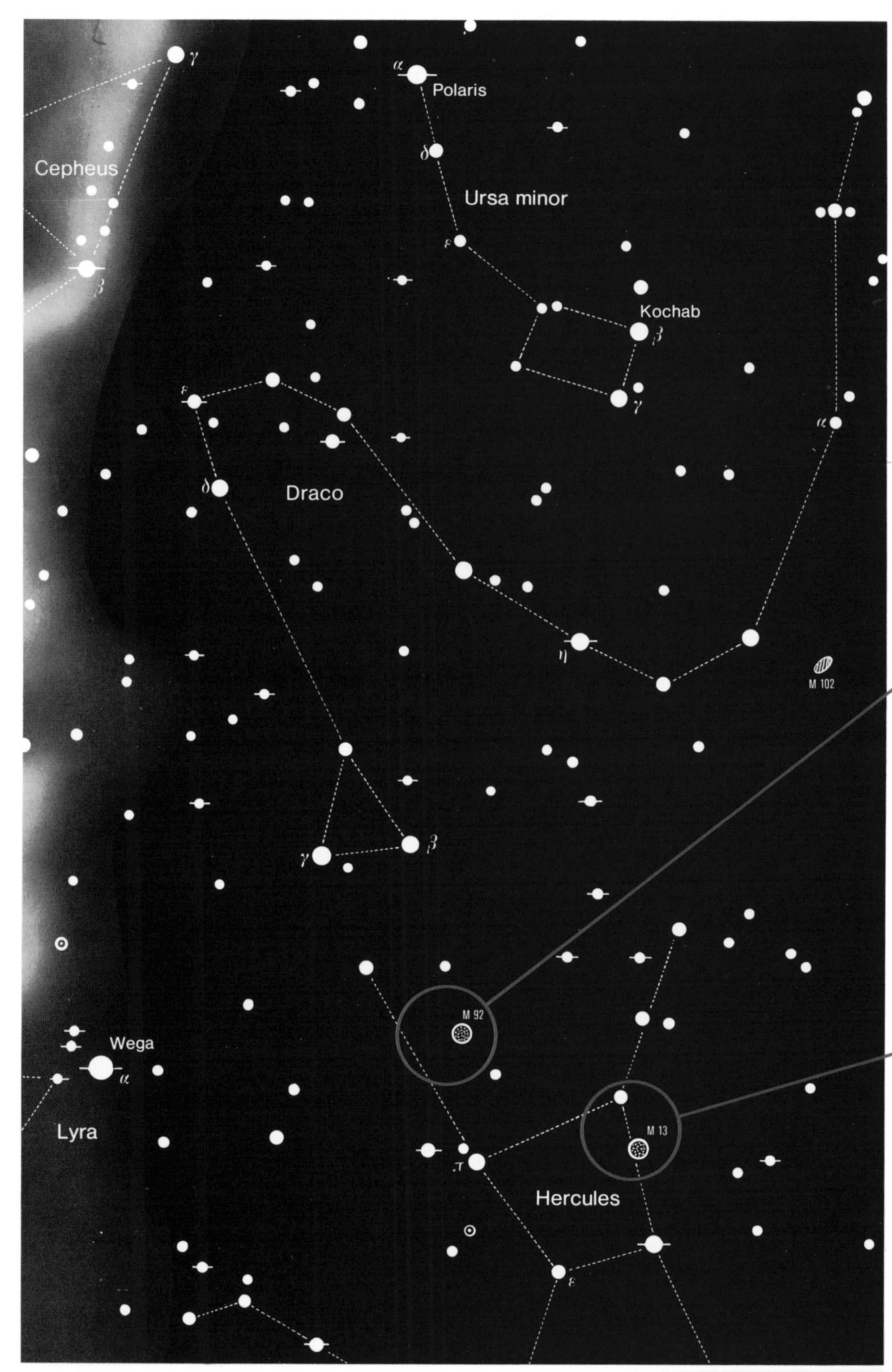

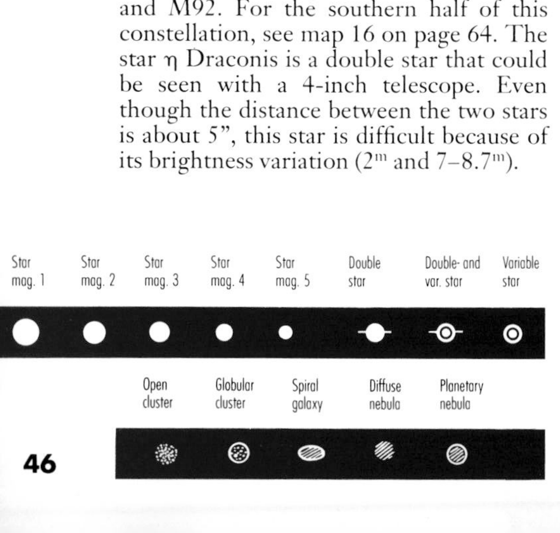

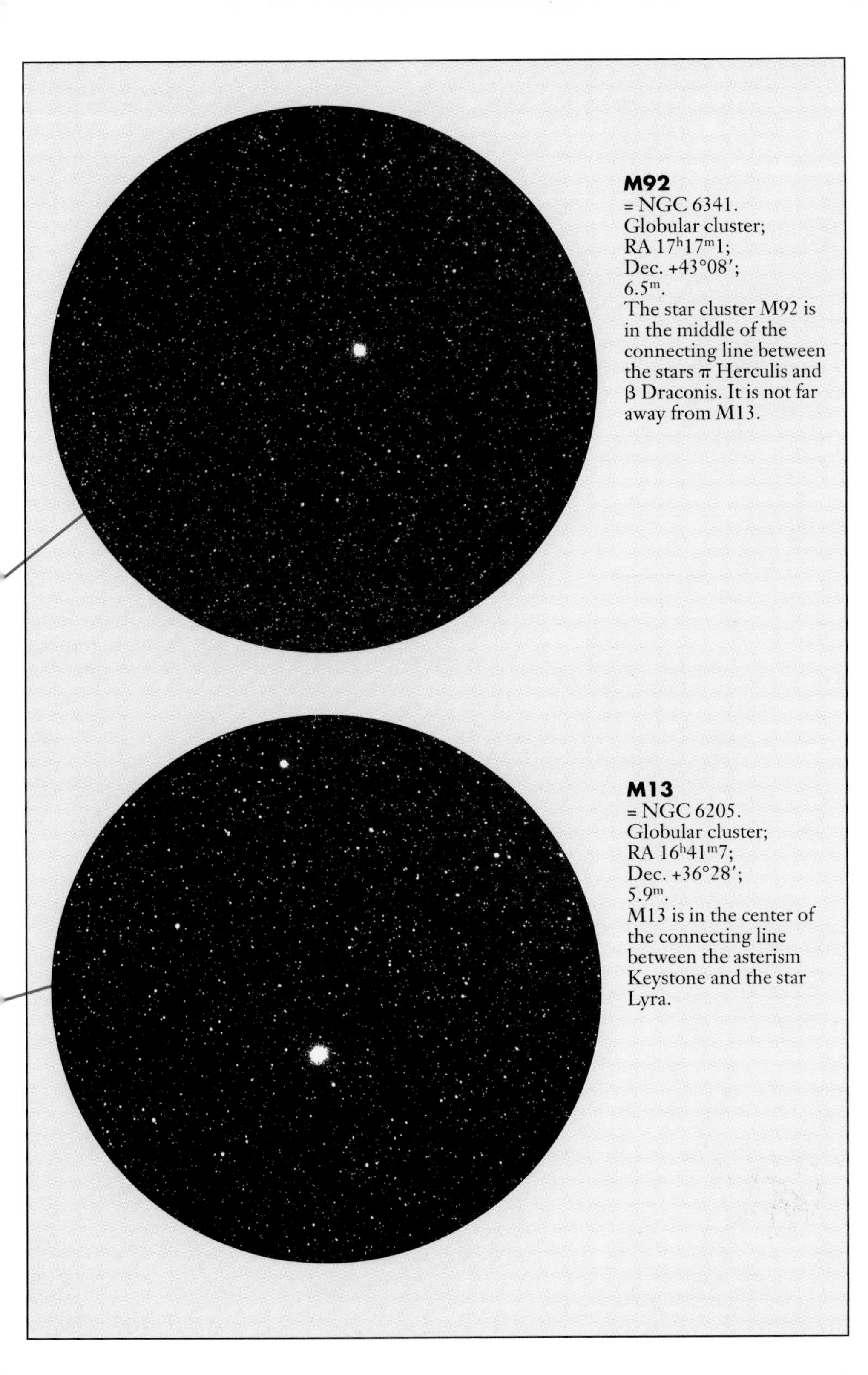

M92
= NGC 6341.
Globular cluster;
RA 17^h17^m1;
Dec. +43°08′;
6.5^m.
The star cluster M92 is in the middle of the connecting line between the stars π Herculis and β Draconis. It is not far away from M13.

M13
= NGC 6205.
Globular cluster;
RA 16^h41^m7;
Dec. +36°28′;
5.9^m.
M13 is in the center of the connecting line between the asterism Keystone and the star Lyra.

Observation Tips for M92

Despite its apparent brightness (6.5^m) and apparent diameter (8 arc-minutes), M92 is still overshadowed by its more famous neighbor, M13. M92 is about 1 magnitude fainter. Since you cannot see anything with the naked eyes, you need at least a pair of binoculars. But in order to resolve the individual stars at the edge, you will need a 4-inch telescope. M92 is smaller and more condensed than M13. Therefore, it is very difficult to resolve the concentration of the stars in the center of this globular cluster.

Observation Tips for M13

This globular cluster is only a few degrees away from the star η Herculis. You can see M13 with the naked eye, although it may appear as a fuzzy star of the apparent 6th magnitude.

M13 has the considerable apparent diameter of 23 arc-minutes and, thus, is not only the most beautiful, but also the brightest globular cluster of the northern sky. Looking at it through a pair of binoculars will show an intense group of stars.

A 3–4-inch telescope (with enlargements between 30- and 50-fold) can resolve the outer edges of M13 into its member stars. The "fuzzy star" becomes more like a crown in the sky, and the number of the individual stars begins to grow. Over 40,000 individual stars have been counted from outside of the center (see photo on p. 146).

The Bright Stars

β Draconis (Rastaban);
RA 17^h30^m4;
Dec. +52°18′;
2.8^m.

γ Draconis (Eltanin);
RA 17^h56^m6;
Dec. +51°29′;
2.2^m.

8 Lyra and Cygnus

The bright star Vega (α Lyrae) in the constellation Lyra might be the birthplace of the solar system's planets. Infrared observations show that the star is brighter than expected and it is surrounded by a ring of cool material, out of which planets could have come about.

Visibility At midnight over the southern horizon (mean local time): July. The stars in the map section stand each month about 2 hours earlier above the southern horizon: August 22^{h}, September 20^{h}.

Lyra (The Lyre) A very impressive, small constellation, which is partially circumpolar. The main star Vega, the fourth brightest star of the sky, creates, together with Deneb in the constellation Cygnus and Altair in the constellation Aquila, the Summer Triangle—the striking figure of an almost isosceles triangle at the northern summer sky.

Cygnus (The Swan) This striking, partially circumpolar constellation is between Lyra and Pegasus. The arrangement of the bright stars forms a swan flying. The bill of the swan is the star β Cygni (Albireo), which is a double star that can be seen through large binoculars.

Star mag. 1 · Star mag. 2 · Star mag. 3 · Star mag. 4 · Star mag. 5 · Double star · Double- and var. star · Variable star

Open cluster · Globular cluster · Spiral galaxy · Diffuse nebula · Planetary nebula

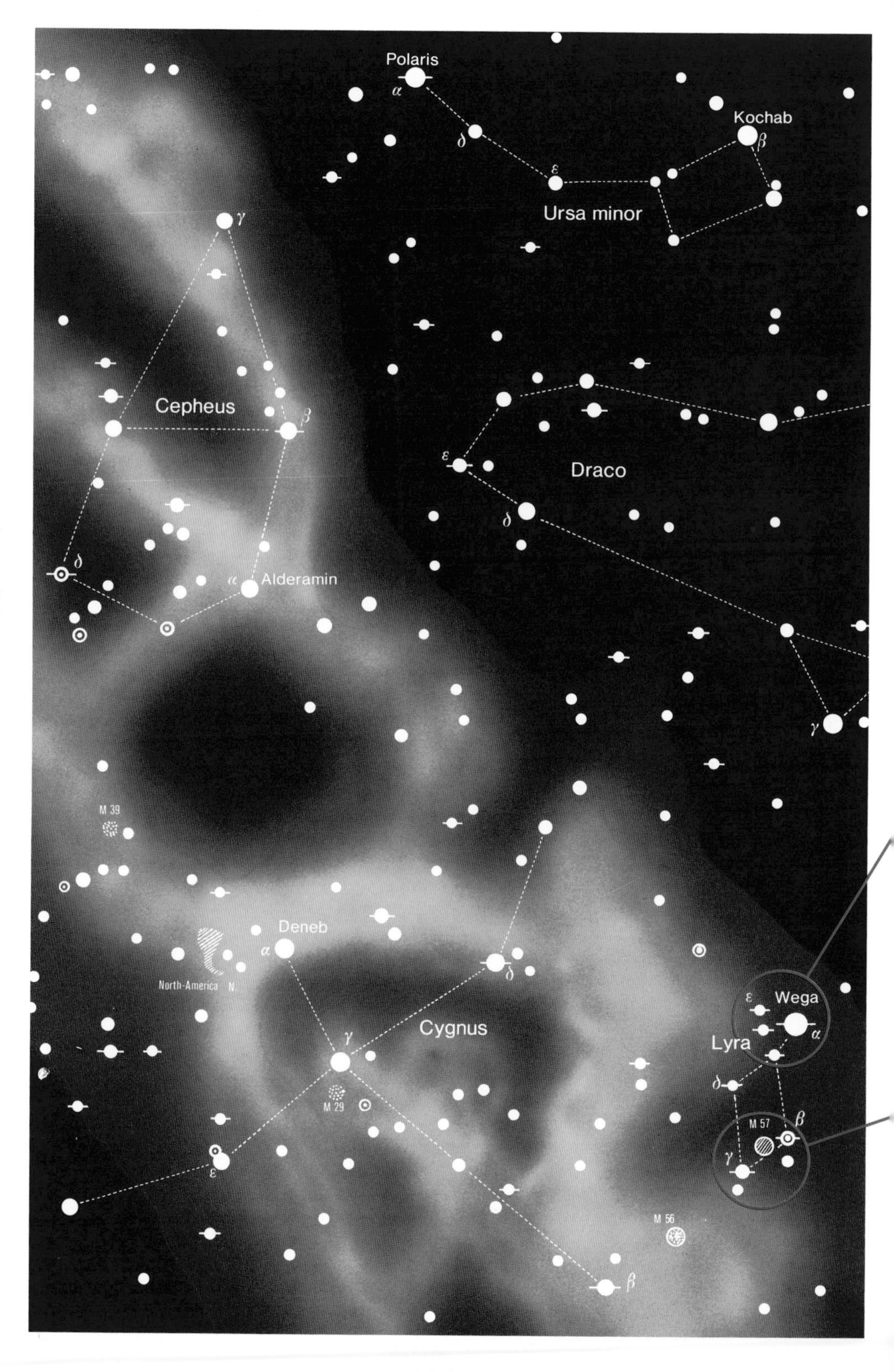

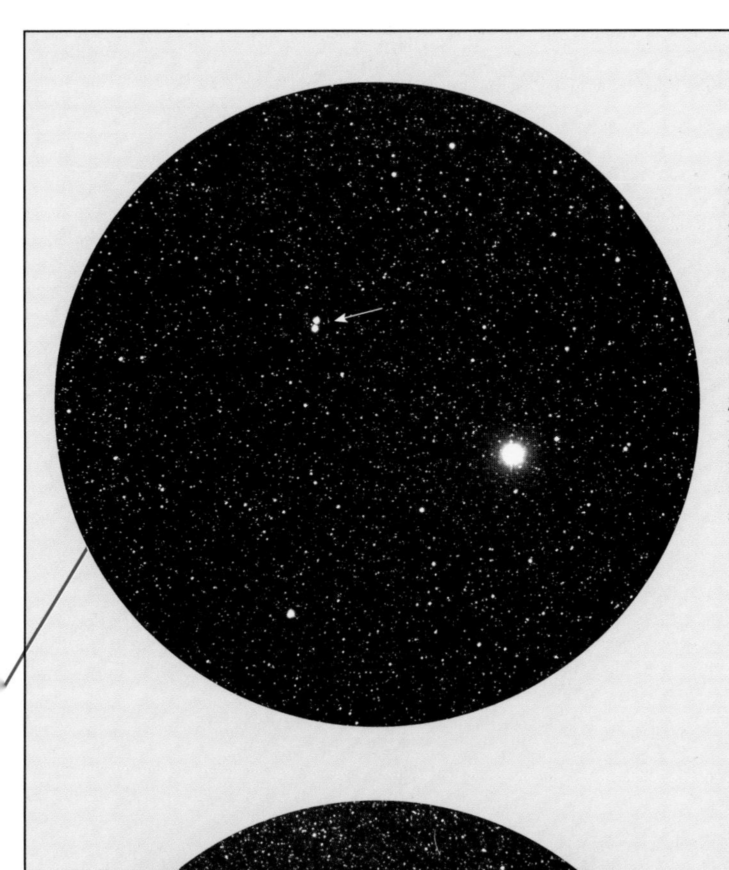

ε Lyrae
Double star;
ε^1: RA 18^h44^m3;
Dec. +39°40′;
5.00^m and/or 6.1^m;
2.6″.
ϵ^2: RA 18^h44^m4;
Dec. +39°37′;
5.2^m and/or 5.5^m;
2.3″.
This double star of about the 4th magnitude can be found 2° to the left of Vega (see arrow).

Ring Nebula in Lyra
= M57 = NGC 6720.
Planetary nebula;
RA 18^h53^m6;
Dec. +33°02′;
9.7^m.
Two bright stars stand out near the center of the picture and at the edge on the top right: γ Lyrae and β Lyrae. They are at about 2.5° apart. In the middle of the distance, closer to β Lyrae, is the planetary nebula M57 (see arrow).

Observation Tips for ε Lyrae

Sharp eyes can separate about 200 arc-seconds distance without the help of an optical instrument. In this section, the double star ϵ Lyrae is to the left of the center of the picture. ϵ Lyrae is special in that each star of the double star is made up of a double star. In order to see that, you need a 3-inch telescope. Binoculars are not enough.

The brighter star to the bottom left in the picture is ζ Lyrae. The star forms with ϵ Lyrae and Vega an almost equilateral triangle in the sky. ζ Lyrae is also a double star with the distance of 44 arc-seconds. You can resolve the individual stars with small binoculars of 8-fold enlargement. By the way, if you look closely at the photo, you will recognize that the little star dot is egg-shaped. This indicates the duplicity.

Observation Tips for the Ring Nebula in Lyra

M57 is better known as the Ring Nebula in Lyra. With binoculars, you do not see much more than a hazy patch of the 9th magnitude. The photo in this section does not offer us much more than that. Through a 3-inch telescope, the "little star" changes into a tiny, pale-looking disk. The typical ring-form only becomes visible with a 6–8 inch telescope. Very large telescopes show the central star (15th mag.), which makes the gas of the nebula shine with the help of its ultraviolet radiation.

By the way, the star β Lyrae is the prototype of a class of eclipsing variable stars (period: 12.9 days; max. 3.3^m, min. 4.2^m). It has three companions with 7m (distance 46″), 9^m (67″), and 9^m (86″).

The Bright Stars

α Cygni (Deneb);
RA 20^h41^m7;
Dec. +45°17′;
1.3m.

γ Cygni (Sadir);
RA 20^h22^m2;
Dec. +40°15′;
2.2^m.

α Lyrae (Vega);
RA 18^h36^m9;
Dec. +38°47′;
$0.0–0.03^m$.

9 Lacerta and Cepheus

In 1784, John Goodricke discovered the pulsating variable star δ Cephei. Since then, astronomers know over five hundred stars of this type, whose pulsation period is so precise that you can measure distances in outer space in relationship to their magnitude and pulsation period.

Visibility At midnight over the southern horizon (mean local time): August. The stars in the map section stand each month about 2 hours earlier above the southern horizon: September 22^{h}, October 20^{h}.

Lacerta (The Lizard) The constellation corresponds to a star chain pointing north-south between the constellations Cygnus and Andromeda.

Cepheus (The King) This constellation represents the mythological King Cepheus of Ethiopia, who is the husband of Cassiopeia and the father of Andromeda, both of which are well-known neighboring constellations. Cepheus can be found in the middle between the constellations Cygnus, Cassiopeia, and Ursa Minor. This constellation became well known in the astronomical science through the star Delta, a classical variable star. Some parts of the Milky Way, rich in stars, belong to this constellation.

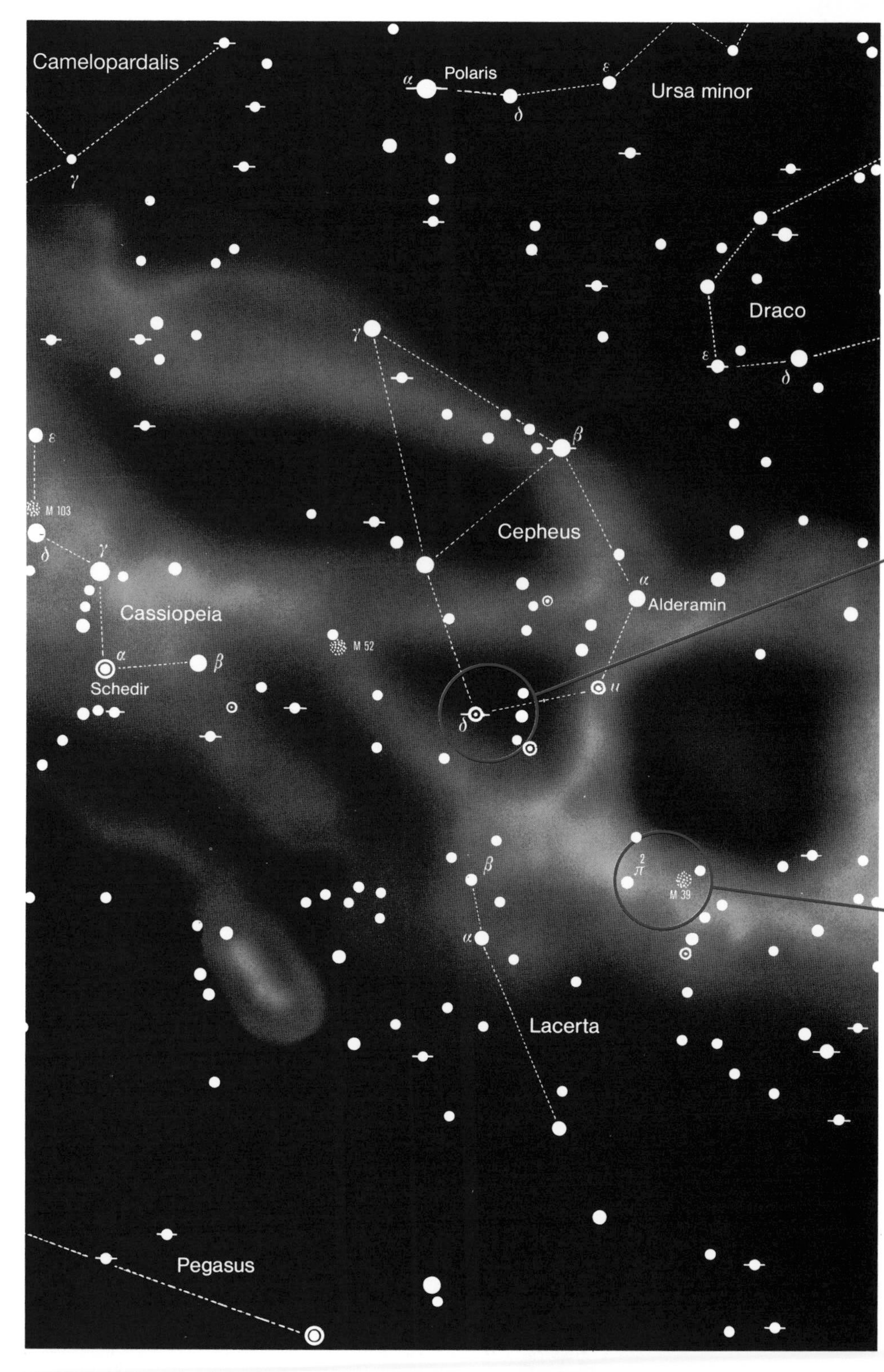

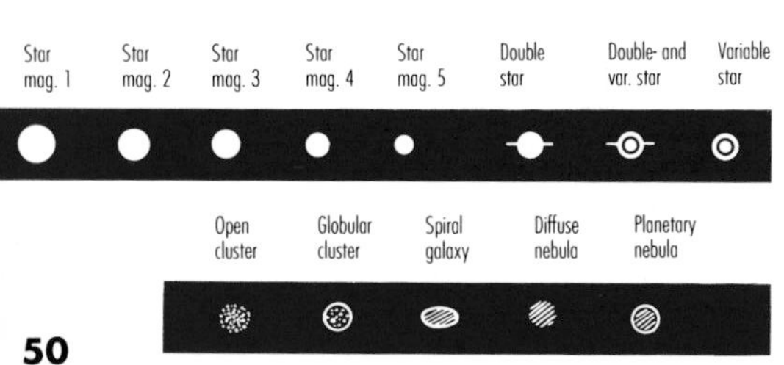

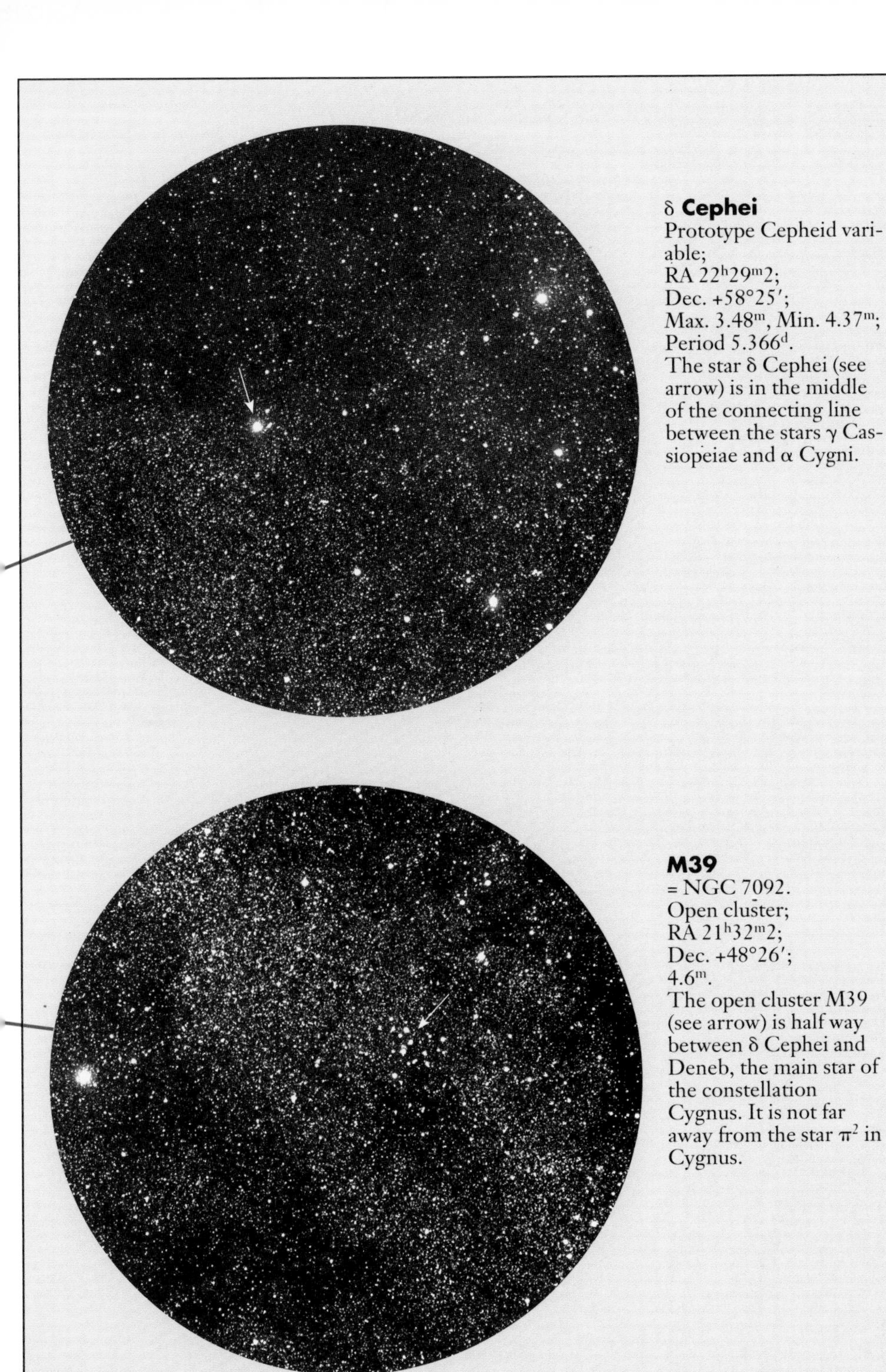

δ Cephei
Prototype Cepheid variable;
RA $22^{h}29^{m}2$;
Dec. +58°25′;
Max. 3.48^{m}, Min. 4.37^{m};
Period 5.366^{d}.
The star δ Cephei (see arrow) is in the middle of the connecting line between the stars γ Cassiopeiae and α Cygni.

M39
= NGC 7092.
Open cluster;
RA $21^{h}32^{m}2$;
Dec. +48°26′;
4.6^{m}.
The open cluster M39 (see arrow) is half way between δ Cephei and Deneb, the main star of the constellation Cygnus. It is not far away from the star π^2 in Cygnus.

Observation Tips for δ Cephei

δ Cephei is surrounded almost in a semicircle by the stars λ Cephei, ζ Cephei, and ϵ Cephei. δ Cephei is variable between 3.5^{m} and 4.4^{m}. The pulsation period lasts 5 days and 6.75 hours.

The star is the prototype of a class of variable stars called Cepheids, which enable scientists to learn about the construction of outer space. By noting a Cepheid star's period of pulsation, apparent magnitude, and absolute magnitude, astronomers can compute a star's distance. They are often called celestial yardsticks. It is very instructive to once observe for oneself one or several pulsation periods. Characteristics for the Cepheids—the name for the class named after Delta—is the difference in brightness of one magnitude, a quick rise in brightness, a slow fall in brightness, and the constancy of the period. Cepheids have already been discovered in other Milky Way systems (galaxies).

Observation Tips for M39

Stars that are pulled relatively far apart characterize this open cluster. M39 has the apparent magnitude of 4.6 and it contains about twenty-five individual stars. You may be able to see this cluster with binoculars.

The many clouds of the Milky Way are found on the left to the center towards the edge of the photo. Also striking is the division of the Milky Way in the constellations Cygnus and Sagittarius-Scorpius (see page 64), which is caused by clouds that absorb interstellar matter.

The Bright Stars

α Cephei (Alderamin);
RA $21^{h}18^{m}6$;
Dec. +62°35′;
2.5^{m}.

β Cephei (Alfirk; double star);
RA $21^{h}28^{m}7$;
Dec. +70°34′;
$3.3–8.0^{m}$.

10 Pisces, Cetus, and Aries

In 150 B.C., the star β Arietis, called Sheratan or Sharatan, marked the vernal equinox. Where the sun lies at that instance is known as the First Point of Aries.

Visibility At midnight over the southern horizon (mean local time): October. The stars in the map section stand each month about 2 hours earlier over the southern horizon: November 22^h, December 20^h.

Pisces (The Fishes) Around March 21, the ecliptic and the celestial equator intersect in this constellation. It is called the vernal equinox, the zero point, from which the right ascension (RA) is counted.

Cetus (The Whale) This is the fourth largest constellation, most of whose stars lie to the south of the celestial equator. The variable star Mira (see pages 139 and 141) is a famous prototype of the class called long-period variables.

Aries (The Ram) A constellation of the zodiac. The star Gamma (γ) is a double star, which can easily be resolved with a 2-inch astronomical telescope. The vernal equinox fell into this constellation 2000 years ago. Due to the precession, the apparent shifting of the sky, it is today in the constellation Pisces.

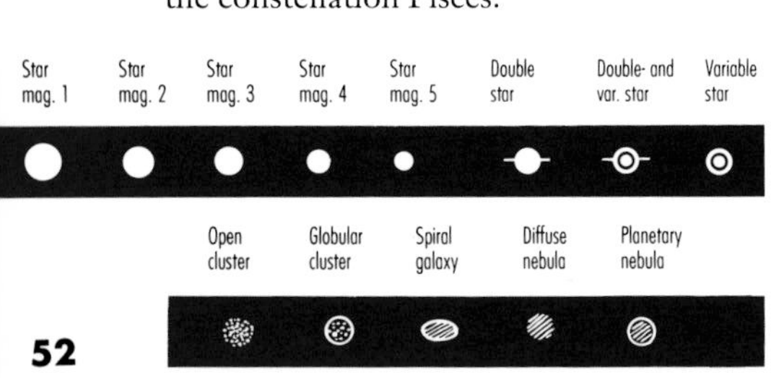

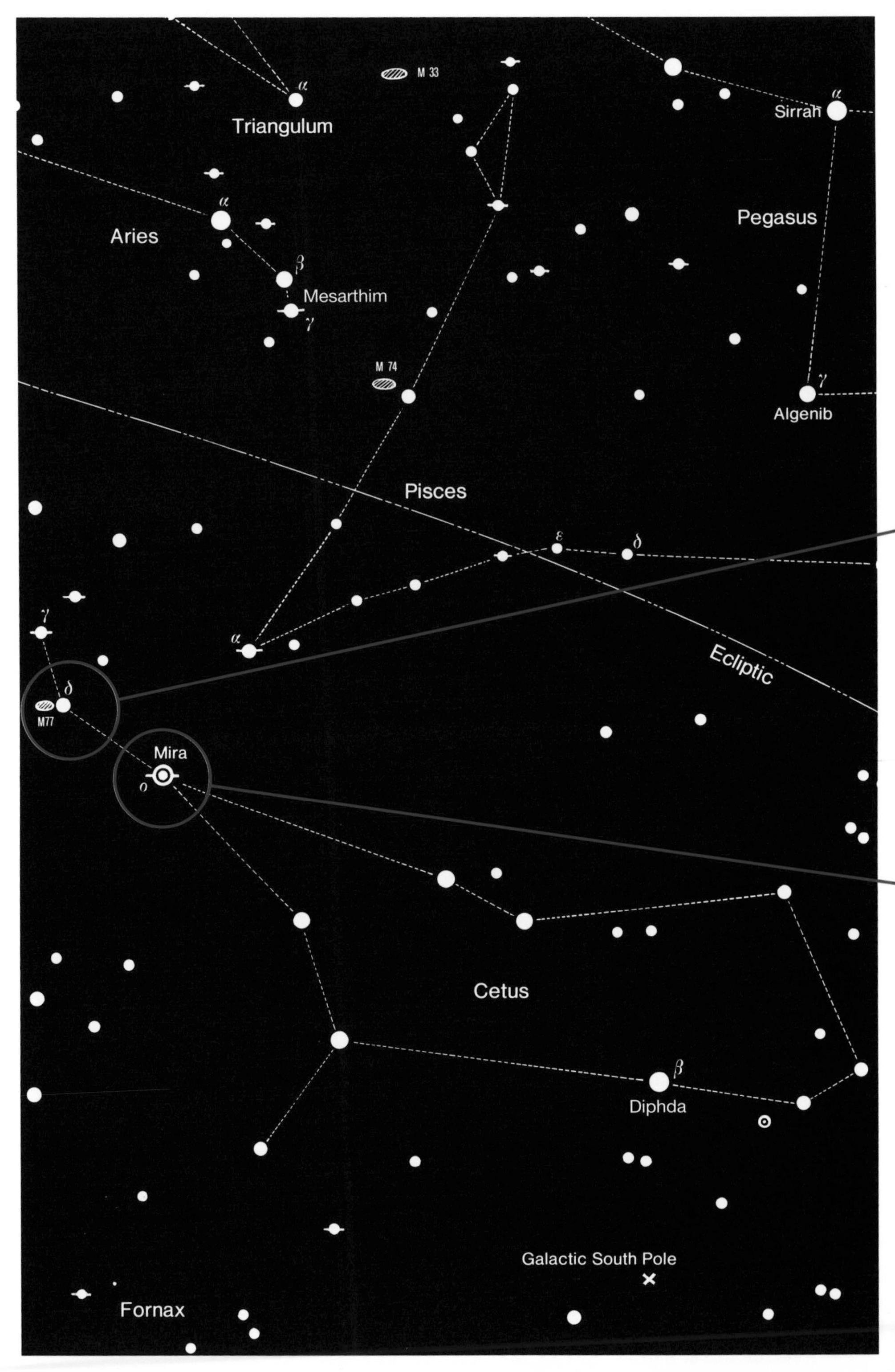

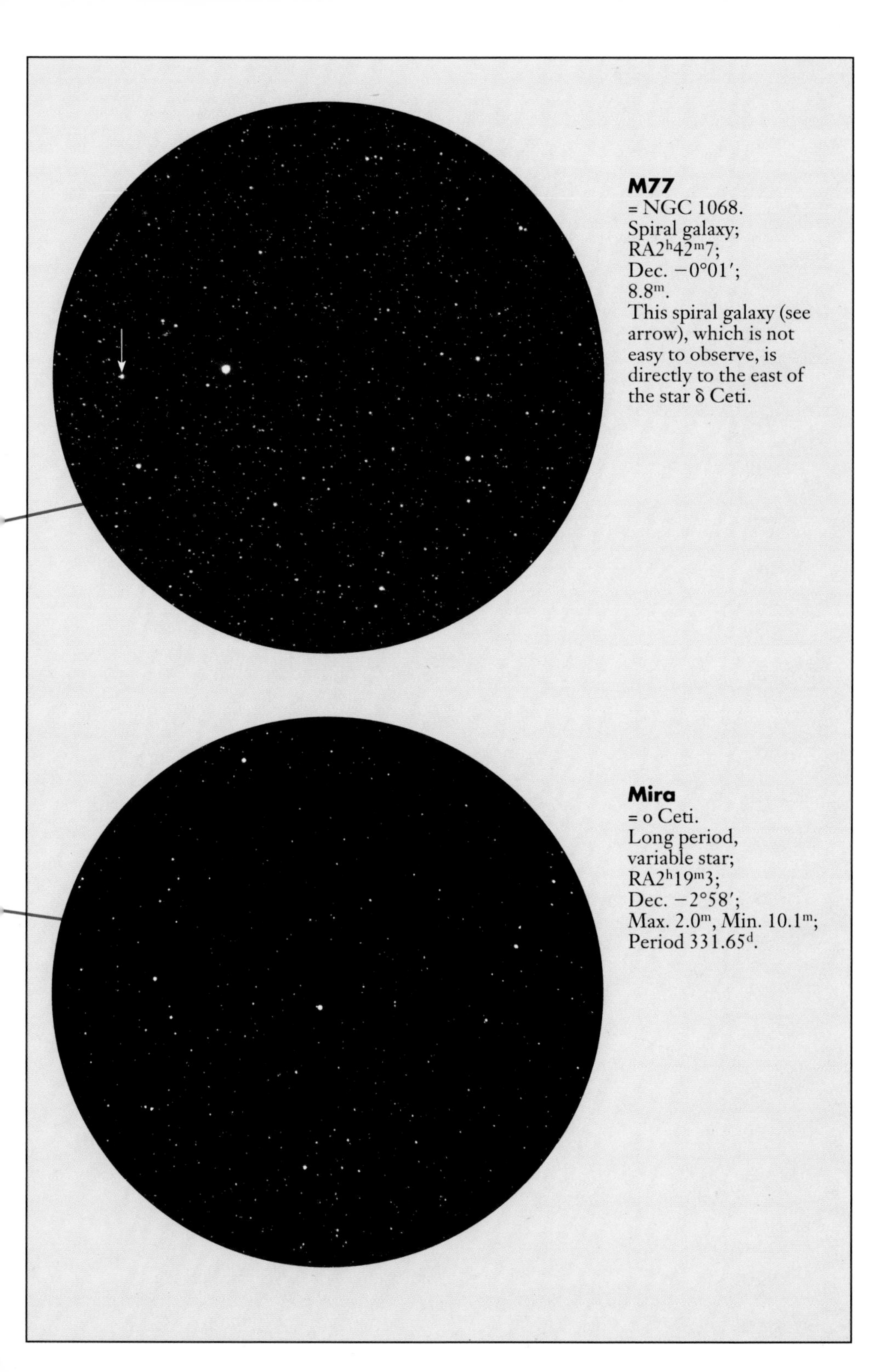

M77
= NGC 1068.
Spiral galaxy;
RA2^h42^m7;
Dec. $-0°01'$;
8.8^m.
This spiral galaxy (see arrow), which is not easy to observe, is directly to the east of the star δ Ceti.

Mira
= o Ceti.
Long period, variable star;
RA2^h19^m3;
Dec. $-2°58'$;
Max. 2.0^m, Min. 10.1^m;
Period 331.65^d.

Observation Tips for M77

It is the brightest spiral galaxy of a larger group in this area (altogether over 40). Astronomers speak of a galaxy cluster, which we are able to prove on an even larger scale in the area of the constellation Virgo (see page 60).

In order to find M77, you need very strong binoculars (14×100 or 22×80) or a 3-inch telescope, as well as a clear, dark night far away from the city lights. A good starting point for the search is the star δ Ceti. Just a degree to the east are two stars that are about of the same brightness (see section to the left), which form an equilateral triangle with the spiral galaxy. Spiral galaxies are, with a few exceptions, difficult to see. They appear as faint little stars (see arrow).

Observation Tips for Mira

This star is the first variable star (other than novae) ever discovered. The light pulsations were first noted in 1596. The reddish color of the star (center of the picture) is striking. By the way, these long-periodic variables are mostly called red giant stars.

The change in brightness does not happen evenly, and the maximum brightness of Mira can vary between the 1st and 4th magnitude. That holds true also for the period, for which there are deviations between 320 and 370 days.

In order to find Mira, locate the stars α Piscium and ζ Ceti (Baten Kaitos). Mira forms with the two stars approximately a right-angled triangle (see also p. 139).

The Bright Stars

β Ceti (Diphda); RA 0^h43^m4; Dec. $-17°59'$; 2.0^m.

α Arietis (Hamal); RA 2^h07^m2; Dec. $+23°28'$; 2.0^m.

β Arietis (Sheratan); RA 1^h54^m6; Dec. $+20°48'$; 2.7^m.

α Piscium (Alrescha); RA 2^h02^m0; Dec. $+2°46'$; 4.18^m or 5.21^m; $1.8''$.

11 Eridanus and Taurus

The name of the main star in the constellation Taurus, Aldebaran, is derived from the Arabic *Al Dabaran*, which means "the follower." In ancient times, the starting point for counting of the moon cycle was in the Pleiades (the Seven Stars). The next counting point was Aldebaran.

Visibility At midnight over the southern horizon (mean local time): November. The stars in the map section stand each month about 2 hours earlier above the southern horizon: December 22^h, January 20^h.

Eridanus (The River) On this map, we see only the northern part of this constellation, which stretches far over the southern sky (see also map 20, p. 72). This constellation also has the name "River Eridanus." To see the northern part of Eridanus, start at the bright Orion star, Rigel.

Taurus (Taurus) Constellation of the zodiac. The bright, reddish shining main star Aldebaran is very striking. It marks the glinting red eye of the bull. To find it, look for the leading constellation Orion, which is adjacent to the constellation Taurus. By extending the three belt stars of Orion along an imaginary straight-line diagonally upwards, you will quickly see Aldebaran and the Hyades.

Star mag. 1 | Star mag. 2 | Star mag. 3 | Star mag. 4 | Star mag. 5 | Double star | Double- and var. star | Variable star

Open cluster | Globular cluster | Spiral galaxy | Diffuse nebula | Planetary nebula

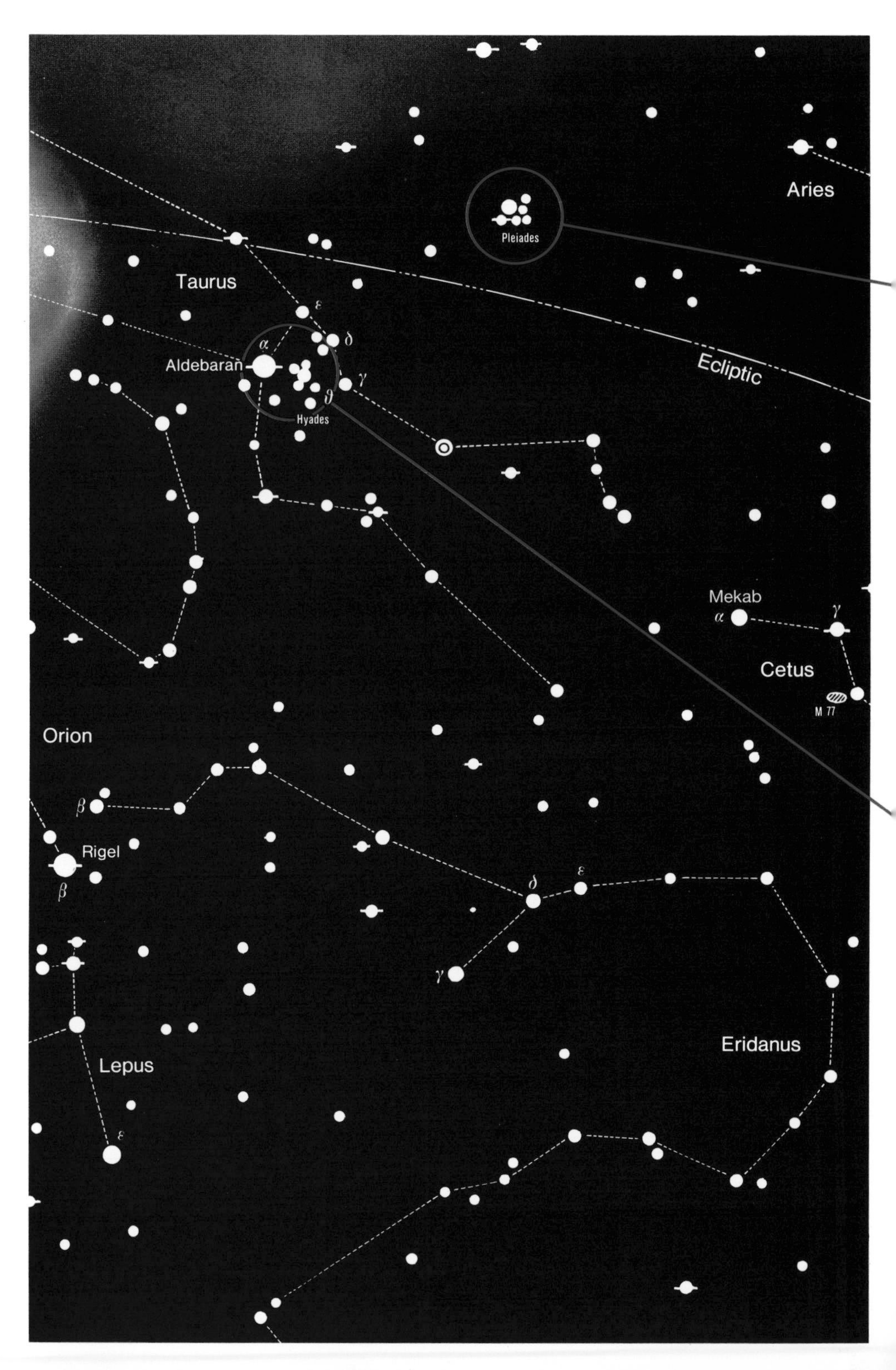

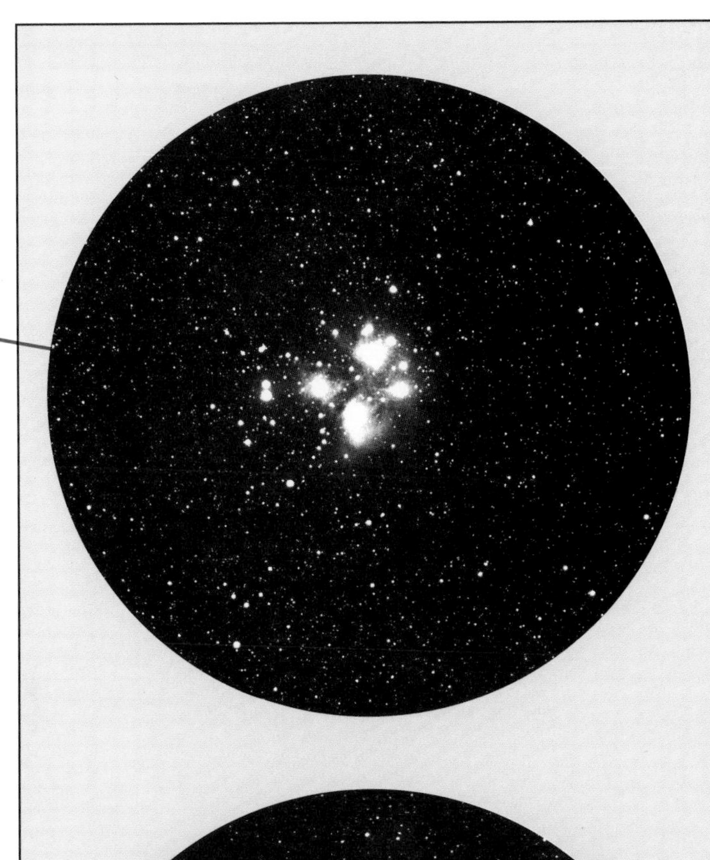

Pleiades
= Seven Sisters = M45.
Open cluster;
RA 3^h47^m0;
Dec. +24°07′;
$2.9-5.6^m$.
Here, you can test your eyes. About six to eight of the brightest stars can be seen with the naked eye. With binoculars, you can easily see more than thirty stars.

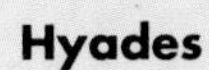

Hyades
= Rainers.
Open cluster;
RA 4^h27^m;
Dec. +16°;
$3.4-4.7^m$.
Compared to the Seven Sisters (Pleiades), the Hyades does not look like an open cluster at first sight. The field looks like it was pulled apart. Because it is so large, it is best to observe it with binoculars.

Observation Tips for the Pleiades

The Pleiades is the most striking and most brilliant open cluster of the entire northern sky. It is popularly known as the Seven Sisters, which is a group of mythological nymphs that are the daughters of Atlas and Pleione. The brightest member of the Pleiades is η Tauri with a magnitude of 2.9.

The Pleiades stars are embedded in a faint nebula, which is noticeable on long-exposure photographs. Under very clear conditions, the brightest part of the nebula, around Merope, can be seen with binoculars or a small telescope.

Observation Tips for the Hyades

Look for the bright, reddish star Aldebaran (α Tauri) to find the Hyades. It is to the left in the photo. To the right of it begins the Hyades stars with the double star θ Tauri (see arrow), which can be resolved with the naked eye.

Overall, two hundred stars, which all travel at the same speed of 43 kilometers per second (27 miles per second) through space, belong to the open cluster of the Hyades. The brightest form a V-shaped pattern on the head of Taurus.

In Greek mythology, the Hyades were daughters of Atlas and Aethra. They were changed into stars by Zeus and called the Hyades. In Greek, the Hyades means "the Rainers" because of their rising and setting in October and April, which coincided with the rainy seasons.

The Bright Stars

β Eridani (Cursa); RA 5^h07^m8; Dec. −5°05′ 2.8^m.

γ Eridani (Zaurak); RA 3^h58^m0; Dec. −13°31′; 2.9^m.

α Tauri (Aldebaran); RA 4^h35^m9; Dec. +16°31′ 0.9^m.

β Tauri (Elnath); RA 5^h26^m3; Dec. +28°36′; 1.7^m.

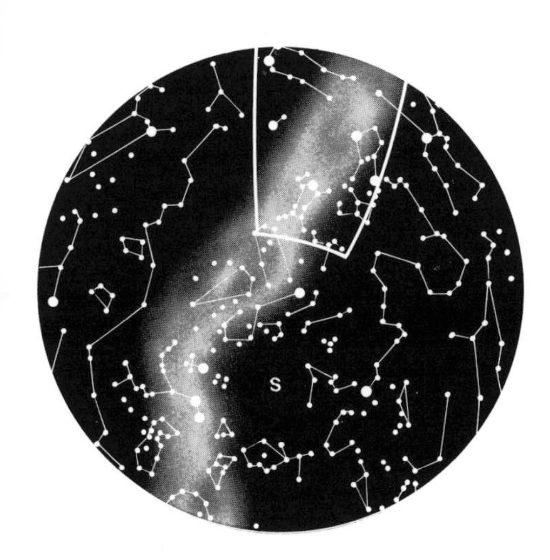

12 Canis Minor, Canis Major, Lepus, Monoceros, Orion, and Gemini

This section shows the magnificent constellations of the northern winter sky.

Visibility At midnight over the southern horizon (mean local time): December. The stars in the map section stand each month about 2 hours earlier above the southern horizon: January 22^h, February 20^h.

Monoceros (Unicorn) Because it lies along the winter Milky Way, it contains many interesting deep-sky objects.

Canis Major (Great Dog) The main star, Sirius, is the brightest star in the entire sky. The belt stars of the constellation Orion can help find it. The straight line they form points in a southeast direction to Sirius.

Lepus (Hare) This constellation to the south of Orion is about the same height as Sirius.

Canis Minor (Little Dog) This is in the middle between the stars Pollus (of Gemini) and Sirius (of Canis Major).

Orion (The Hunter) This constellation is crammed with objects of interest for observatory instruments of all sizes.

Star mag. 1 · Star mag. 2 · Star mag. 3 · Star mag. 4 · Star mag. 5 · Double star · Double- and var. star · Variable star

Open cluster · Globular cluster · Spiral galaxy · Diffuse nebula · Planetary nebula

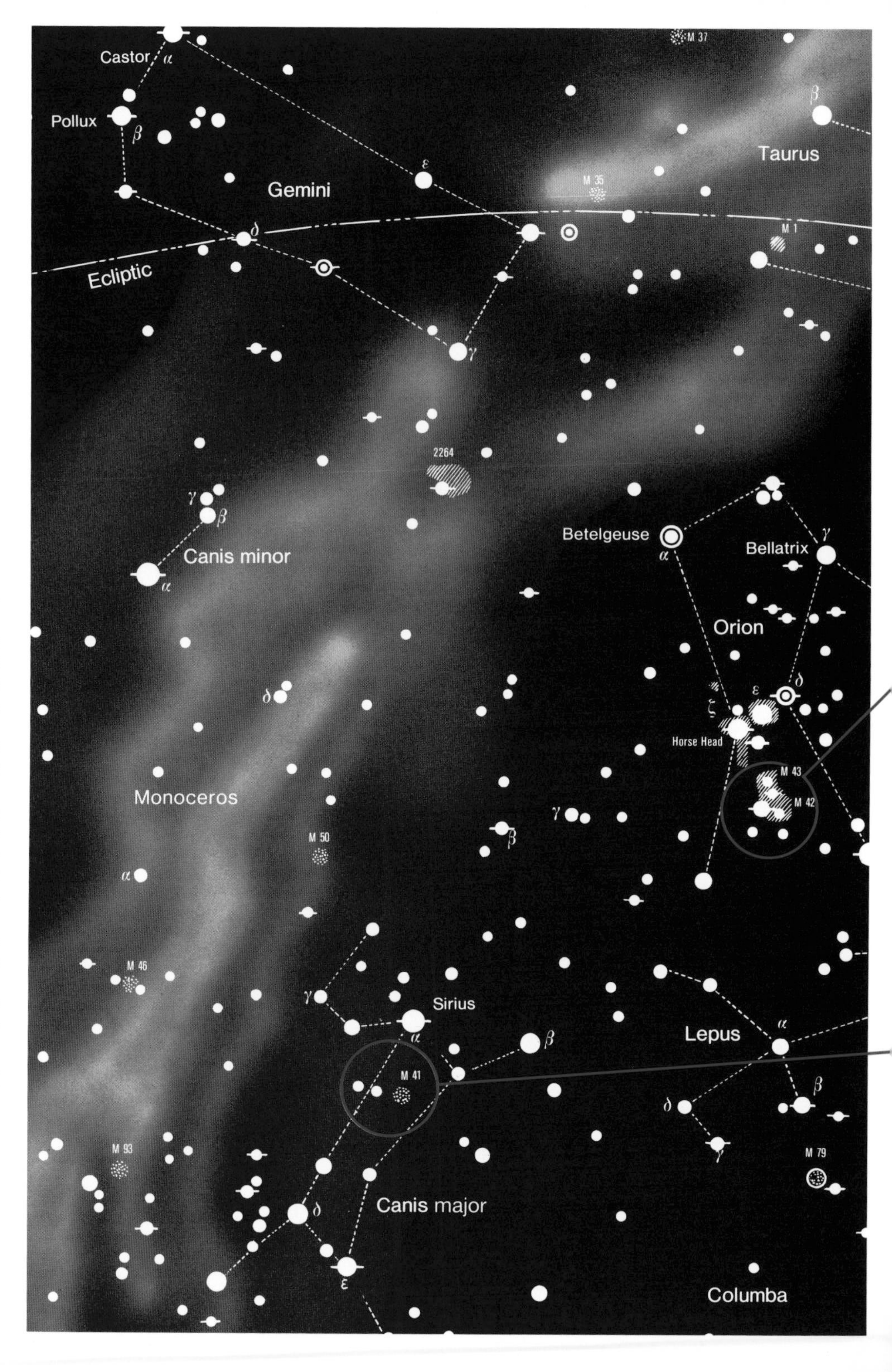

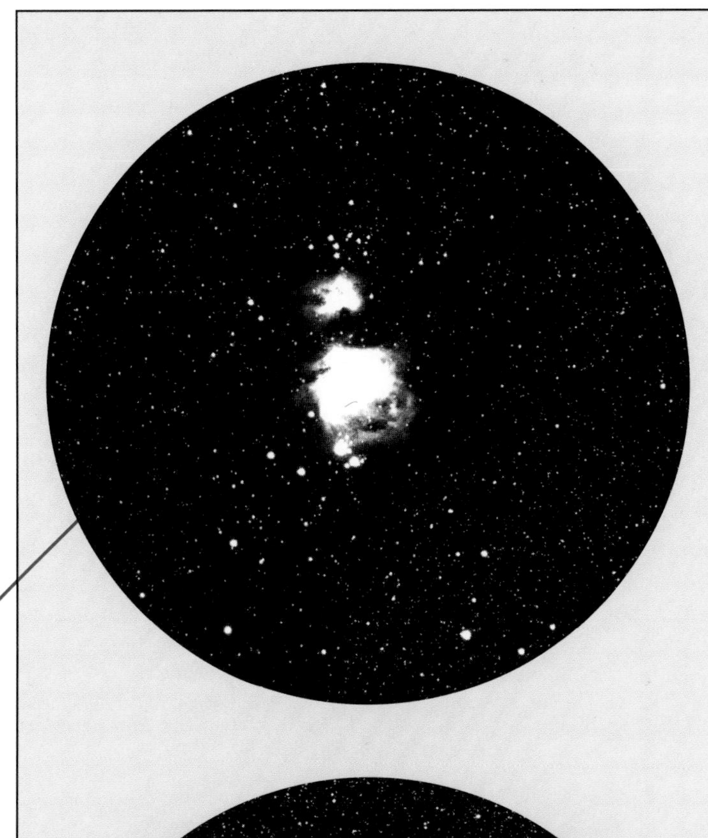

Orion Nebula
= M 42 = NGC 1976.
Nebula;
RA 5^h35^m4;
Dec. $-5°27'$;
2.9^m.
To find the nebula, create a guiding line between the stars Rigel and ζ Orionis (one of the three belt stars). In the middle of this line is the object M42, which, together with its companion nebula, has the double name M42–M43.

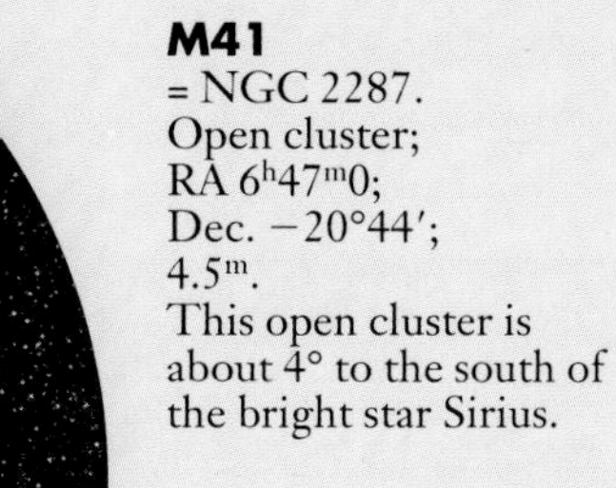

M41
= NGC 2287.
Open cluster;
RA 6^h47^m0;
Dec. $-20°44'$;
4.5^m.
This open cluster is about 4° to the south of the bright star Sirius.

α Orionis (Betelgeuse) is a red giant star about 500 times the sun's diameter. It is so large that it is unstable.
Gemini (Gemini) A constellation of the zodiac (the most northern one) with the bright stars Castor and Pollux.

Observation Tips for the Orion Nebula

This nebula is an object of our Milky Way. The greenish-white clouds of the nebula reflect light of surrounding stars. The very thin gas substance itself does not have its own light.

With binoculars, you can discover clouds and the four striking stars of θ Orionis, the Trapezium, at the heart of Orion Nebula. Striking dark spots (low on stars) let us see the absorption through the gas.

Capturing this in a photo is not so easy. The short exposure catches only the center. Although the long exposure shows the considerable expansion of the nebula, it leads to an overexposure in the center (see photo on p. 139).

Observation Tips for M41

Both Sirius and M41 (4° south of Sirius) show up in the same visual field of a pair of binoculars. Since M41 (an open cluster; see arrow) has the brightness of 6^m, you can try to look for this object with naked eyes on a dark, clear night. It is a fairly rich cluster with about eighty stars. H. Vehrenberg mentions in *Messier Book*, "In the center of the cluster, one discovers a star with a strikingly red light."

Three stars form a distinct triangle (to the left towards the edge in the photo). The two outer stars (top, π Canis majoris; bottom, 17 Canis majoris) are double stars with the distances 12" and/or 44". The former can be resolved with a 2–3-inch telescope; the latter can be resolved with 15 × 60 or 22 × 80 binoculars.

The Bright Stars

α Canis majoris (Sirius); RA 6^h45^m1; Dec. $-16°43'$; -1.4^m.

α Canis minoris (Procyon); RA 7^h39^m3; Dec. $+5°14'$; 0.4^m.

α Geminorum (Castor); RA 7^h34^m6; Dec. $+31°53'$; 1.6^m.

α Orionis (Betelgeuse); RA 5^h55^m2; Dec. $+7°24'$; $0.4\text{-}1.3^m$; Period 2110^d.

13 Cancer, Leo, Sextans, and Hydra

The largest constellation in the sky is Hydra. It stretches from the constellations Cancer, Leo, and Virgo to Libra.

Visibility At midnight over the southern horizon (mean local time): February. The stars in the map section stand each month about 2 hours earlier over the southern horizon: March 22^{h}, April 20^{h}.

Cancer (Cancer) This is the faintest of the twelve constellations of the zodiac through which the sun passes each year. In the middle on the connecting line between the bright stars Prokyon (Canis minor) and Regulus (Leo) is the open cluster M67, which can be seen with binoculars.

Leo (Leo) The map shows only the head of the lion with the bright star Regulus. The star γ Leonis is a double star that can be seen with a smaller (2-inch) telescope.

Sextans (The Sextant) This constellation is between the stars Regulus (Leo) and Alfard (Hydra).

Hydra (Water Snake) The brightest star Alfard forms an almost isoceles triangle with the Leo stars Regulus and Denebola.

Star mag. 1	Star mag. 2	Star mag. 3	Star mag. 4	Star mag. 5	Double star	Double- and var. star	Variable star

Open cluster	Globular cluster	Spiral galaxy	Diffuse nebula	Planetary nebula

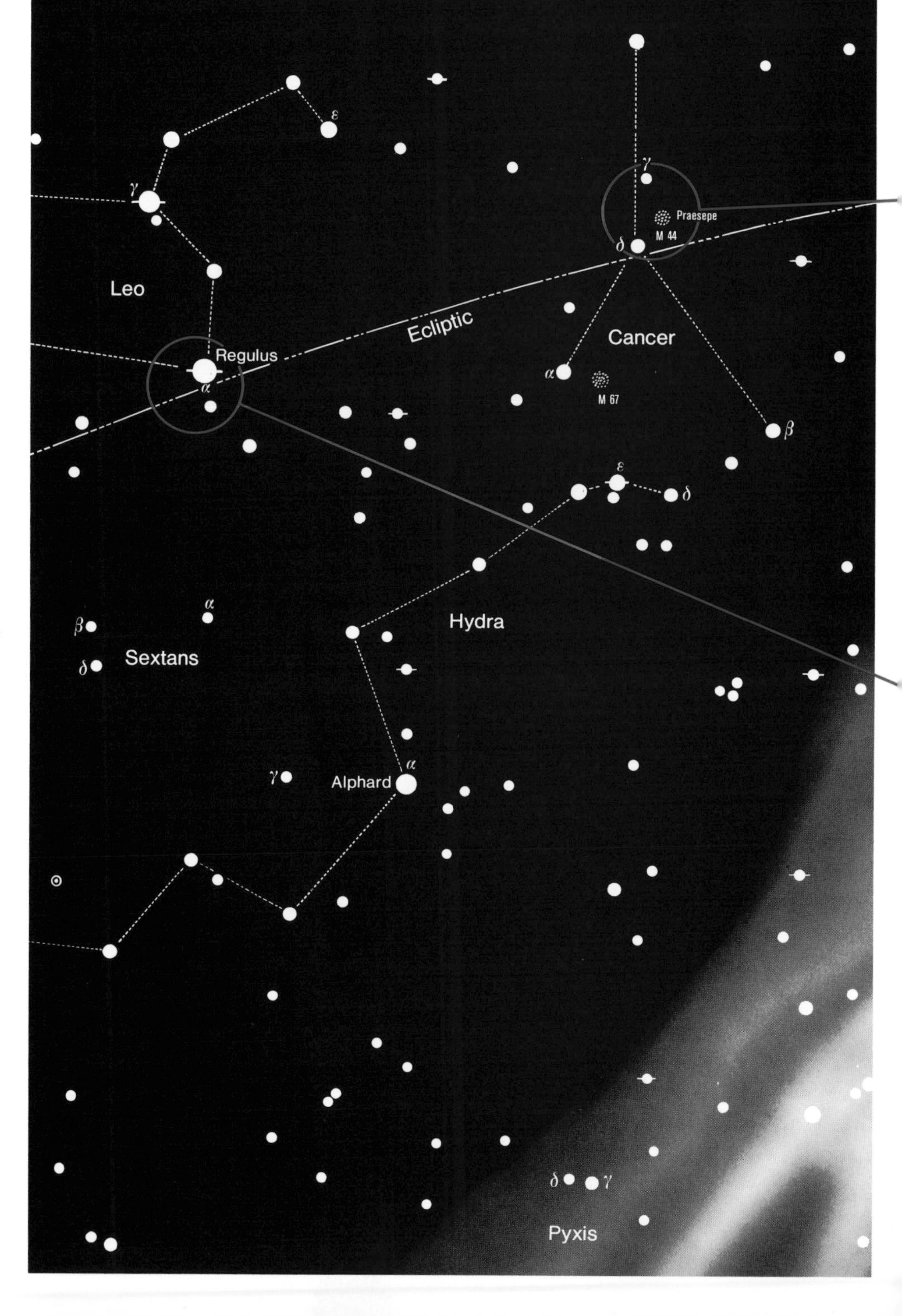

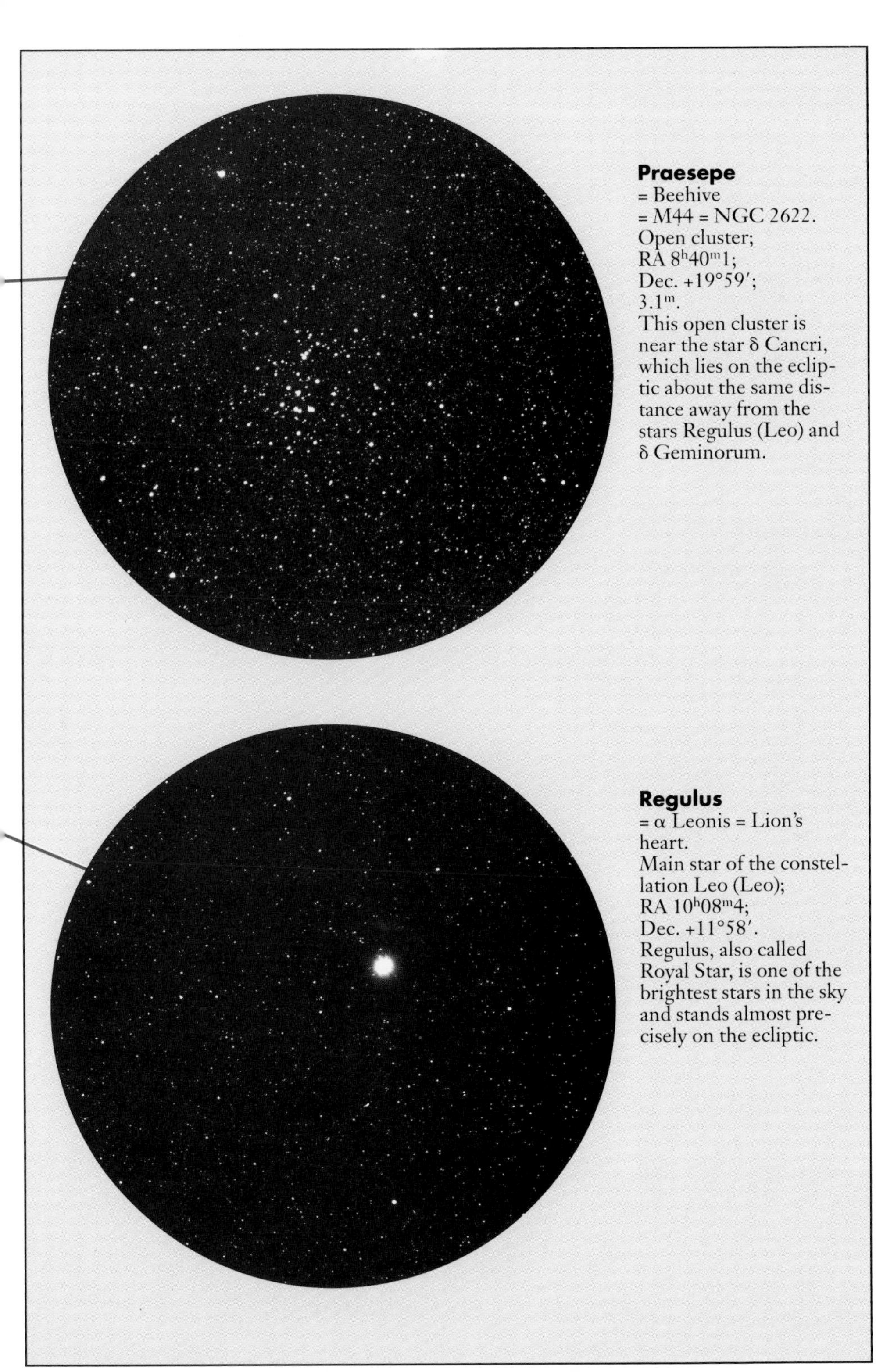

Praesepe
= Beehive
= M44 = NGC 2622.
Open cluster;
RA 8^h40^m1;
Dec. +19°59′;
3.1^m.
This open cluster is near the star δ Cancri, which lies on the ecliptic about the same distance away from the stars Regulus (Leo) and δ Geminorum.

Regulus
= α Leonis = Lion's heart.
Main star of the constellation Leo (Leo);
RA 10^h08^m4;
Dec. +11°58′.
Regulus, also called Royal Star, is one of the brightest stars in the sky and stands almost precisely on the ecliptic.

Observation Tips for the Praesepe

Praesepe is seen as a misty patch to the naked eye. Binoculars will give you a better resolution to the individual stars. Since the stars are spread out comparatively far apart, it is an object for little enlargements.

Galileo saw with his telescope thirty individual stars, which you can also see with binoculars. Scientists found out that over seventy-five visible stars belong to this open cluster. They also see a certain resemblance to the Hyades (see p. 55), because of the loose distribution of the stars.

By the way, you can test how pollution in the environment affects the viewing of this open cluster. On a clear night, the misty patch that you see with the naked eye is pale. But through a polluted atmosphere, it visibly becomes blurred.

Observation Tips for Regulus

The sun, each year in the course of its apparent orbit through the twelve constellations of the zodiac, passes once in front of Regulus around the time of August 23. Of course, we cannot see this, since the sun outshines everything. Regulus lies directly on the ecliptic and is capable of being covered by the moon and planets. Of course there are many stars that lie on the ecliptic and can be covered by the moon and by the planets (for example, Spica, in the constellation Virgo, and Antares, in the constellation Scorpius, which lie close to the ecliptic). It is just that the observation is especially impressive and easier with such a bright star as Regulus.

The Bright Stars

α Leonis (Regulus);
RA 10^h08^m4;
Dec. +11°58′;
1.4^m.

γ Leonis (Algieba);
RA 10^h20^m0;
Dec. +19°51′;
2.0^m.

α Hydrae (Alfard);
RA 9^h27^m6;
Dec. −8°40′;
2.0^m.

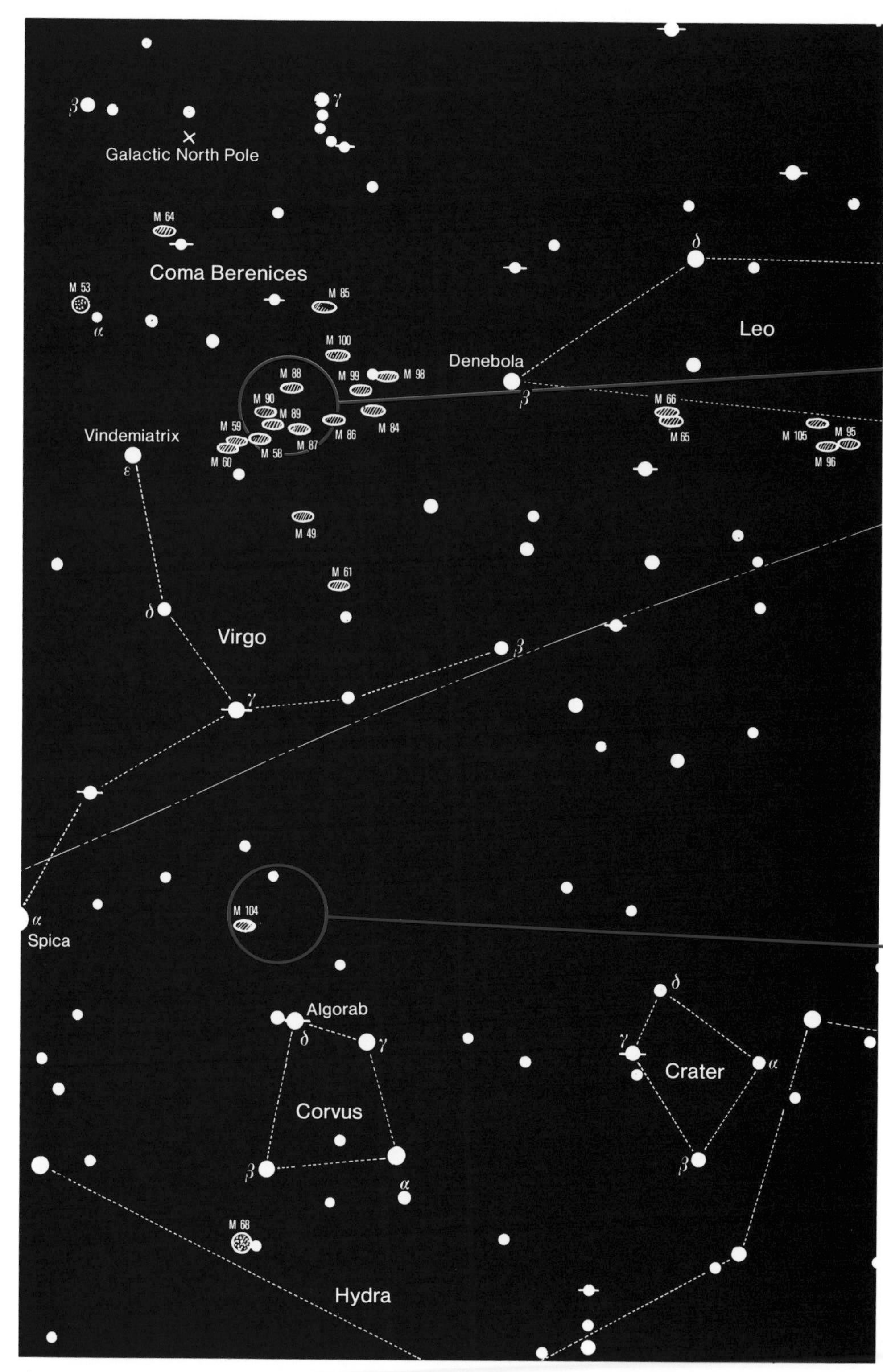

14 Crater, Coma Berenices, Virgo, and Corvus

Galaxies or extragalactic star systems appear especially frequently in certain celestial areas. In the constellation Coma Berenices, you can observe over one thousand galaxies with large telescopes.

Visibility At midnight over the southern horizon (mean local time): March. The stars in the map section stand each month about 2 hours earlier over the southern horizon: April 22^{h}, May 20^{h}.

Crater (The Cup) This constellation of the southern sky contains no objects of particular interest. It is in the middle between the bright stars Spica (constellation Virgo) and Alphard (constellation Hydra).

Coma Berenices (Berenice's Hair) This constellation is tween Boötes and Leo. It contains clusters of galaxies that can be seen only with larger telescopes.

Virgo (Virgo) The second largest constellation of zodiac. It has bright star Spica (ear of wheat), which stands very close to the ecliptic. There are numerous extragalactic systems in Virgo. One speaks of a so-called Virgo cluster, an expanded clus-

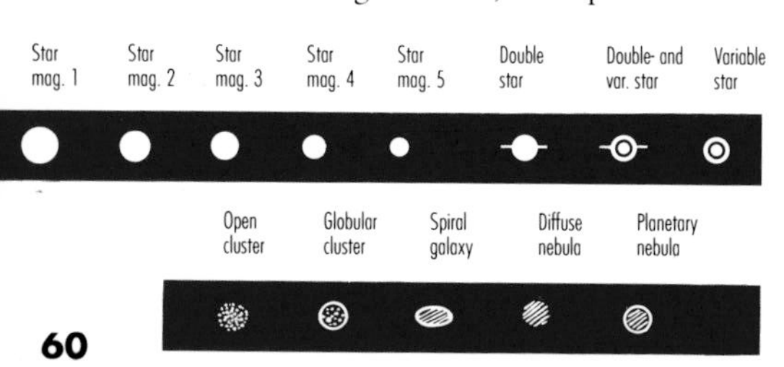

ter of galaxies. γ Virginis (distance 3″) is an easy-to-see multiple star for the 3-inch telescope.

Corvus (The Raven) This constellation is very close to the constellation Virgo. As in the constellation Crater, the brighter stars form a square.

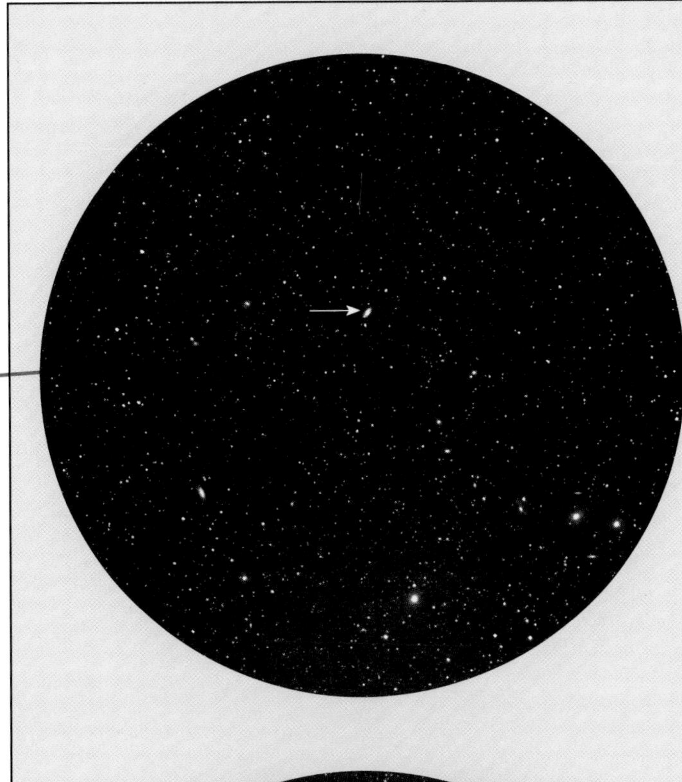

M88
= NGC 4501.
Spiral galaxy;
RA 12^h32^m0;
Dec. +14°25′;
10.3^m.
M88 is directly above the center of the picture of the section (see arrow) with a spiral structure, which resembles the one of the Andromeda Galaxy.

Observation Tips for M88

To look for M88, point the telescope to the center of the connecting line between the stars ε Virginis and β Leonis. Since M88 has a brightness of only 10^m, it is not easy to find the galaxy. A lot depends on the transparency of the atmosphere and on an optics that is strong in light.

Look closely into the lower part of the picture and you'll see other galaxies, which belong to the Virgo galaxy cluster. On the left bottom is M90, which has a brightness of 10^m and looks similar to M88. Collisions of galaxies have probably taken place in such a dense cluster. The lack of spiral arms suggests there was a collision before the spiral structure developed.

Sombrero Galaxy
= M104 = NGC 4594.
Spiral galaxy;
RA 12^h40^m0;
Dec. −11°37′;
9.3^m.
This spiral galaxy is to the left of the center of the picture towards the edge (see arrow). Its name was given because of its distinctive large center and the dark lane of dust that runs along the equatorial plane.

Observation Tips for the Sombrero Galaxy

Close-ups of this spiral galaxy show a certain resemblance to Saturn, even though the two have nothing in common. M104, not far away from the star δ Corvi, belongs to the constellation Virgo and the so-called Virgo cluster—that accumulation of galaxies in the constellation Virgo to which M88 also belongs. Hans Vehrenberg mentions in *Messier Book:* "There are signs that the cluster stretches over the large area of the sky between 10^h and 15^h45^m in RA and −3 and +20 in dec. It might count about 13,500 galaxies up to a magnitude of 15.7^m."

The Bright Stars

γ Corvi (Gienah); RA 12^h15^m8; Dec. −17°33′; 2.6^m.

α Virginis (Spica); RA 13^h25^m2; Dec. −11°10′; 1.0^m.

γ Virginis (Porrima); RA 12^h41^m7; Dec. −1°27′; 2.7^m; Multiple star.

15 Boötes, Corona Borealis, Serpens, and Libra

Arcturus, the main star of Boötes, is one of the oldest star names. It belonged in antiquity to the stars that indicated to the farmers the time of harvest. Arcturus in the morning sky was the sign for the wine harvest.

Visibility At midnight over the southern horizon (mean local time): May. The stars in the map section stand each month about 2 hours earlier over the southern horizon: June 22^{h}, July 20^{h}.

Boötes (The Herdsman) The southern part of the constellation is shown here with Arcturus—a bright, striking, reddish-yellow star. Its name, going back to the Greek world of legends, means "Ox-Driver." The northern part of this constellation is shown on page 44.

Corona Borealis (The Northern Crown) This constellation is between Boötes and Hercules.

Serpens (The Serpent) This constellation is "actually split into two halves." The part shown here is Serpens Caput, the head of the snake, that is directly to the south of the Corona Borealis.

Star mag. 1	Star mag. 2	Star mag. 3	Star mag. 4	Star mag. 5	Double star	Double- and var. star	Variable star

Open cluster	Globular cluster	Spiral galaxy	Diffuse nebula	Planetary nebula

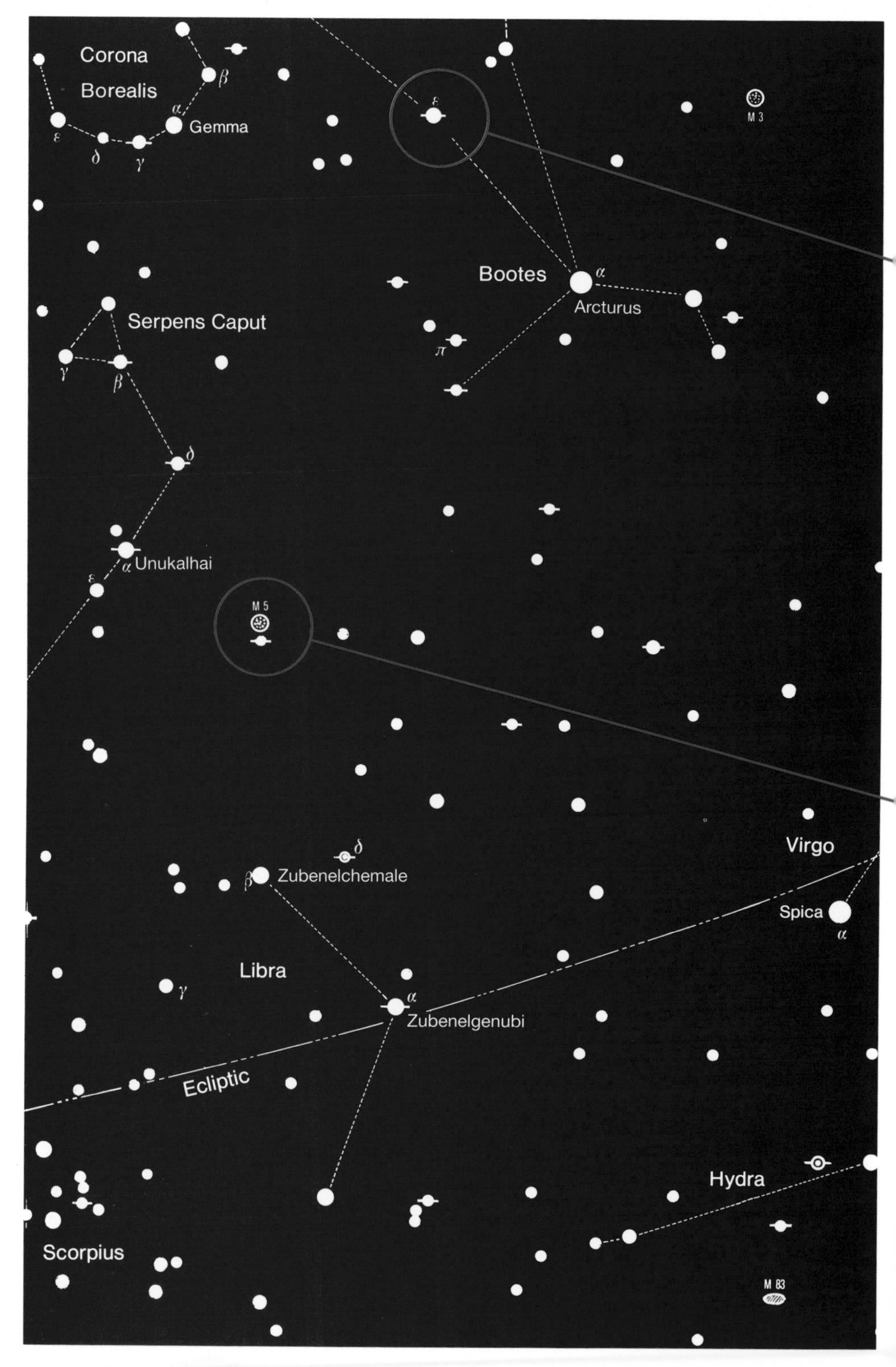

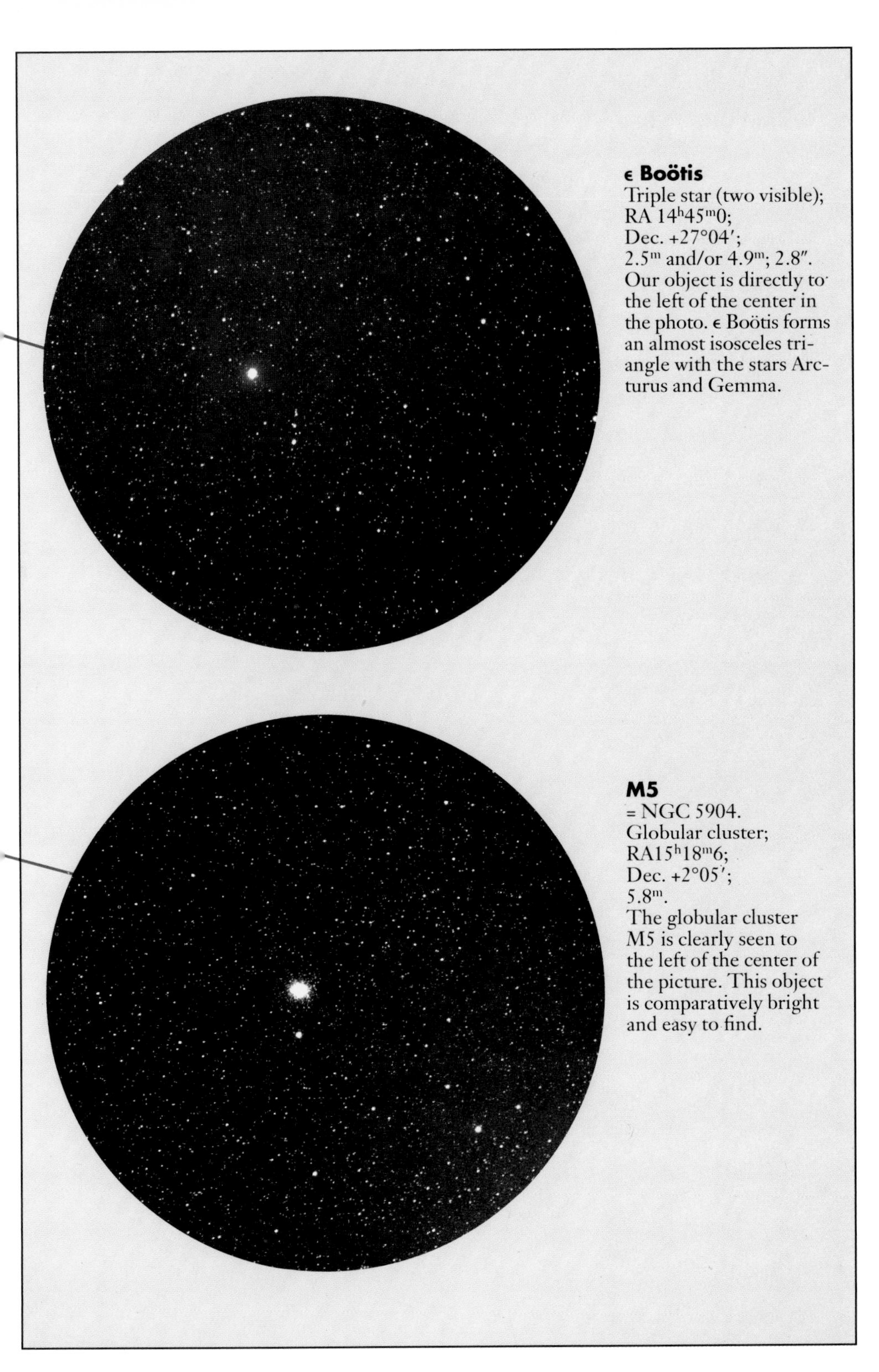

ϵ Boötis
Triple star (two visible); RA 14^h45^m0; Dec. +27°04′; 2.5^m and/or 4.9^m; 2.8″. Our object is directly to the left of the center in the photo. ϵ Boötis forms an almost isosceles triangle with the stars Arcturus and Gemma.

M5
= NGC 5904. Globular cluster; RA15^h18^m6; Dec. +2°05′; 5.8^m. The globular cluster M5 is clearly seen to the left of the center of the picture. This object is comparatively bright and easy to find.

Libra (The Scales) This is a faint constellation of the zodiac. Its main star α Librae stands practically on the ecliptic. Look for it between Spica and Antares.

Observation Tips for ϵ Boötis

Although ϵ Boötis is actually a triple star, only two are visible. Of the two, the brighter star has a magnitude of 2.5 and a yellow color; the darker companion has a magnitude of 4.9 and a bluish color. The distance between them is 2.8 arc-seconds (″). You can try to resolve these two stars with 3- and 4-inch telescopes. Anything less will be more difficult, even though the indicated distance would be far enough for a resolution. Theoretically, in order to see a resolution, the stars must be of the same brightness. A difference of 1 magnitude makes the resolution harder. In our photo, ϵ Boötis appears egg shaped.

Observation Tips for M5

To look for it, start at the star δ Serpentis or β Librae. M5 forms an almost isosceles triangle with both stars. Very close to M5 is 5 Serpentis, a star of the 5th magnitude. Since the area does not have many brighter stars, this cluster stands out quite well. M5 is an easy positive object for binoculars. With a 3- to 4-inch telescope, at about 40-fold enlargement, the sight intensifies; parts of the edge resolve into individual stars. The view is impressive, because numerous stars of the 12th to 14th magnitude are present in the globular cluster. M5 is over 10 billion years old, which makes it one of the old cosmic objects. 5 Serpentis, below M5, and the star 6 Serpentis, about 1° away from it in the extension of the straight line, are double stars.

The Bright Stars

α Bootis (Arcturus); RA 14^h15^m7; Dec. +19°11′; 0.1^m.

α Corona Borealis (Gemma); RA 15^h34^m7; Dec. +26°43′; 2.3^m.

α Serpentis (Unukalhai); RA 15^h44^m3; Dec. +6°26′; 2.7^m.

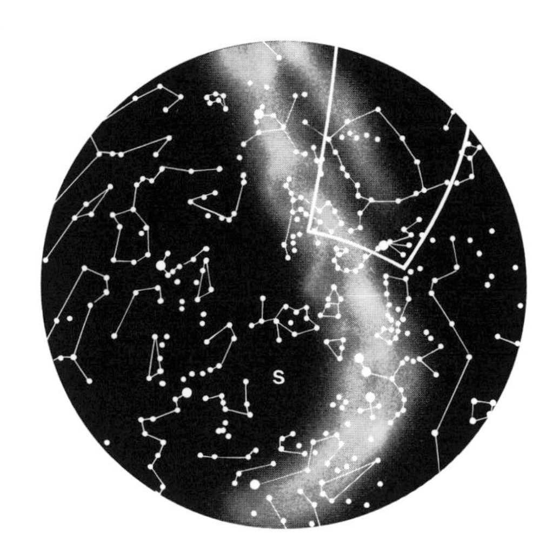

16 Hercules, Ophiuchus, Sagittarius, and Scorpius

The Ionian nature-philosopher Demokrit von Abdera said in 400 B.C. that the Milky Way consists of stars. Galileo confirmed this with his telescope, which he invented in 1609. In the constellations Sagittarius and Ophiuchus, the Milky Way can be seen especially impressively.

Visibility At midnight over the southern horizon (mean local time): June. The stars in the map section stand each month about 2 hours earlier over the southern horizon: July 22^{h}, August 20^{h}.

Hercules (Hercules) The map here shows the southern part of this constellation. Rasalgethi (α Herculis) is a red giant star some six hundred times the diameter of the sun, making it one of the largest stars known. It is erratically variable and it has a bluish companion 4.7″ distance away with a magnitude of 5.4. Rasalgethi is not far away from Rasalhague, the brighter main star of the constellation Ophiuchus. The northern part of Hercules is shown on page 46.

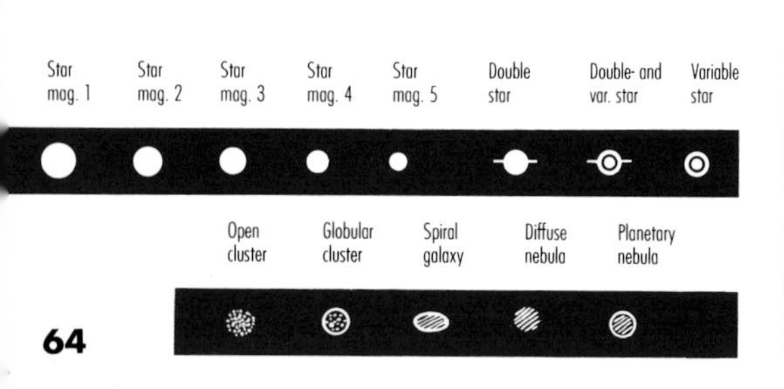

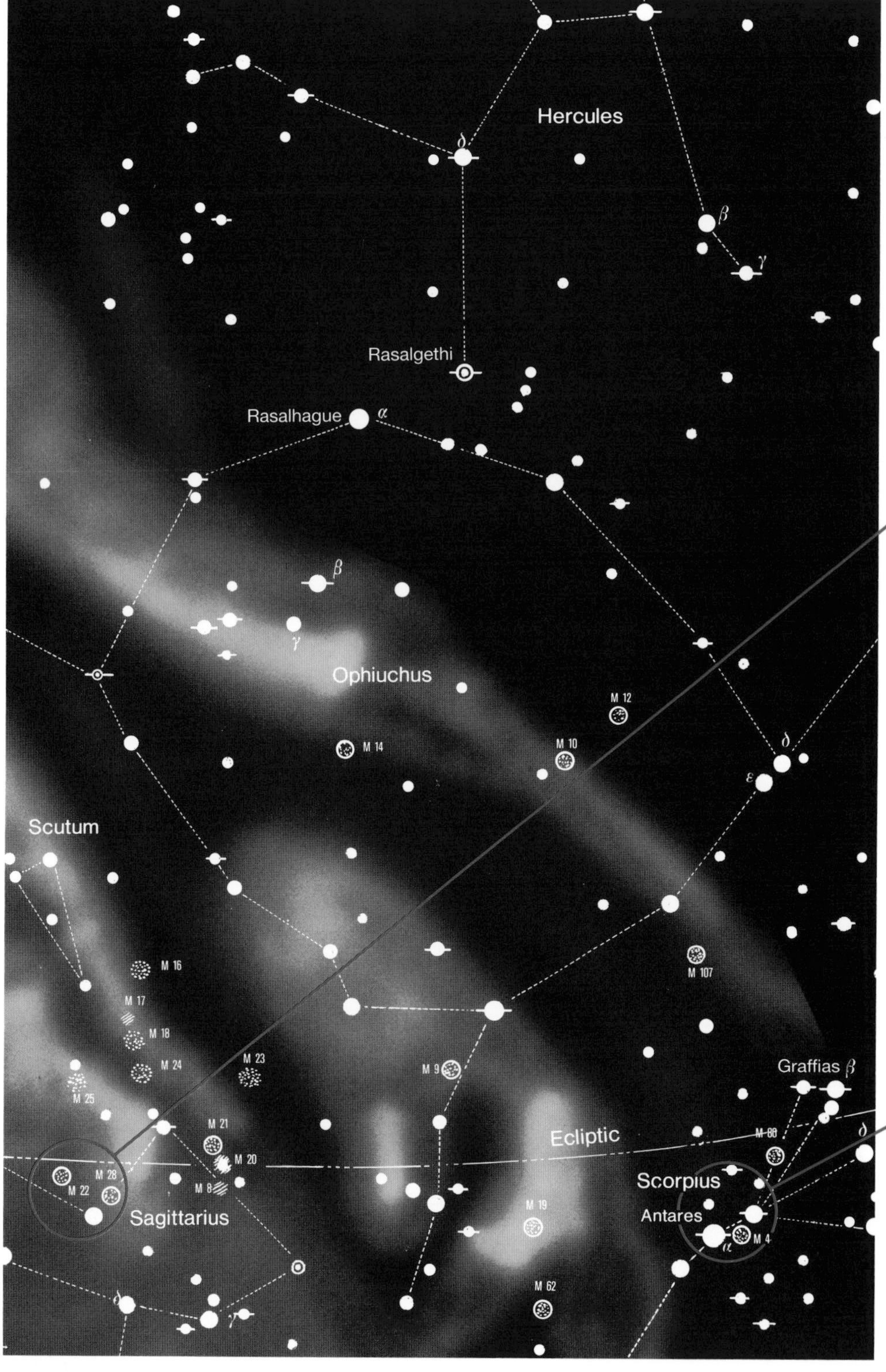

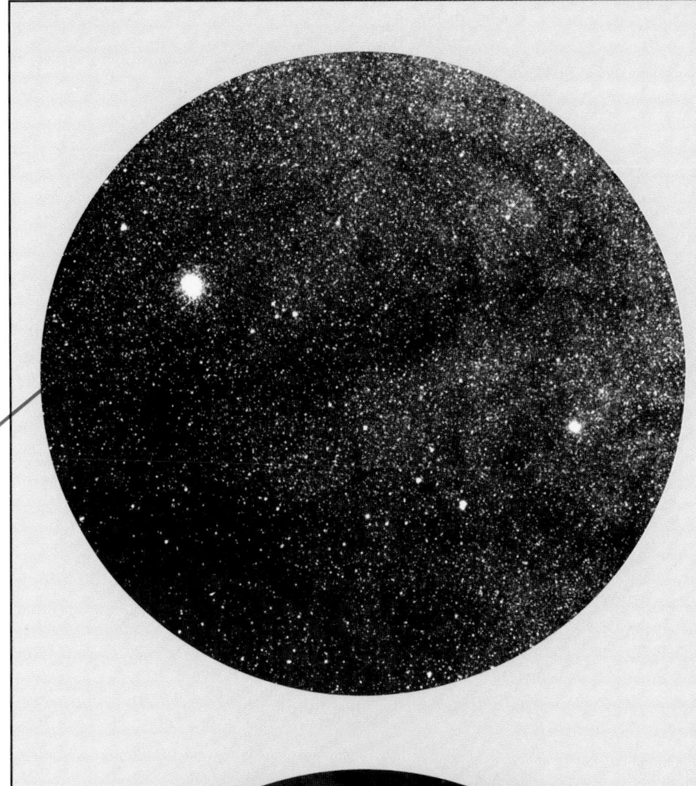

M22
= NGC 6656.
Globular cluster;
RA 18^h36^m4;
Dec. −23°54′;
5.1^m.
To the left of the center of the picture is the globular cluster M22. Its closeness to the Milky Way, which is rich in stars, cannot be overlooked.

Antares
= α Scorpii.
RA 16^h29^m4;
Dec. −26°26′.
This section shows, from the center of the picture to the left downward, the constellation σ Scorpii, M4, and α Scorpii.

Ophiuchus (The Serpent) This constellation is between Hercules and Scorpius, with its southernmost regions extending into the Milky Way.
Sagittarius (The Archer) The main attraction of Sagittarius is its clusters and nebulae. There are about fifteen objects in Sagittarius, more than in any other constellation (see also p. 66).
Scorpius (The Scorpion) Constellation of of the zodiac (see also p. 82). The bright reddish star Antares "rival of Mars" cannot be overlooked at the horizon.

Observation Tips for M22

Look for M22 in the midsummer in central northern latitudes. It stands to the south of the most southern spot of the ecliptic, and it reaches only at the latitude of the northern tropic (Mexico, Cuba, East Bengal, Arabia) a height like the Summer Triangle in northern latitudes.

M22 is a brilliant object to see with binoculars. It is rightfully compared to the most beautiful globular cluster of the northern sky, M13 (see p. 47). To find it, use binoculars to locate the star Atair in the constellation Aquila. Then go along the Milky Way to the bright star σ Sagittarii (Nunki; see also map on p. 66). The bright, star-like spot to the right towards the edge is the globular cluster M28.

Observation Tips for Antares and Its Surroundings

Antares (α Scorpii) is a red supergiant that is three hundred times the diameter of the sun. It has a brightness of 1^m and it is related to Arcturus (in the constellation Boötes) and Betelgeuse (in the constellation Orion). The reddish color gave it the name Antares.

M4 is a globular cluster that is loosely scattered. Many stars can be resolved with a 3-inch telescope. Brighter stars are surrounded by a "veil" of interstellar gas and dust accumulations, which hinder the view of the stars that are lower in brightness.

The Bright Stars

α Herculis (Ras Algethi); RA 17^h14^m6; Dec. +14°23′; Max. 3^m, Min. 4^m; Semiregular variable; Period: 100^d.

α Ophiuchi (Ras Alhague); RA 17^h34^m9; Dec. +12°34′; 2.1^m

α Scorpii (Antares); RA 16^h29^m4; Dec. −26°26′; Max. 2.3^m; Min. 1.8^m; Period: 1733^d.

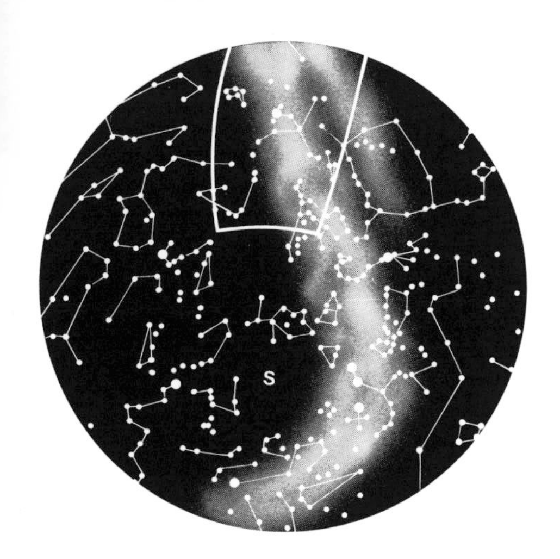

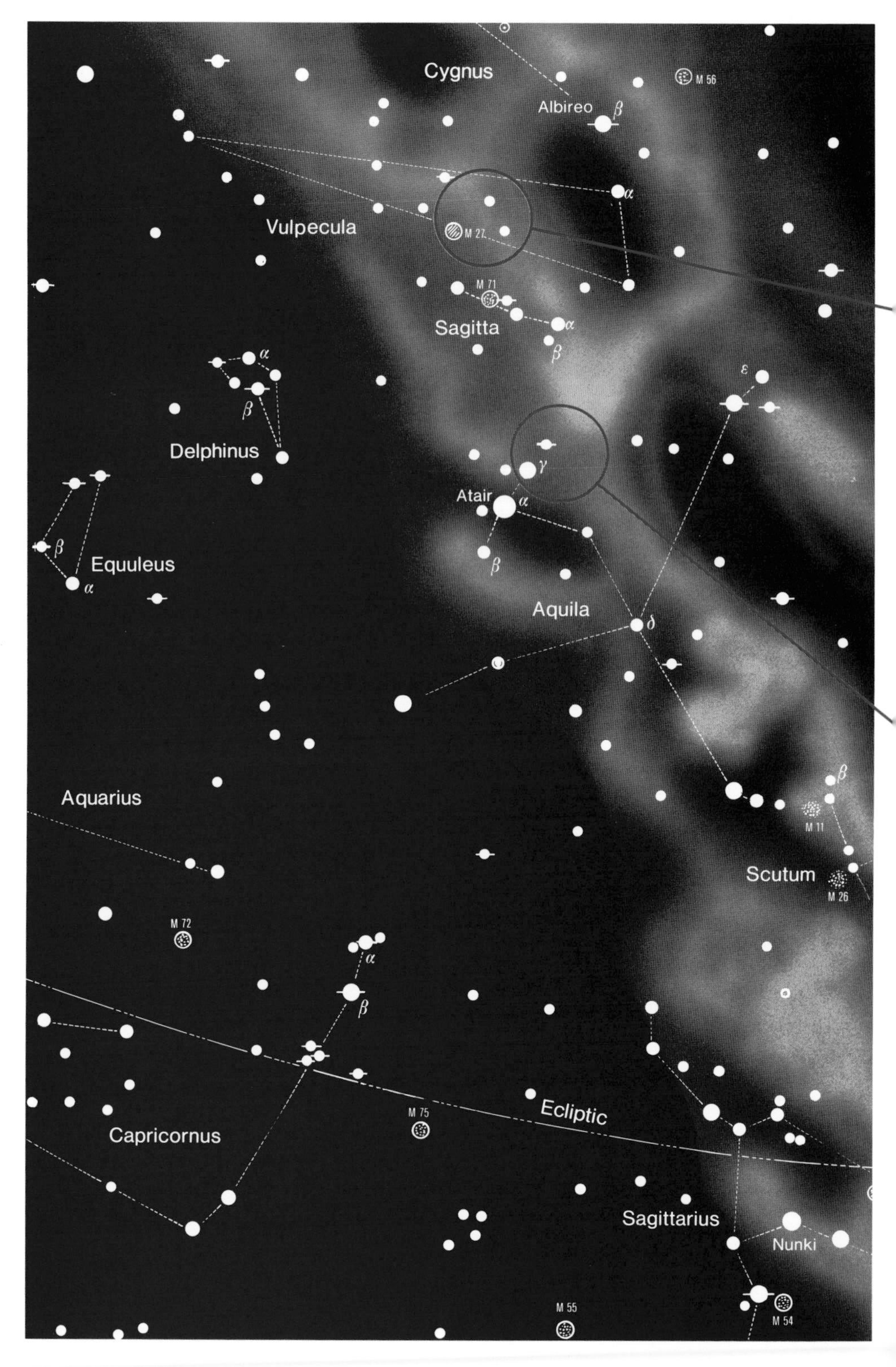

17 Aquila, Capricornus, Delphinus, Equuleus, Sagitta, Sagittarius, and Vulpecula

In 1918, the Nova Aquilae star reached, in a few hours, the brightness of Sirius, which is the brightest star in the sky. This outbreak of brightness characterizes the eruptive variable star, which reaches its minimum years later again.

Visibility At midnight over the southern horizon (mean local time): July. The stars in the map section stand each month about 2 hours earlier over the southern horizon: August 22^h, September 20^h.

Aquila (The Eagle) This constellation resembles a flying bird, whose head is depicted by the stars Atair (α), Alshain (β), and Tarazed (γ).

Caprjcornus (The Sea Goat) With binoculars, you can see that β Capricorni is a double star. The brighter star is yellow, the weaker star is bluish.

Delphinus (The Dolphin) This is a small constellation of stars that are equally bright (between 3.5 and 4.7^m).

Equuleus (The Little Horse) This square resembles a horse's head. Since it has no bright stars, it is difficult to find.

Star mag. 1 | Star mag. 2 | Star mag. 3 | Star mag. 4 | Star mag. 5 | Double star | Double- and var. star | Variable star

Open cluster | Globular cluster | Spiral galaxy | Diffuse nebula | Planetary nebula

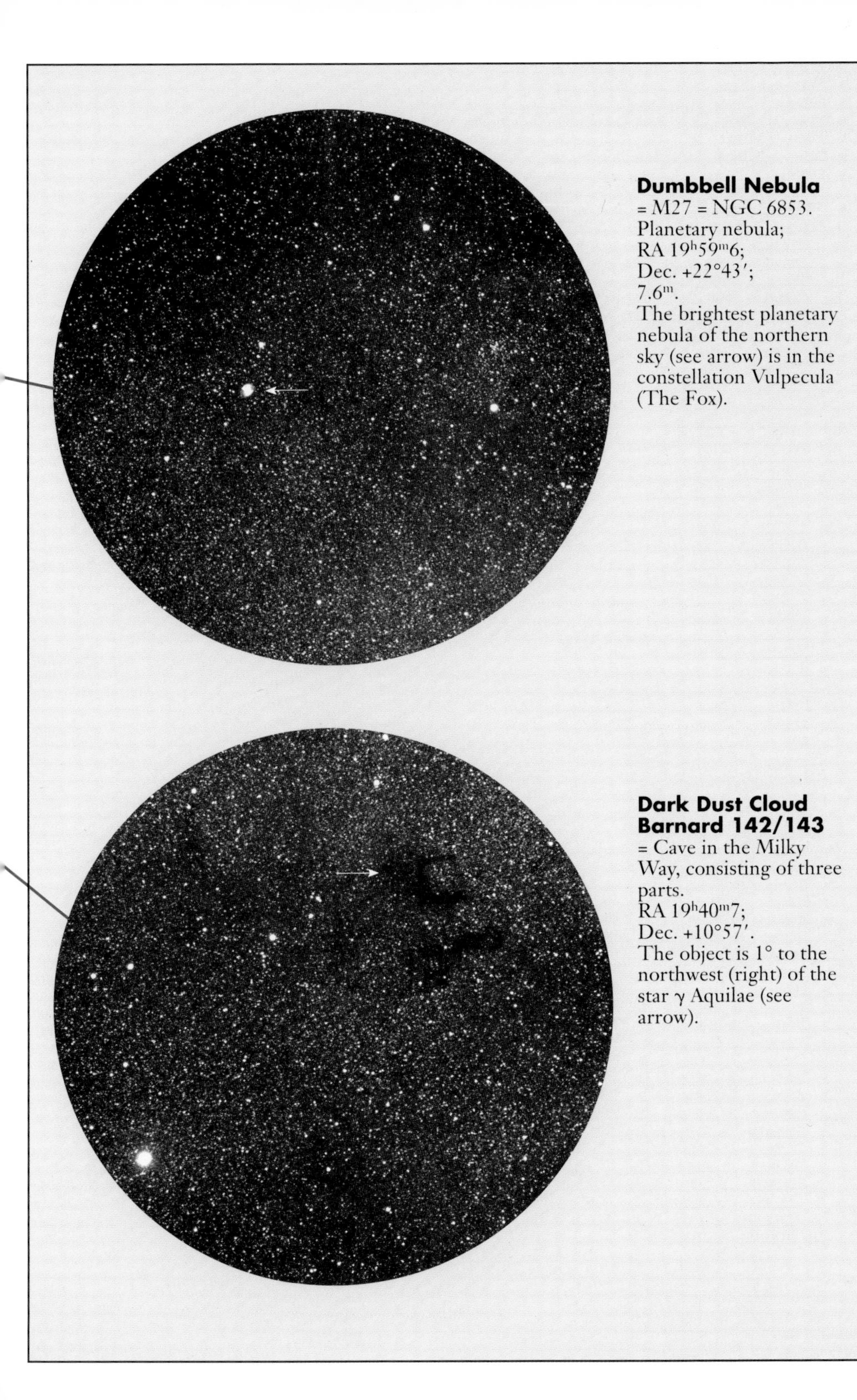

Dumbbell Nebula
= M27 = NGC 6853.
Planetary nebula;
RA $19^{h}59^{m}6$;
Dec. +22°43′;
7.6^{m}.
The brightest planetary nebula of the northern sky (see arrow) is in the constellation Vulpecula (The Fox).

Dark Dust Cloud Barnard 142/143
= Cave in the Milky Way, consisting of three parts.
RA $19^{h}40^{m}7$;
Dec. +10°57′.
The object is 1° to the northwest (right) of the star γ Aquilae (see arrow).

Sagitta (The Arrow) This constellation contains stars of the 4th magnitude, which are in a strikingly linear order.
Sagittarius (The Archer) You can see many objects in this constellation of the zodiac with binoculars. It is the most southern of all constellations of the zodiac. It has numerous nebulae and star clusters.
Vulpecula (The Fox) This constellation is between Aquila and Cygnus.

Observation Tips for the Dumbbell Nebula

The sphere of gas gives the impression of a pale-shining planetary disk through a telescope. From this comes the name planetary nebula for such celestial objects, even though they have nothing to do at all with the planets. The Englishman Lord Rosse found in his telescope a resemblance to a dumbbell and therefore he gave the nebula the name "Dumbbell Nebula." The bright sphere of gs outshines the central star.

The Dumbbell Nebula is recognizable in the star field, at 6–8-fold enlargement, as a distinct little nebulus disk. A good starting point to search for it is the star α Aquilae (Altair). From there, go to the star Gamma in the constellation Sagitta, and then about 3° to the north up to the Dumbbell Nebula (see arrow).

Observation Tips for the Dark Dust Cloud Barnard 142/143

You need a moonless night with a good view to see this object with binoculars. This dark dust cloud is a striking and clear contrast to the surrounding area of rich stars. One of the most impressive appearances in the sky is the Milky Way. The sparkles of the many stars show up magnificently through binoculars. Looking carefully, you will discover that the stars are once in a while disrupted by "black holes" that contain little or no stars. Scientists have found that these huge dark clouds are made from cosmic gas and dust, which cloud the view of stars or make it impossible to see them.

The Bright Stars

α Aquilae (Altair); RA $19^{h}50^{m}8$; Dec. +8°52′; 0.8^{m}.
γ Aquilae (Reda, Tarazed); RA $19^{h}46^{m}3$; Dec. +10°37′; 2.8^{m}.
β Capricorni (Dabih); RA $20^{h}21^{m}0$; Dec. −14°47′; $3.1-6.2^{m}$ (double star).
σ Sagittarii (Nunki); RA $18^{h}55^{m}3$; Dec. −26°18′; 2.1^{m}.

18 Pegasus, Aquarius, and Piscis Austrinus

The constellation Pegasus appears in the nothern evening sky in the fall. In Greek mythology, Pegasus is the winged horse born from the blood of Medusa after she was slain by Perseus.

Visibility At midnight over the southern horizon (mean local time): September. The stars in the map section stand each month about 2 hours earlier over the southern horizon: October 22^h, November 20^h.

Pegasus (Pegasus) The bright stars form a large square in the northern sky (leading constellation). In our section of the map, γ Pegasi and α Andromedae (see also map on p. 52) are missing in order to complete the square. The star Sirrah (α Andromedae) is added to the Pegasus square, even though it belongs to the constellation Andromeda.

Piscis Austrinus (Southern Fish) The Arabic name means "fish's mouth." The bright star Fomalhaut, which is the eighteenth brightest star of the entire sky, stands out.

Aquarius (Aquarius) This constellation of the zodiac is between Pegasus and Piscis Austrinus.

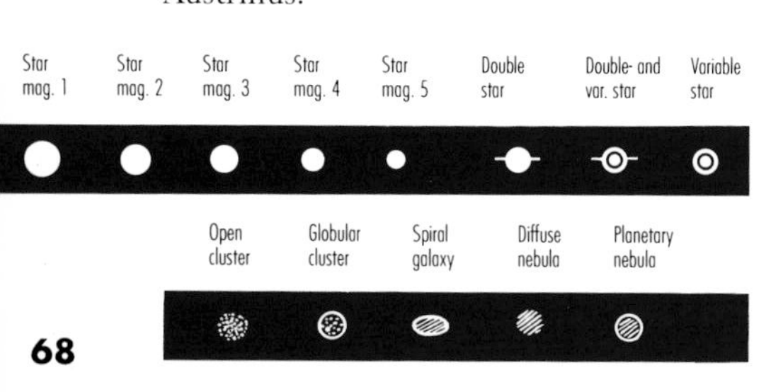

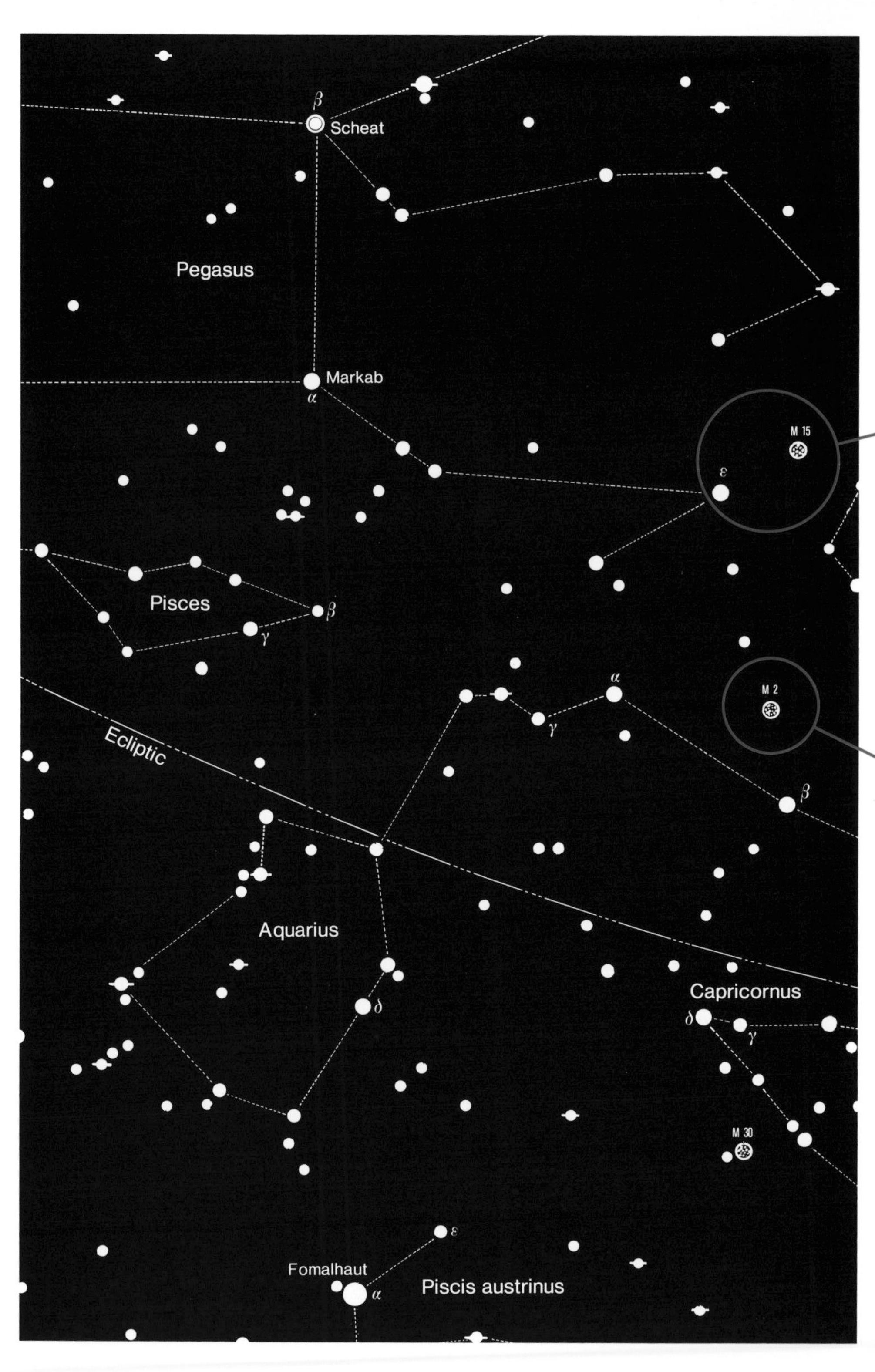

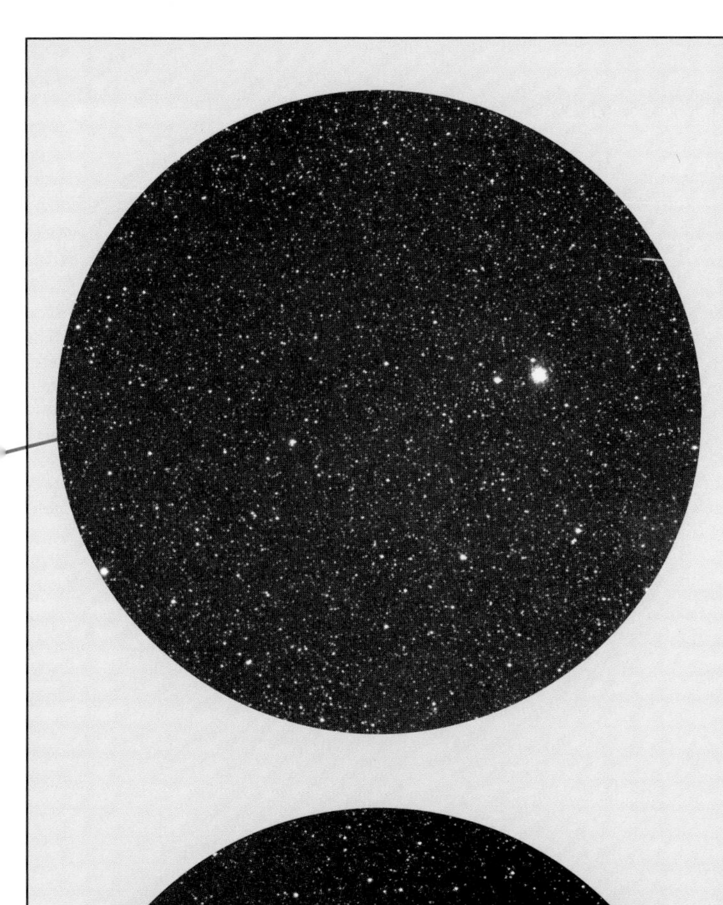

M15
= NGC 7078.
Globular cluster;
RA 21^h30^m0;
Dec. +12°10′;
6.3^m.
The globular cluster M15 is near the star ϵ Pegasi. Both objects are brought into the visual field by wide-angle binoculars.

M2
= NGC 7089.
Globular cluster;
RA 21^h33^m5;
Dec. −0°49′;
6.5^m.
M2 is directly in the center of the section of the picture. It can be seen with the naked eye under favorable circumstances.

Observation Tips for M15

The apparent brightness of M15 is 6.3, which makes it similar to M2—a star system that is easy to find. In the photo, M15, a ball-shaped cluster, is to the right of the center. A weak meteor trace is at the edge of the picture on the top right.

M15 in Pegasus forms with M3, M13, and M92 a group of globular clusters in the northern summer sky, which ends towards the east. The constellations Sagittarius, Scorpius, and, partly, Ophiuchus are rich in globular clusters, which account for the southern sky.

M15 has a limited naked-eye visibility, but it is easiliy seen with binoculars. Its distance is 10 parsec. The brightest individual stars have apparent magnitudes of around 14.5. About twenty-five of these stars are known.

Observation Tips for M2

To find M2, begin at the stars α Aquarii and β Aquarii. It appears in the visual field of wide-angle binoculars. β Aquarii is about a 5° distance from M2.

In contrast to M4 (see p. 65), M2 is a very compact globular cluster, and therefore it is difficult to resolve the individual stars, even along the edges. M2 is extremely evenly interspersed with weaker stars. It is hard to recognize any concentration of stars or dark clouds. Indeed, this sky region appears, in comparison to the overcrowded regions in the Cygnus or Sagittarius, rather uniform.

The Bright Stars

β Aquarii (Sadalsud); RA 21^h31^m6; Dec. −5°34′; 2.8^m.

α Pegasi (Markab); RA 23^h04^m8; Dec. +15°12′; 2.5^m.

α Piscis Austrini (Fomalhaut); RA 22^h57^m6; Dec. −29°37′; 1.2^m.

19 Phoenix and Tucana

The star atlas *Uranometria omnium asterismorum* by Johann Bayer, published in 1603, is said to be one of the most famous star atlases. It contains 1706 stars of the 1st to the 6th magnitude, which are distributed over forty-eight constellations of the northern and southern sky.

Visibility At midnight over the northern horizon (mean local time): October. The stars in the map section stand each month about 2 hours earlier over the northern horizon: November 22^h, December 20^h.

Phoenix (The Phoenix) Constellations of the southern sky, insofar as they cannot be observed from the northern hemisphere of the Earth, were depicted for the first time on maps in *Uranometria* by J. Bayer. The observations of navigators, who had sailed to Java and Sumatra, supplied the basic facts. They had arbitrarily formed new constellations and had named them after the tropical world of the birds. Look for the constellation Phoenix between the bright stars Fomalhaut (Piscis Austrinus) and Achernar (Eridanus).

Tucana (The Toucan) This constellation borders on the constellation Phoenix, which is directly to the south.

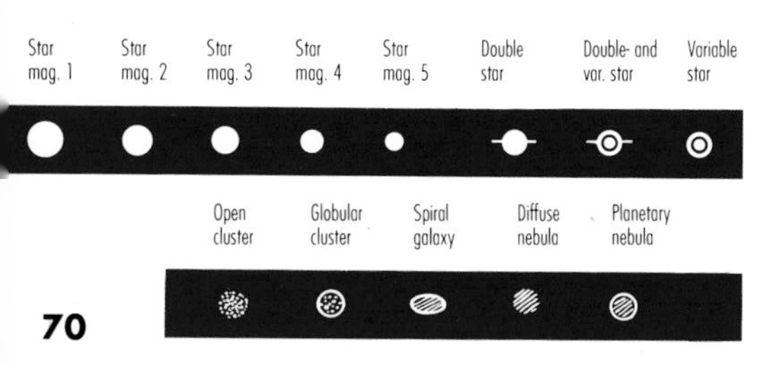

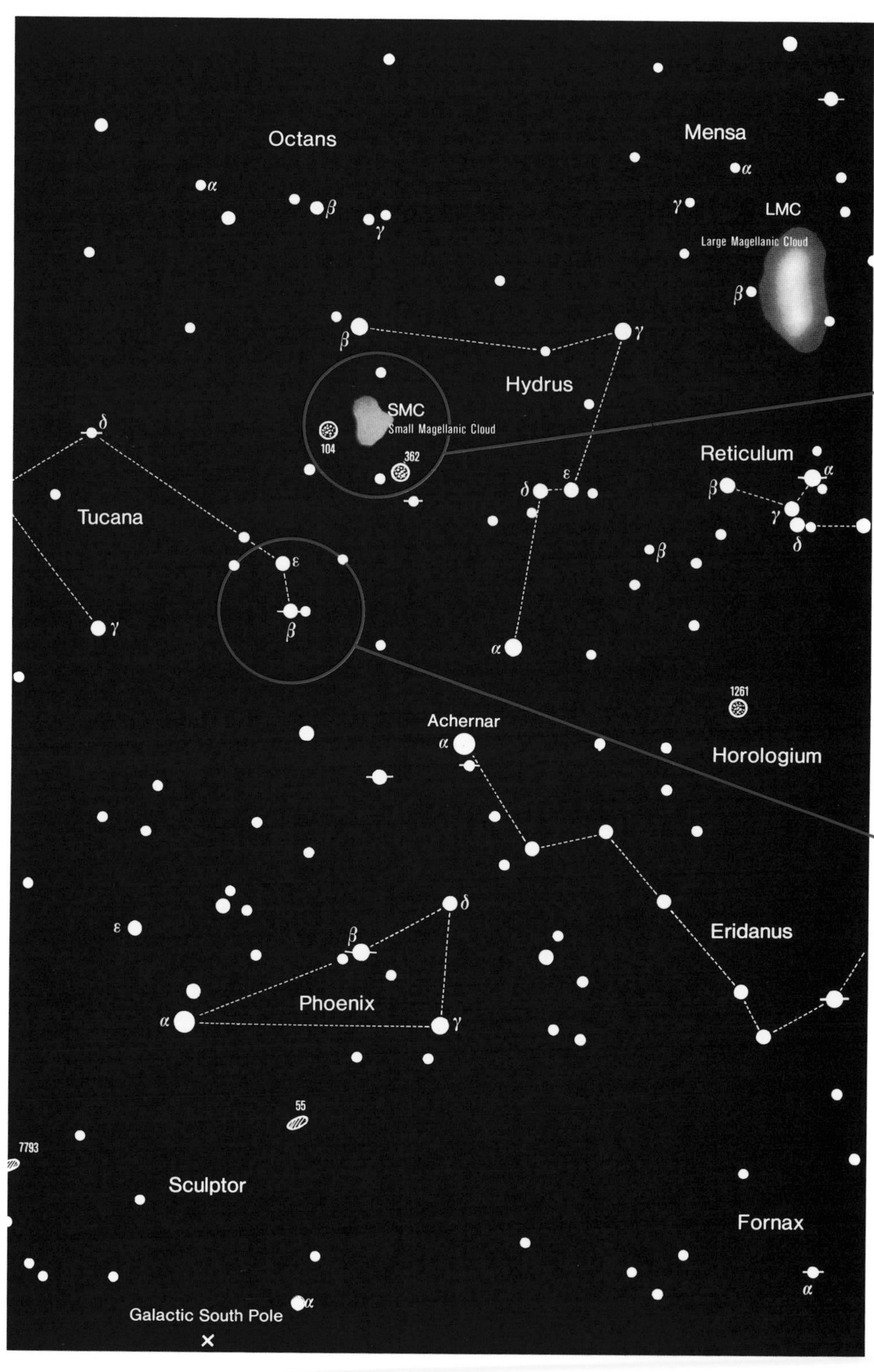

Small Magellanic Cloud
= SMC = Small Magellanic Cloud. Irregular galaxy; RA 0^h52^m7; Dec. −72°50′; 2.2^m. Astronomers in southern latitudes can see the next two extragalactic star systems: the two Magellanic Clouds (for the Large Magellanic Cloud, see p. 75).

β Tucanae
Double star; RA 0^h31^m5; Dec. −62°58′; 4.4^m and/or 4.5^m; 27″. Our object is in the center of the picture. Both stars are about equally bright (4.5 mag) and they are 27 arcseconds apart.

Observation Tips for the Small Magellanic Cloud

The cloud in the picture reminds us of an open cluster. Viewing it through a telescope will also find that resemblance. But the Small Magellanic Cloud is about ten times closer than the famous Andromeda Galaxy, which is why telescopes and cameras show so much detail.

The Small Magellanic Cloud is a striking object in the southern sky. The bright star Achernar (of the constellation Eridanus) may serve as an orientation guide. Starting from there, search for the object southwards towards the pole. The high southern declination makes the galaxy only in central southern latitudes a worthwhile object to see. There, the star system is circumpolar.

The Small Magellanic Cloud is almost free of absorbing dust; through it you can observe galaxies farther away. There are numerous globular clusters and clusters of variables. At the left, towards the edge of the picture, is the beautiful globular cluster 47 Tucanae (=NGC 104).

Observation Tips for β Tucanae

This is a double star for a small telescope (or binoculars from 12-fold enlargement or higher). It has a companion: The third star in the system has an apparent brightness of 14.0^m. At a distance of 2.2 arcseconds, it can only be resolved with large telescopes.

Tucanae forms an isosceles triangle with the bright star Achernar (of the constellation Eridanus) and the Small Magellanic Cloud. Brighter stars in the field are, for the most part, missing or low. This is because it is near the galactic South Pole in the constellation Sculptor (The Sculptor). With the declination −63°, β Tucanae can never be observed from Europe. From Central America, it can be seen rising a little above the horizon. In Australia and New Zealand, it can be observed almost the entire year.

The Bright Stars

α Phoenicis (Ankaa); RA 0^h26^m3; Dec. −42°18′; 2.4^m.

β Tucanae: RA 0^h31^m5; Dec. −62°58′; 4.5^m.

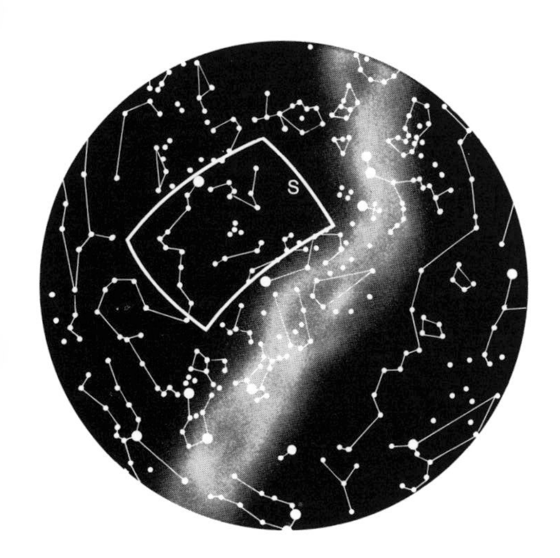

20 Eridanus, Hydrus, Horologium, Caelum, and Reticulum

The Babylonians saw the Euphrates in the constellation Eridanus, the Egyptians saw the Nile in it, while the Greeks saw the Eridanos, which flows far up in the north of Greece.

Visibility At midnight over the northern horizon (mean local time): November. The stars in the map section stand each month about 2 hours earlier above the northern horizon: December 22^h, January 20^h.

Eridanus (The River) This constellation begins at the constellation Orion (see p. 54) and reaches the southern declination with its main star Achernar.

Caelum (The Chisel) This constellation is between Eridanus and Columba (The Dove).

Hydrus (Lesser Water Snake) This constellation lies between the two Magellanic Clouds.

Reticulum (The Net) In the 1750s, Abbé Nicolas-Louis de Lacaille completed a list of the names of the constellations of

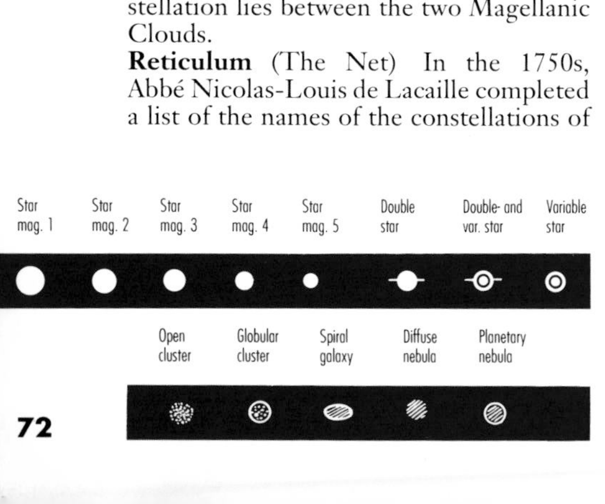

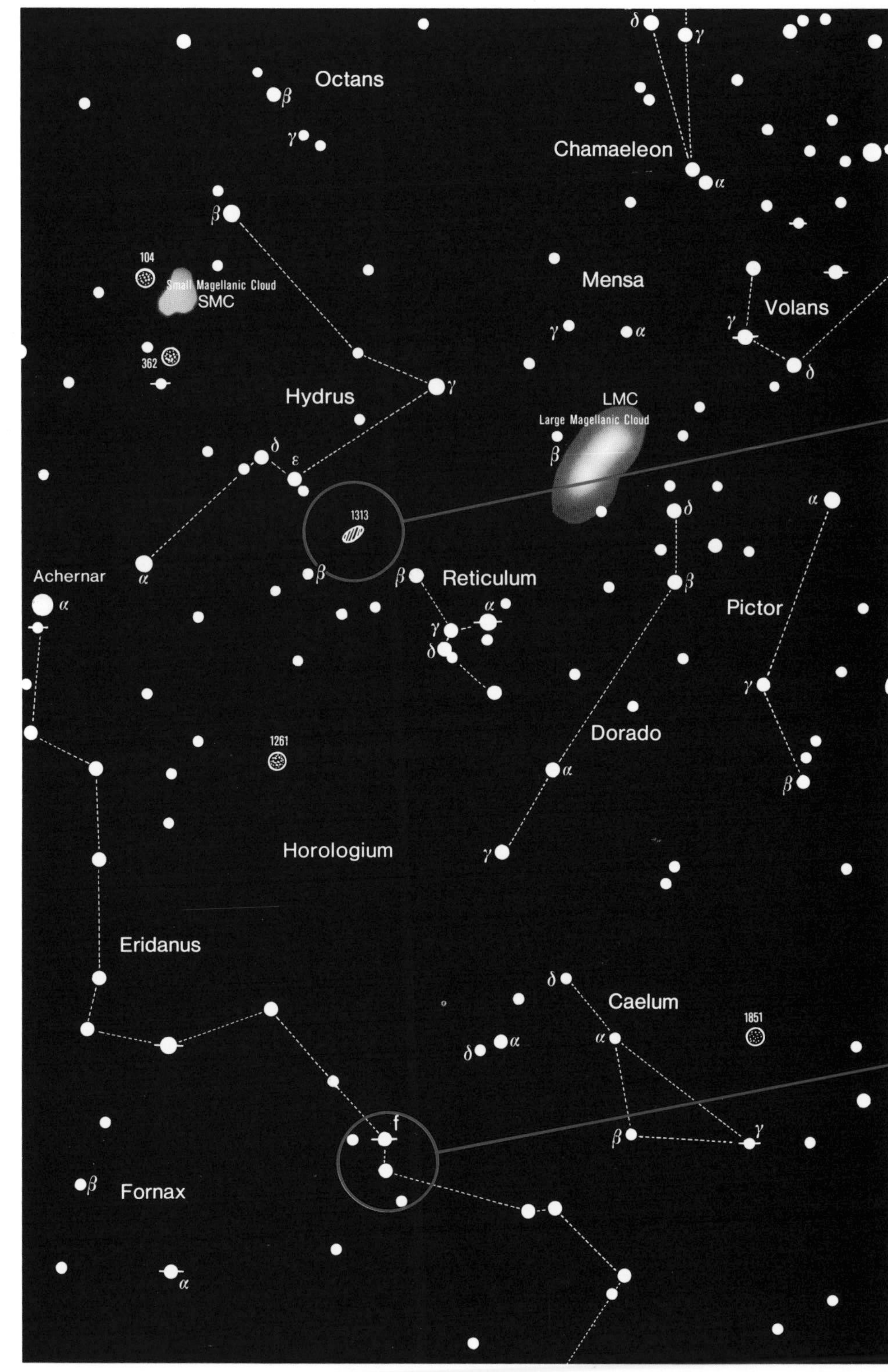

NGC 1313
Galaxy;
RA $3^{h}18^{m}3$;
Dec. $-66°30'$;
9.4^{m}.
The extragalactic system NGC 1313 stands in the center of the photo (see arrow). With an apparent magnitude of 9, it is not easy to see.

f Eridani
Double star;
RA $3^{h}48^{m}6$;
Dec. $-37°37'$;
4.8^{m} and/or 5.3^{m}; $7.9''$.
The double star f Eridani is in the middle of the connecting line between the stars β Fornacis and β Caeli.

the southern sky. He observed the southern sky from 1751 until 1754 from the Cape of Good Hope. He gave the name Reticulum to this constellation of tightly arranged stars in the middle between Achernar and Canopus.
Horologium (The Pendulum Clock) This constellation is between Reticulum and Eridanus constellations.

Observation Tips for NGC 1313

To find it, choose as a starting point the bright star Achernar (constellation Eridanus). From there, continue via α Hydri to β Horologii. NGC 1313 forms an isosceles triangle with β Horologii and the somewhat brighter star β Reticuli. Despite the moderate brightness and a small boundary, you will find NGC 1313 on a dark, moonless night with a telescope that is strong in light, since the object is neither embedded in Milky Way clouds, nor does it have a dense surrounding of stars.

Observation Tips for f Eridani

The stars have a brightness of 5.3^{m} and 4.8^{m}. The distance between them is 7.9 arc-seconds. The double star can more easily be found by first locating the arrangment of three stars, which are about equally bright and appear effortlessly in the visual field (see photo). f Eridani stands a little above the center of the picture (see arrow), about 2° above the three stars. Binoculars bring this distance into the visual field.

The Bright Stars

α Eridani (Achernar); RA $1^{h}37^{m}7$; Dec. $-57°14'$; 0.6^{m}.

α Hydri; RA $1^{h}58^{m}8$; Dec. $-61°34'$; 2.9^{m}.

α Reticuli; RA $4^{h}14^{m}4$; Dec. $-62°28'$; 3.4^{m}.

21 Puppis, Dorado, Volans, Carina, Pictor, Mensa, and Columba

The two Magellanic Clouds in the southern sky belong to what is called a Local Group of a combination of galaxies, which includes our Milky Way and the Andromeda Galaxy. They are named after Ferdinand Magellan, the Portuguese navigator.
Visibility At midnight over the northern horizon (mean local time): December. The stars in the map section stand each month about 2 hours earlier over the northern horizon: January 22h, Februay 20h.
Puppis (The Stern) This constellation lies between the constellation Canis Major and Carina. Only the southern part is shown here.
Volans (The Flying Fish) This small constellation lies directly next to the bright stars β Carinae and ϵ Carinae.
Dorado (The Swordfish) This constellation is with the Large Magellanic Cloud.
Pictor (The Painter) This constellation is between Dorado and Carina.

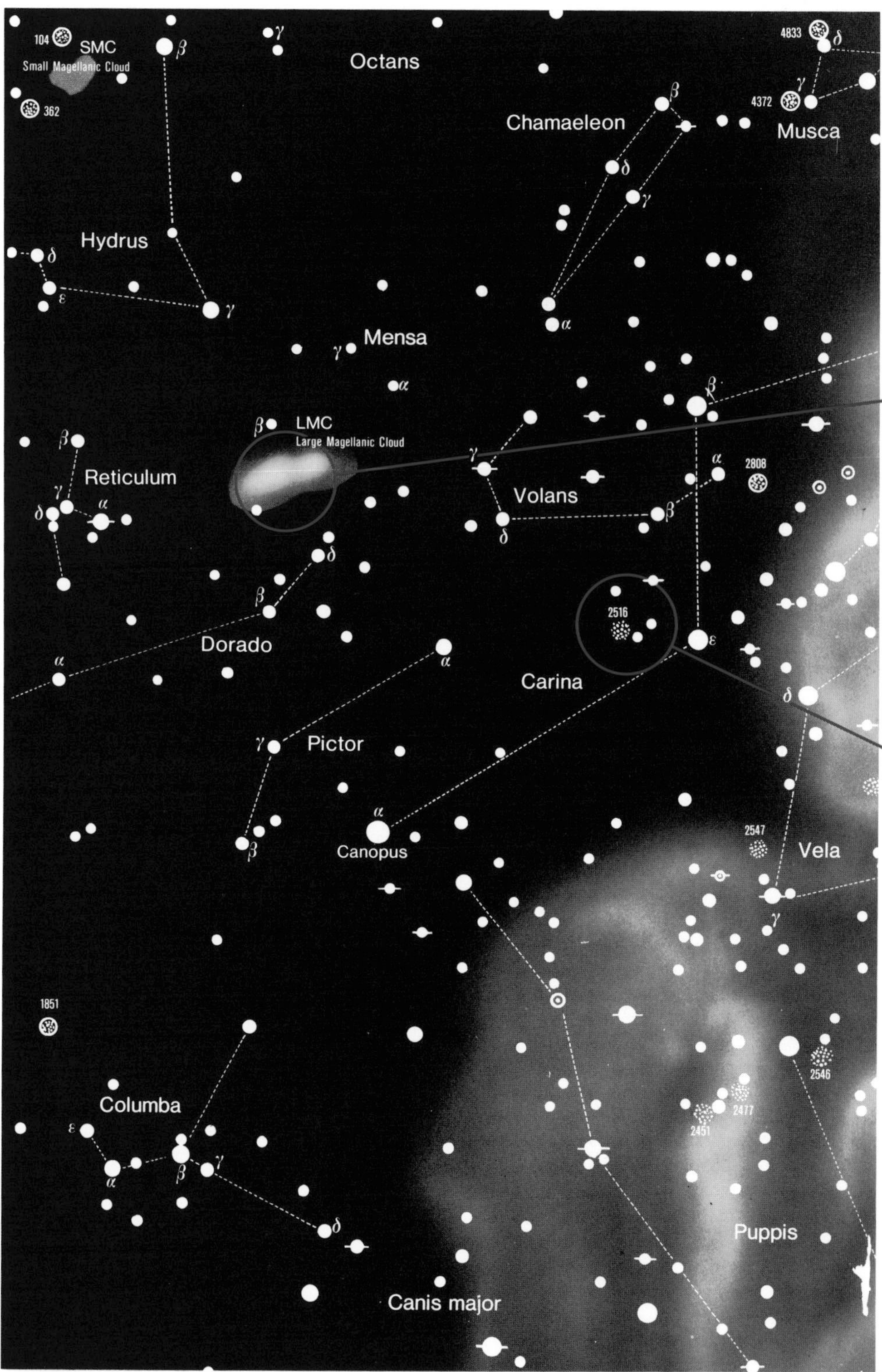

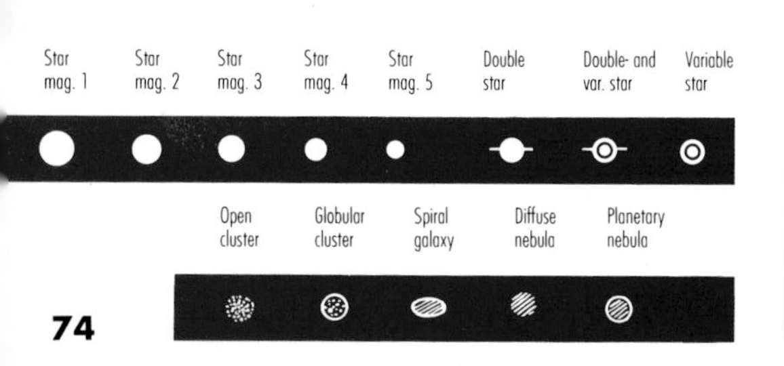

Carina (The Keel) This constellation in the southern sky lies in the Milky Way. It has rich star fields, especially three very bright stars.
Mensa (Table Mountain) This constellation has branches of the Large Magellanic Cloud.
Columba (The Dove) This is below the constellation Lepus.

Large Magellanic Cloud
= LMC = Large Magellanic Cloud.
Galaxy;
RA $5^{h}23^{m}6$;
Dec. $-69°45'$;
0.6^{m}.
The Large Magellanic Cloud is the only extragalactic system that can be seen with the naked eye during a full moon.

Observation Tips for the Large Magellanic Cloud

The galaxy almost forms an isosceles triangle with the bright Carina stars, Alpha and Beta. You have to be to the south of the Earth's equator in order to see the magnificent object high above the horizon. In central southern latitudes, the Large Magellanic Cloud is circumpolar. It covers an $8 \times 8°$ sky area. All the objects, already known from the Milky Way, have been found there: variable stars, open clusters, and globular clusters, planetary nebulae, nebulae, dark clouds. The brightest nebula in the Large Magellanic Cloud is NGC 2070; it is at the right edge of the photo and it is called the Tarantula Nebula. It can be well seen with the naked eye and it is the same type of nebula as the Orion Nebula.

NGC 2516
Open cluster;
RA $7^{h}58^{m}3$;
Dec. $-60°52'$;
3.8^{m}.
The bright open cluster NGC 2516 in the constellation Carina is in the center of the picture.

Observation Tips for NGC 2516

It is next to the very bright star ϵ Carinae. Wide-angle binoculars can bring NGC 2516 and ϵ Carinae into the same visual field. This open cluster can be seen with the naked eye; its apparent magnitude is 3.0. The environment of NGC 2516 is an outskirts area of the Milky Way, not comparable to the star swarm in the constellation Puppis and in areas of Carina, which are close to Crux (the Southern Cross). All open clusters belong to our Milky Way system and they are concentrated towards the galactic plane. That's why the open clusters occur in the star clouds of the Milky Way and in the outskirts of it. Our object is a matter of a concentrated open cluster, whose details become nicely visible in the small telescope.

The Bright Stars

α Carinae (Canopus); RA $6^{h}24^{m}0$; Dec. $-52°42'$; -0.72^{m}.

β Carinae (Miaplacidus); RA $9^{h}13^{m}2$; Dec. $-69°34'$; 1.7^{m}.

ϵ Carinae; RA $8^{h}22^{m}5$; Dec. $-59°31'$; 1.9^{m}.

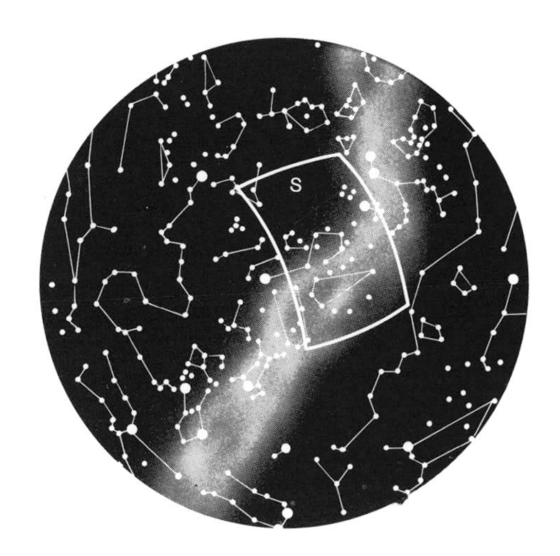

22 Antlia, Carina, Pyxis, and Vela

The galactic coordinates orient themselves according to the main plane of the Milky Way. Their intersection with the celestial sphere marks the galactic equator, which runs through the constellation Vela. It is inclined towards the celestial equator 62°36′.

Visibility At midnight over the northern horizon (mean local time): February. The stars in the map section stand each month about 2 hours earlier above the northern horizon: March 22^{h}, April 20^{h}.

Antlia (Air Pump) This constellation is to the south of the star Alphard (Hydra).

Carina (The Keel) The magniificent star clouds of the Milky Way characterize this part of the constellation (see also p. 74).

Pyxis (The Compass) This small constellation is between Puppis and Antlia.

Vela (The Sails) The galactic equator runs through this constellation. The bright star γ Velorum A is said to be one of the hottest stars known. It is bluish and the surface temperature reaches 30,000° (sun 6000°). It is too far south for most northern observers to see.

Star mag. 1 | Star mag. 2 | Star mag. 3 | Star mag. 4 | Star mag. 5 | Double star | Double- and var. star | Variable star

Open cluster | Globular cluster | Spiral galaxy | Diffuse nebula | Planetary nebula

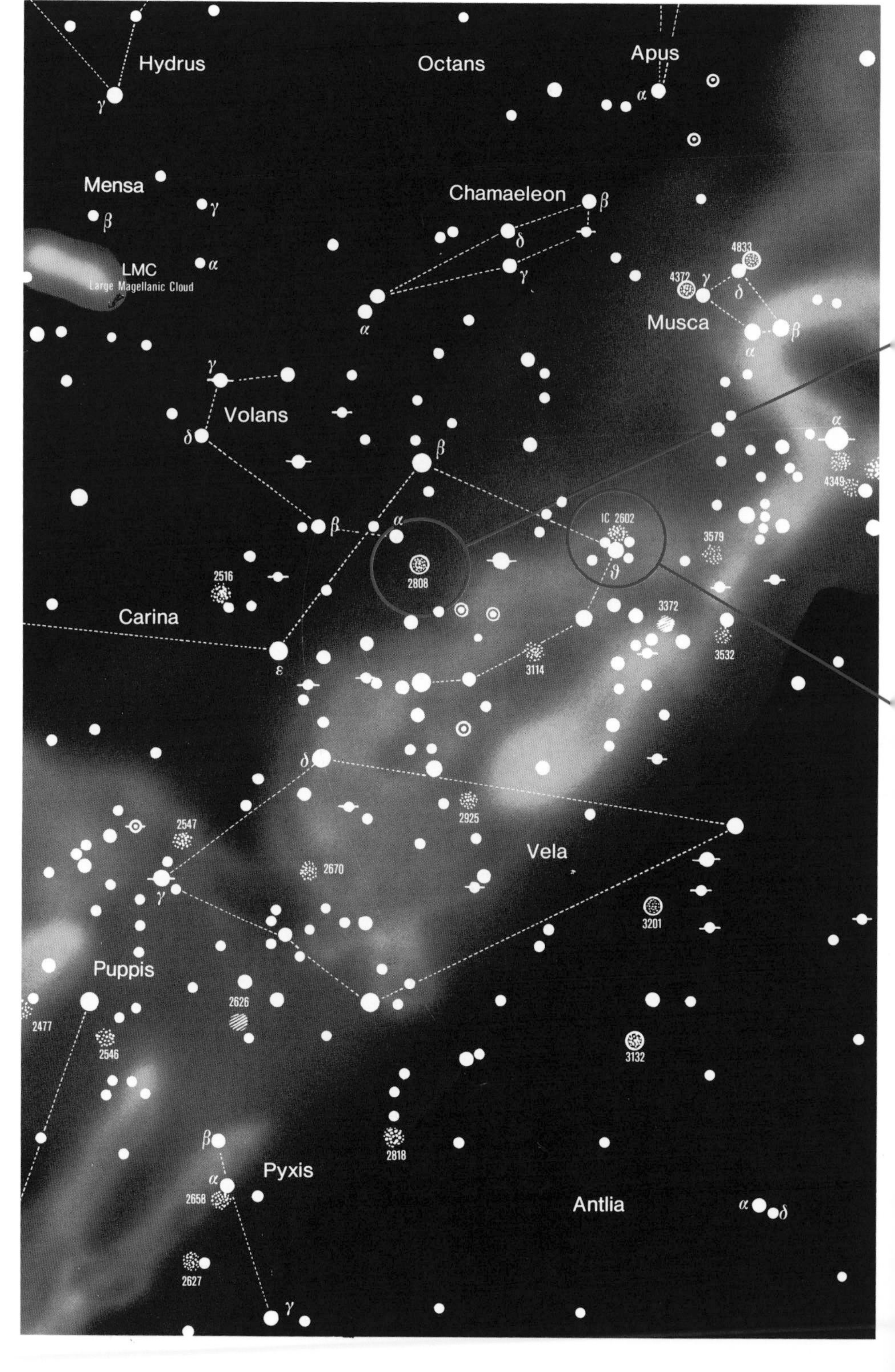

NGC 2808
Globular cluster;
RA 9^h12^m0;
Dec. $-64°52'$;
6.3^m.
The globular cluster NGC 2808 is in the middle of the photo. At the center of NGC 2808 is a group of loosely packed stars.

IC 2602
Open cluster;
RA 10^h43^m2;
Dec. $-64°24'$; 1.9^m.
This is an open cluster in the clouds of the Milky Way. IC 2602 is directly in the center of the photo (see arrow). The little cluster to the left above is the object Melotte 101.

Observation Tips for NGC 2808

Even though the area is full of stars, this globular cluster stands out, since there are no brighter star fields directly close by. In order to find it, use binoculars and start at the bright star β Carinae. The next step is α Volantis, which is in the top left of the photo. NGC 2080 is in the same visual field. (As a general rule, the visual field of the binoculars is usually larger—between 3° and 8°—than that of an astronomical telescope.)

With the declination −64°, NGC 2808 is only in Central America—and in South India to some extent—high enough above the horizon. In Brazil, South Africa, and Australia, the cluster reaches the proximity of the zenith.

Observation Tips for IC 2602

The objects preceded with an "M" go back to the nebula list of the French astronomer Charles Messier; they are brighter objects between the declination +70° and −35° (see also p. 144–145). A new catalog for nebulae and clusters has the name NGC (*New General Catalogue* by Dreyer). Also M-objects have an NGC number; there are over six thousand NGC objects. There is also a supplement called the *Index Catalogue*, with the abbreviation IC. This abbreviation is also used to identify nebulae and clusters in connection with a number—for example: the open cluster IC 2602.

In order to find IC 2602, it is helpful to know that the star θ Carinae is very close by. Once you have this star in your visual field, then you also automatically have IC 2602 and Melotte 101. θ Carinae lies almost in the middle of the line from ϵ Carinae to α Crucis.

The Bright Stars

γ Velorum;
RA 8^h09^m5;
Dec. $-47°20'$;
1.8^m.

δ Velorum;
RA 8^h44^m7;
Dec. $-54°42'$;
1.9^m.

λ Velorum
(Suhail);
RA 9^h08^m0;
Dec. $-43°26'$
2.2^m.

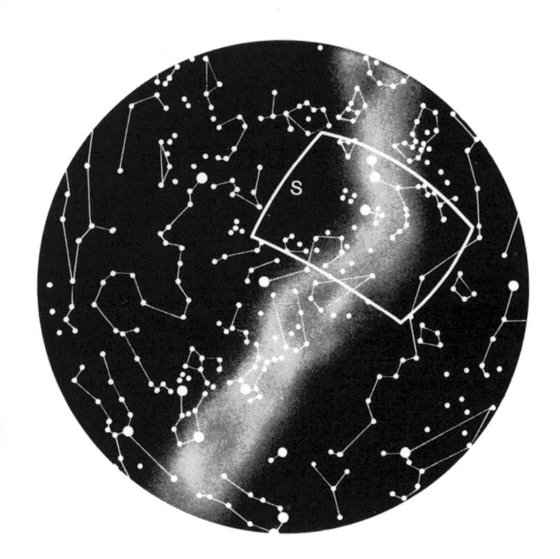

23 Chamaeleon, Musca, Centaurus, and Crux

Crux, the Southern Cross, was in antiquity part of the constellation Centaurus. In 1225, it was pictured on an Arabian celestial globe and in 1574 on a globe of the astronomical clock at the Cathedral in Strassburg, France.

Visibility At midnight over the northern horizon (mean local time): March. The stars in the map section stand each month about 2 hours earlier over the northern horizon: April; 22^h, May 20^h.

Chamaeleon (The Chameleon) The constellation, which is located close to the celestial South Pole, consists of six main stars of the 4th and 5th magnitude.

Musca (The Fly) This constellation lies between Crux and Chamaeleon.

Centaurus (The Centaur) This constellation, with its numerous bright stars, frames Crux. The two main stars, together with Crux, are an orientation help in the southern sky (see also p. 21). Alphae Centauri A is the third brightest star in the sky; Beta Centauri is the tenth brightest.

Crux (The Southern Cross) This the most famous constellation in the southern sky embedded in the Milky Way. The stars

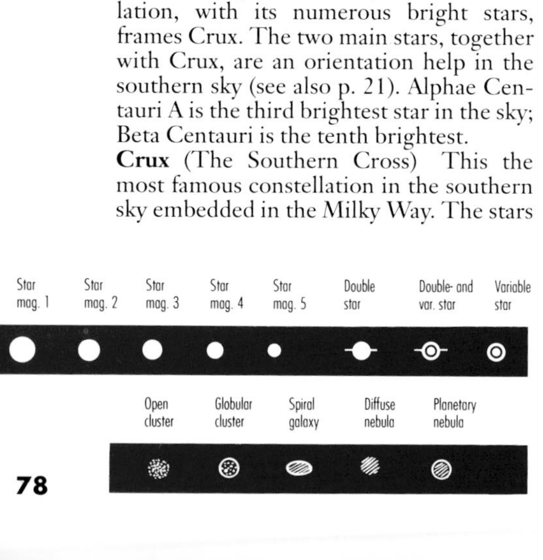

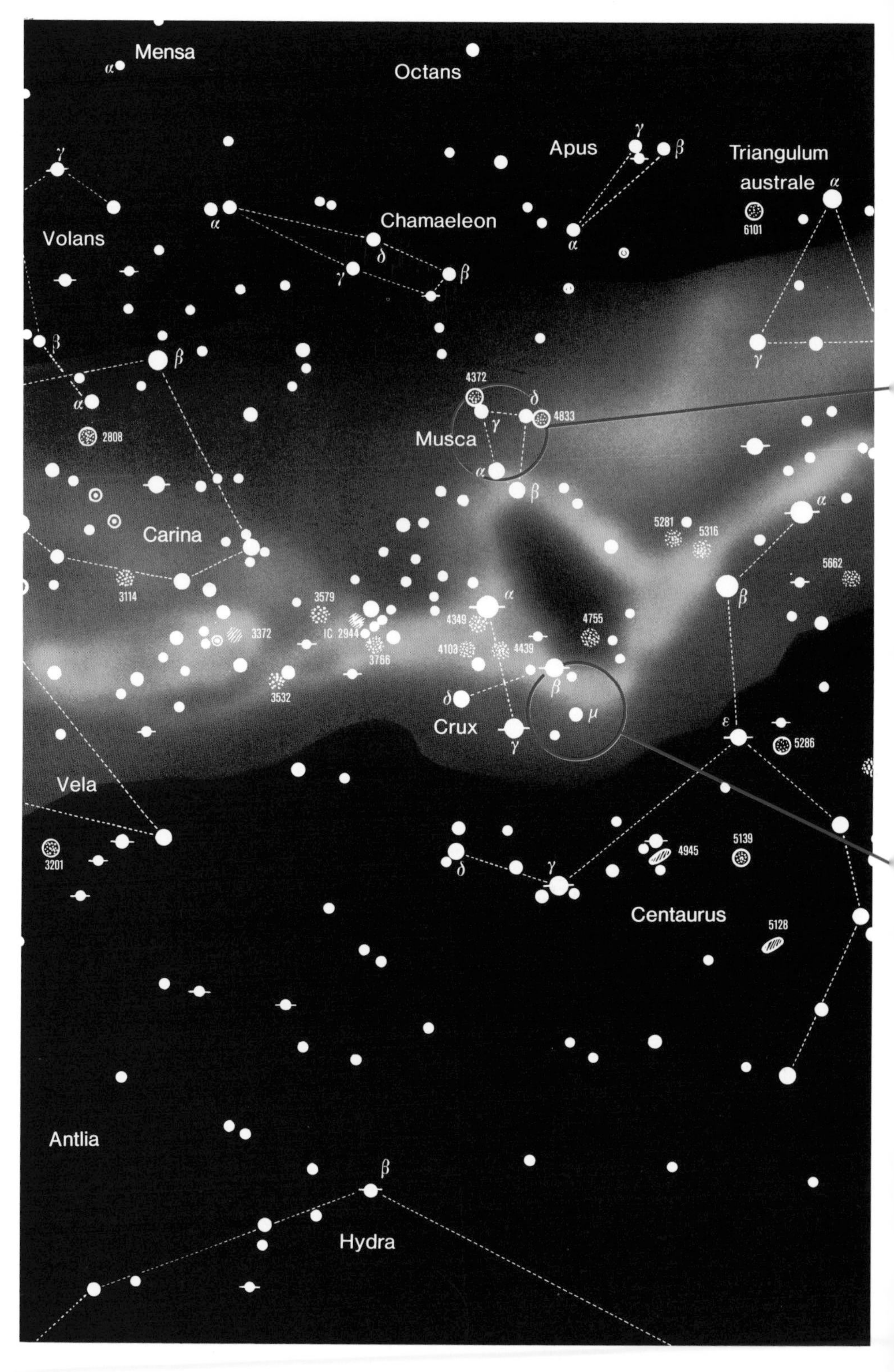

NGC 4372 and 4833
NGC 4372: Globular cluster;
RA 12^h25^m8;
Dec. $-72°40'$;
7.8^m.
NGC 4833: Globular cluster;
RA 12^h59^m6;
Dec. $-70°53'$.
An interesting section in the constellation Musca (The Fly).

μ Crucis
Star pair and double star;
RA 12^h54^m6;
Dec. $-57°11'$;
4.3^m and/or 5.3^m; 35″.
This section shows in the upper part a little bit of the Milky Way in the area of Crux (the Southern Cross). The field is scattered with stars!

α Crucis and γ Crucis form the long beam of the cross, which points directly towards the celestial South Pole.

Observation Tips for NGC 4372 and 4833

α Muscae is imbedded in the Milky Way, close to the lower edge of the photo. Towards the right edge stands NGC 4833 and on the top left NGC 4372 (see arrow). Starting from the center of the picture, in the direction of NGC 4372, is a distinct, dark cloud.

The closeness of Crux is helpful in finding α Muscae. Start at the bright star α Crucis. Search in the direction of the South Pole. In wide-angle binoculars, α Crucis and α Muscae are together in the visual field. Very close to NGC 4372 is γ Muscae, a star that clearly stands out of the field. Near NGC 4833 is also a bright star (a star pair in the telescope). Above α Muscae are three small dark clouds. In the dense star field of the Milky Way, the dark clouds are clearly marked although not as strikingly as the above-described dark cloud.

Observation Tips for μ Crucis

The bright star at the upper edge of the photo is β Crucis; to the right of it is an open cluster, NGC 4852 in the constellation Centaurus. In the lower center of the picture is the double star μ Crucis (see arrow). With the less bright star, which is below it to the right, it gives the appearance of a double star to the naked eye. But here, it is only a matter of a constellation—it has nothing to do with the "real" double star μ Crucis.

The companion of μ Crucis is at a distance of 35 arc-seconds at a magnitude difference of 4.3 to 5.3—i.e., an easy object for binoculars. At a declination of almost −57°, this northernmost brighter star of Crux can be seen only in North Africa and Mexico close to the horizon. It comes close to the zenith in mid-southern latitudes.

The Bright Stars

α Centauri A; (Toliman); RA 14^h39^m6; Dec. $-60°50'$; -0.1^m.
β Centauri (Agena); RA 14^h03^m8; Dec. $-60°22'$; 0.6^m.
α Crucis (Acrux); RA 12^h26^m6; Dec. $-63°06'$; 1.1^m.
β Crucis; RA 12^h47^m7; Dec. $-59°41'$; 1.3^m.
γ Crucis; RA 12^h31^m2; Dec. $-57°07'$; 1.6^m.

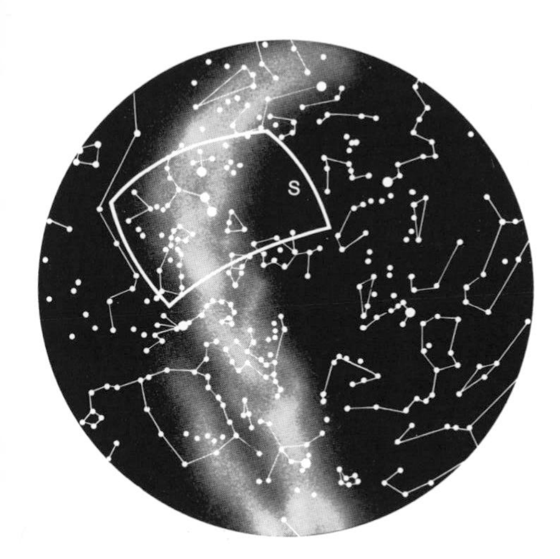

24 Apus, Triangulum Australe, Norma, Lupus, and Circinus

The bright star α Centauri A is the fixed star that, at 4.3 light-years away, is closest to our solar system. It forms with two stars a multiple star system. The third star of the system, α Centauri C, was discovered only in 1915.

Visibility At midnight over the northern sky (mean local time): May. The stars in the map section stand each month about 2 hours earlier over the northern horizon: June 22^{h}, July 20^{h}.

Apus (The Bird of Paradise) This is one of the constellations that is closest to the South Pole.

Triangulum Australe (The Southern Triangle) This triangular constellation does not stand out much in the area of the Milky Way. It is close to the bright stars α Centauri and β Centauri.

Norma (The Level) This constellation is at the borderline of visibility with the naked eye. Globular clusters and a planetary nebula!

Lupus (The Wolf) This constellation was formerly combined with the constellation Centaurus. The spear-throwing Cen-

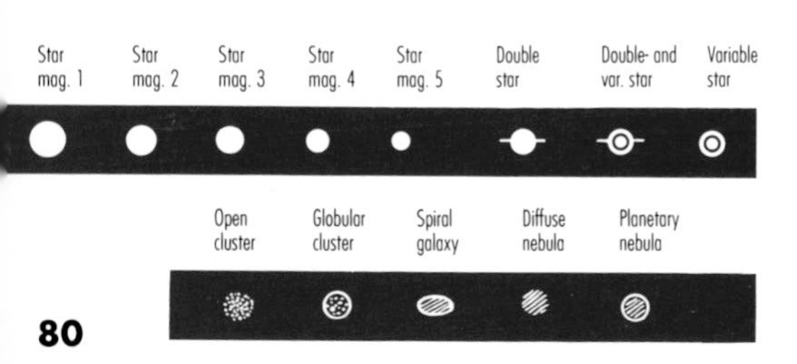

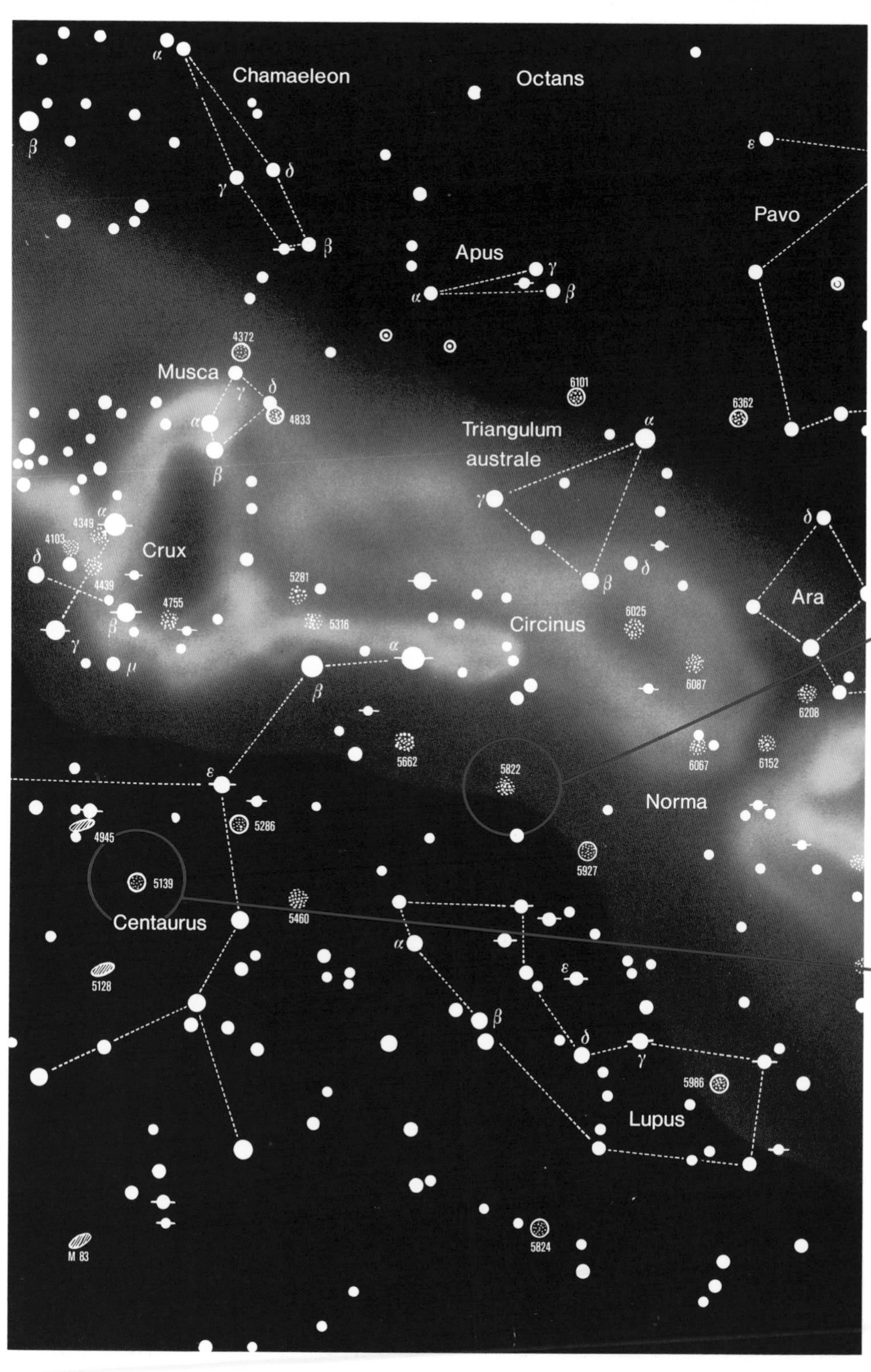

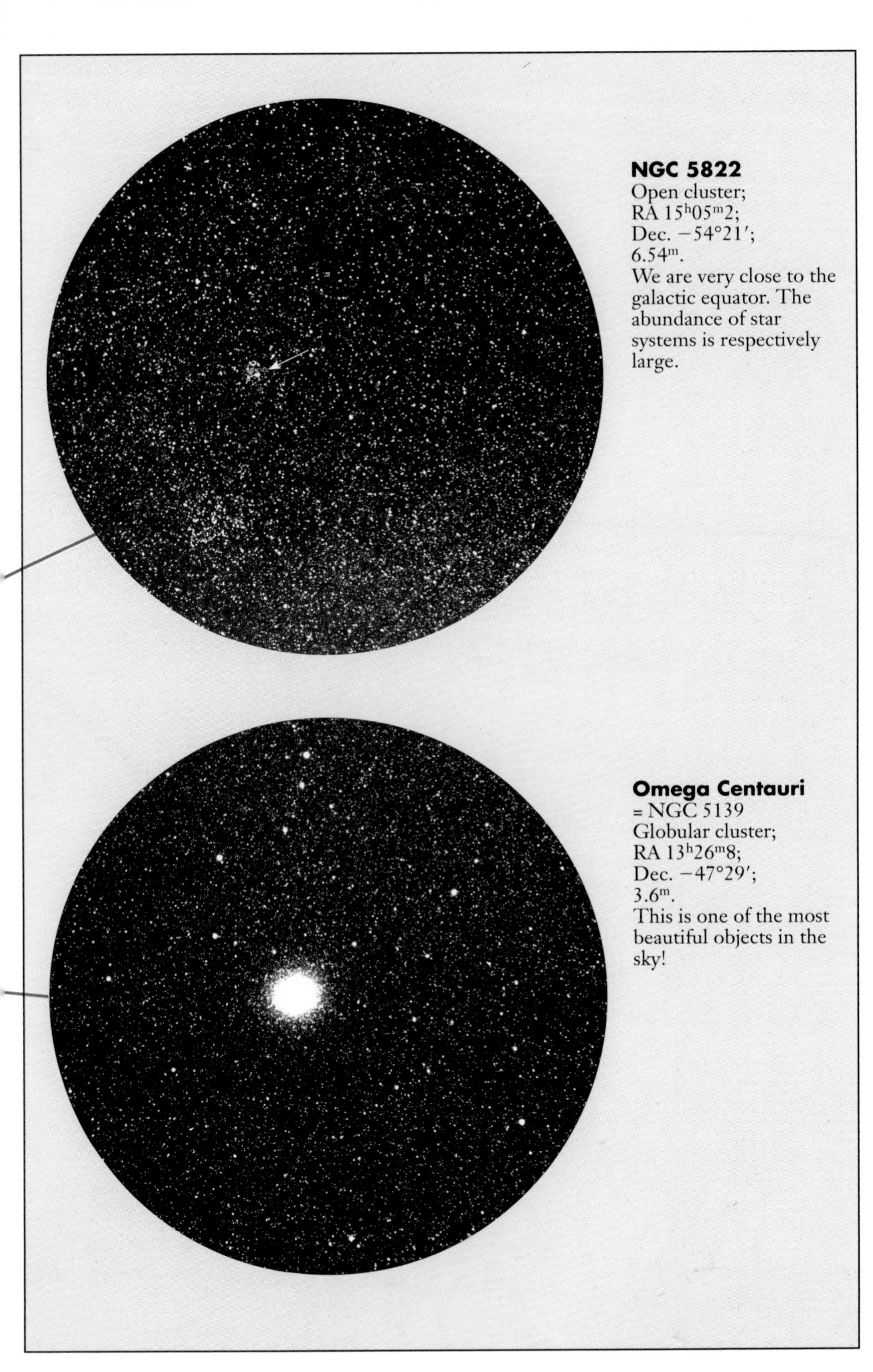

NGC 5822
Open cluster;
RA 15^h05^m2;
Dec. $-54°21'$;
6.54^m.
We are very close to the galactic equator. The abundance of star systems is respectively large.

Omega Centauri
= NGC 5139
Globular cluster;
RA 13^h26^m8;
Dec. $-47°29'$;
3.6^m.
This is one of the most beautiful objects in the sky!

taur, who hunts the wolf, appears on many older illustrations. Most of its stars are not visible from mid-northern latitudes.
Circinus (The Compasses) This constellation is between the Triangulum Australe and Centaurus.

Observation Tips for NGC 5822

This large and loose open cluster (of about 150 stars) is visible with binoculars or a small telescope. In the picture, it is to the left of the center of the picture (see arrow). To find it, draw a connecting line between the stars α Centauri and γ Lupi. NGC 5822 is in the first third of the distance, closer to the star α Centauri.

You cannot recognize a central increase in density in open clusters, which is the case for NGC 5822. This form is irregular. Open clusters are also called galactic clusters, because they lie in the plane of our galaxy. The accumulation near the galactic equator confirms that. You can watch interstellar dark clouds again and again in these regions.

Observation Tips for Omega Centauri

ω Centauri (NGC 5139) forms an equilateral triangle with the stars β Centauri and α Lupi. Its apparent diameter reaches the full moon while its apparent brightness is 4th magnitude.

J. Herschel (from his 1837 observation from South Africa) describes his impression: "This most beautiful of all star clusters fills my 18-inch telescope with its densest part filled with thousands of stars." Binoculars or a small telescope can give you the same impression.

Because it is so far south of the celestial equator, ω Centauri is visible only in the southernmost states of the U.S., such as the Gulf States, New Mexico, Arizona, and Southern California. There, you can detect it low on the horizon on clear spring and summer evenings. To the naked eye, ω Centauri looks like a fuzzy star. Through binoculars or a small telescope, it looks like a hazy patch.

The Bright Stars

α Lupi;
RA 14^h41^m9;
Dec. $-47°23'$; 2.3^m.
β Lupi;
RA 14^h58^m5;
Dec. $-43°08'$; 2.7^m.
α Triangulum Australis;
RA 16^h48^m7;
Dec. $-69°02'$;
1.9^m.

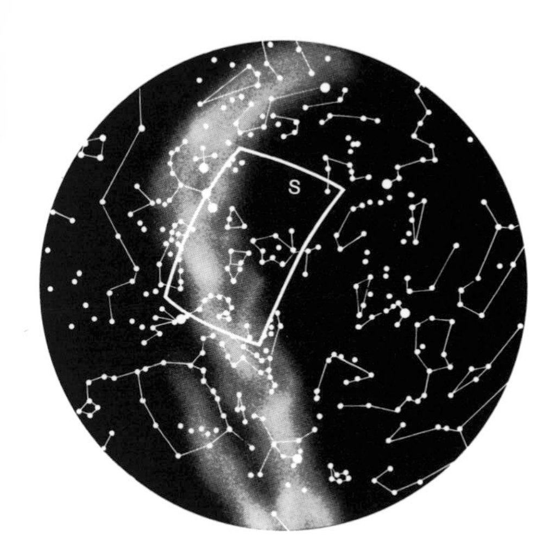

25 Ara, Pavo, and Scorpius

In the 3rd millenium B.C., the people in the Caucasus drew the constellations Scorpius and Sagittarius. The different brightnesses were carved into rocks in circles of different sizes.

Visibility At midnight over the northern horizon (mean local time): June. The stars in the map section stand each month about 2 hours earlier over the northern horizon: July 22^h, August 20^h.

Ara (The Altar) This constellation is to the south of Scorpius in the rich part of the Milky Way.

Pavo (The Peacock) Ten stars with brightnesses between 3rd and 4th magnitude form this constellation. The bright star α Pavonis (2.1 mag), which has also the name Peacock, stands at the border to the constellation Telescopium.

Scorpius (The Scorpion) The southern part is shown in this section. When you have the entire constellation in front of you (it ranges between −10° and −45° declination), it is easy to make out the scorpion (see p. 12). In the U.S., this magnificent constellation is visible in its entirety from the Gulf States, Arizona, and South-

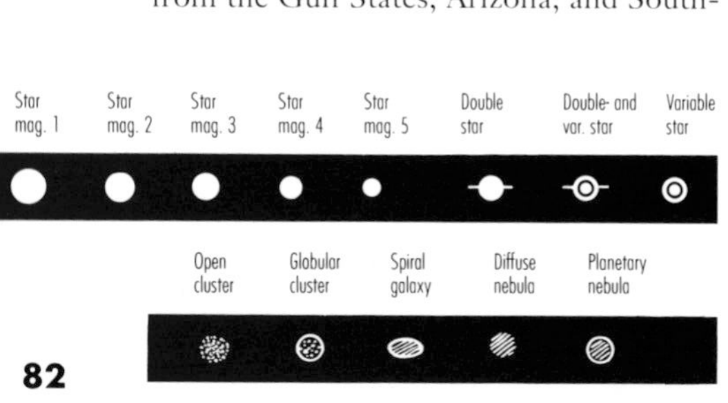

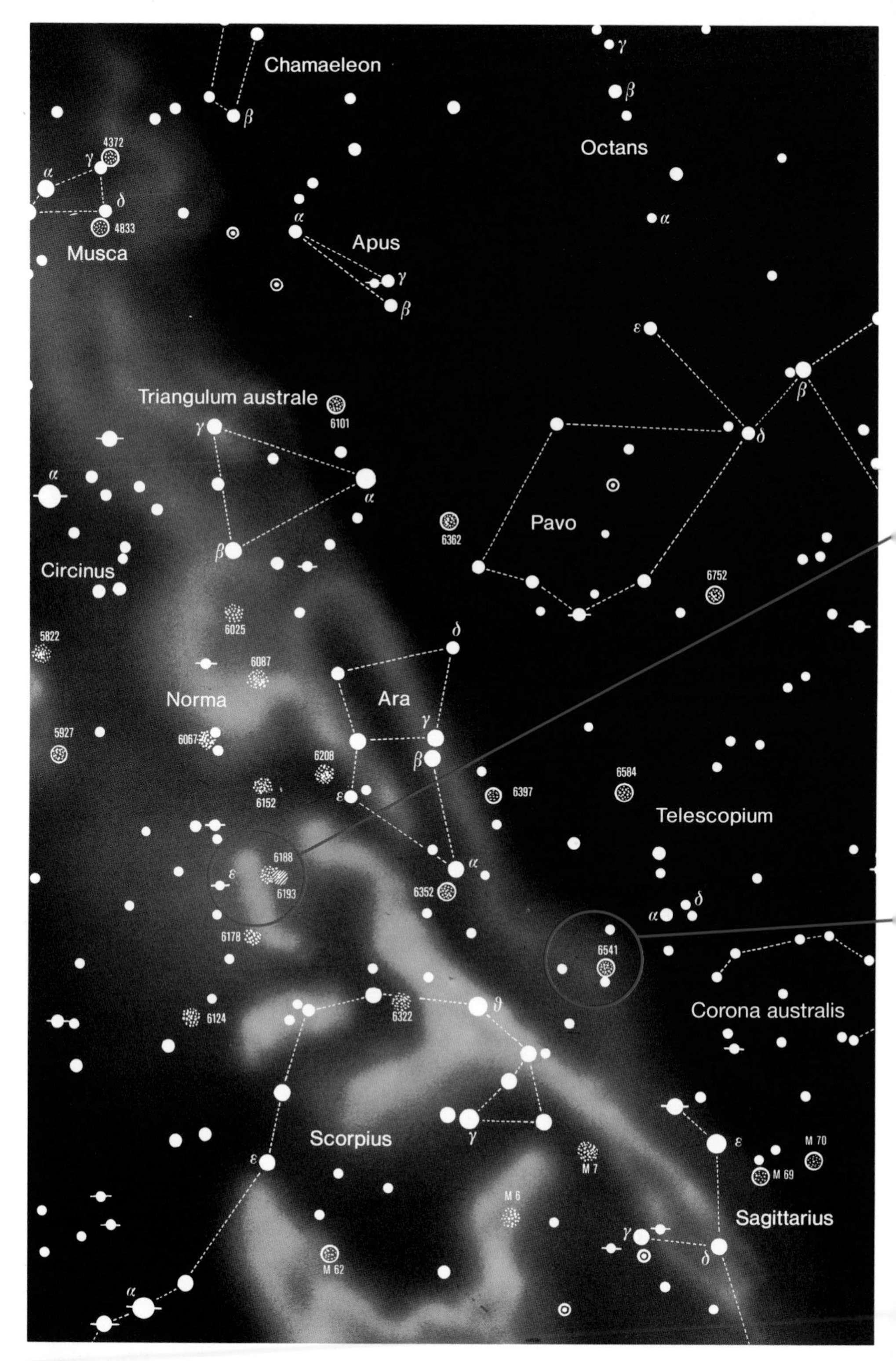

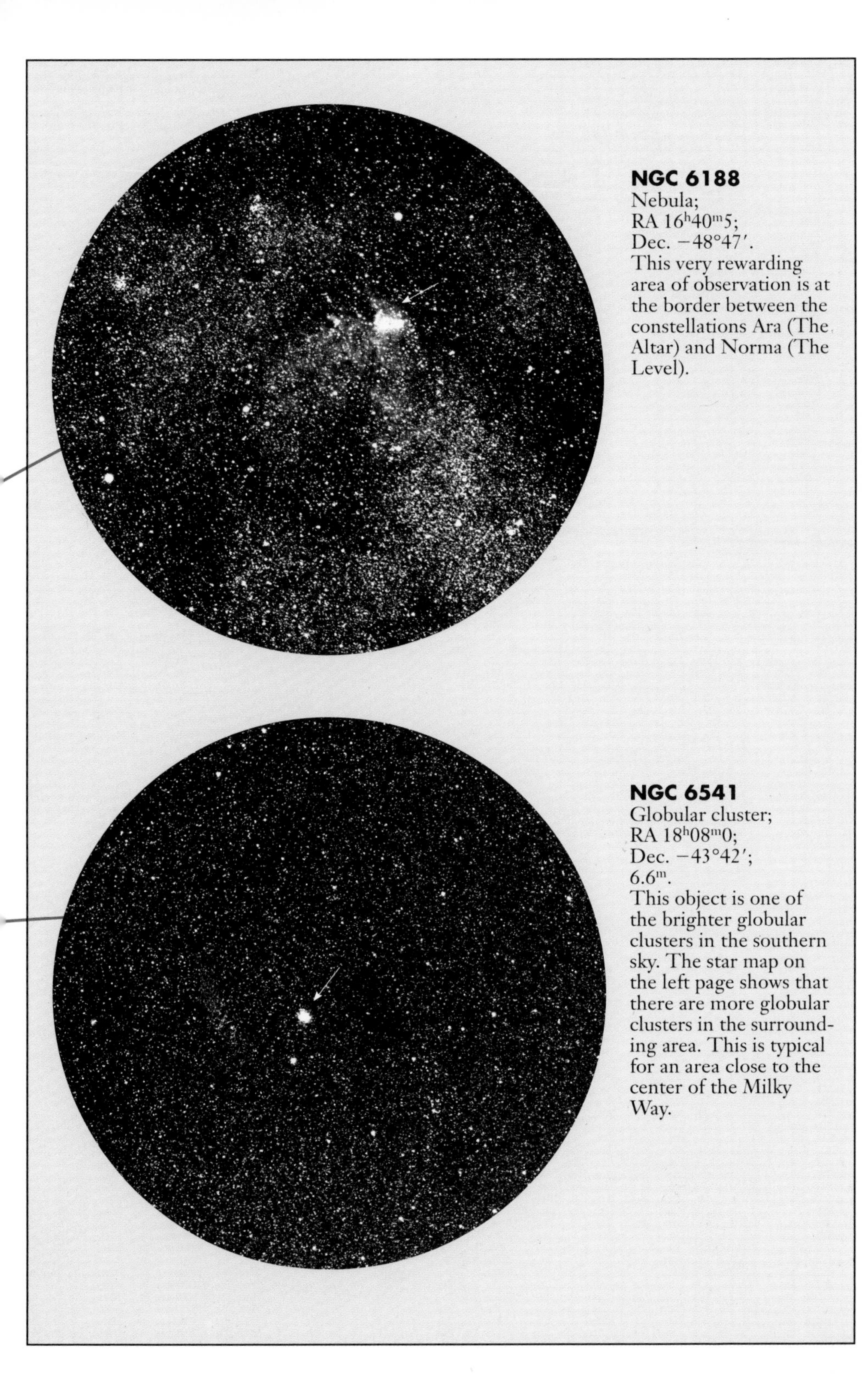

NGC 6188
Nebula;
RA $16^{h}40^{m}5$;
Dec. −48°47′.
This very rewarding area of observation is at the border between the constellations Ara (The Altar) and Norma (The Level).

NGC 6541
Globular cluster;
RA $18^{h}08^{m}0$;
Dec. −43°42′;
6.6^{m}.
This object is one of the brighter globular clusters in the southern sky. The star map on the left page shows that there are more globular clusters in the surrounding area. This is typical for an area close to the center of the Milky Way.

ern California. In the southern part of the constellation, there are impressive Milky Way clouds. (See also page 64.)

Observation Tips for NGC 6188

NGC 6188 is a luminous nebula where dark and bright clouds of gas are intermixed. In the photo, the nebula is directly to the right of the center of the picture (see arrow). NGC 6193, an open cluster, is also there, and it is difficult to keep the two objects apart. To find NGC 6188, draw a line between the bright stars α Centauri and γ Scorpii. It is approximately in the middle of this line, which is shown in the picture. The brightest star on the left bottom edge of the picture is ϵ Normae, a star that is close to the galactic equator. To the south of ϵ Normae is the open cluster NGC 6143 (upper left edge of picture). Another open cluster is NGC 6167, which is above the center of the picture to the left between NGC 6143 and NGC 6188. And at the edge of the picture in the bottom right is the open cluster NGC 6204.

Observation Tips for NGC 6541

This globular cluster (see arrow) is in the constellation Crux, which is described on p. 84. To find it, begin at the bright star θ Scorpii. Then look for the object about in the middle of the line towards α Telescopii. It is visible with binoculars or a small telescope.

The density of stars in this field is substantial. It is also near the Milky Way plane. It is estimated that our Milky Way system has overall 200 billion stars, of which only a couple of billions can be seen through a telescope. About 6000 stars up to the 6th magnitude can be distinguished with the naked eye in the sky. Up to the 7th magnitude are already 10,000 stars, and up to the 8th magnitude are about 20,000. Some 2 million stars are up to the 16th magnitude. And 2 billion stars are up to the 20th magnitude.

The Bright Stars

β Arae;
RA $17^{h}25^{m}3$;
Dec. −55°32′;
2.9^{m}.

α Arae;
RA $17^{h}31^{m}8$;
Dec. −49°53′;
2.9^{m}.

λ Scorpii (Shaula);
RA $17^{h}33^{m}6$;
Dec. −37°06′;
1.6^{m}.

θ Scorpii;
RA $17^{h}37^{m}3$;
Dec. −43°00′;
1.9^{m}.

26 Telescopium, Microscopium, and Corona Australis

In the light clouds of the southern Milky Way, you can often see dark clouds, which are a sign of interstellar matter—i.e., clouds of gas and dust surrounding regions where new stars are forming. You can also see dark accumulations of substance. The Coalsack Nebula in the constellation Crux is famous.

Visibility At midnight over the northern horizon (mean local time): July. The stars in the map section stand each month about 2 hours earlier over the northern horizon: August 22^h, September 20^h.

Telescopium (The Telescope) The name of this constellation comes from Abbé Nicolas-Louis deLacaille to honor the most important instrument for astronomy. This constellation is faint and contains very few interesting objects.

Microscopium (The Microscope) This insignificant constellation is between Sagittarius and Grus (The Crane).

Corona Australis (The Southern Crown) The stars form a distinctly pronounced

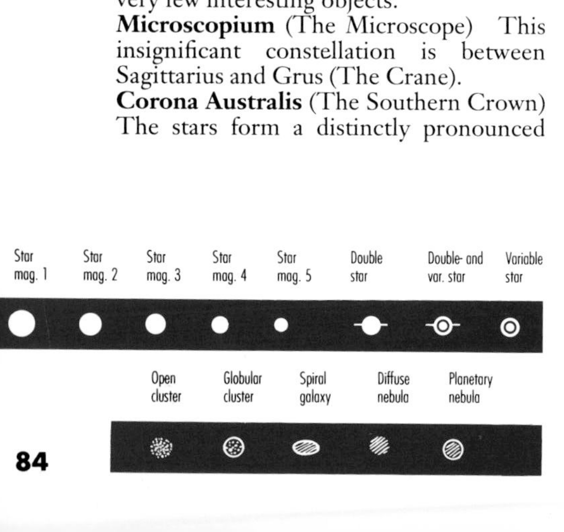

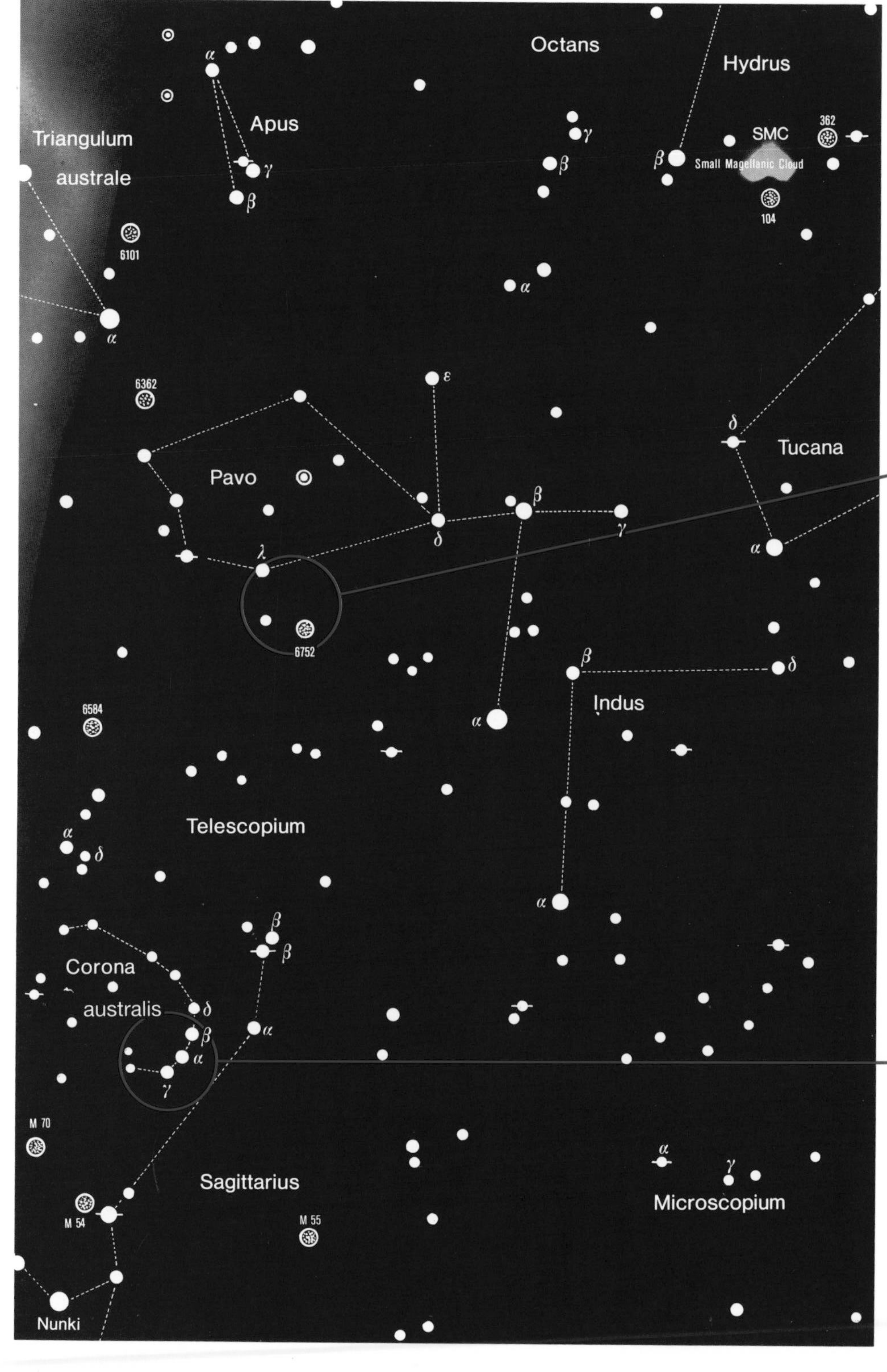

arch, similar to the constellation Corona Borealis between the constellations Hercules and Boötes. Corona Australis lies at the edge of the Milky Way.

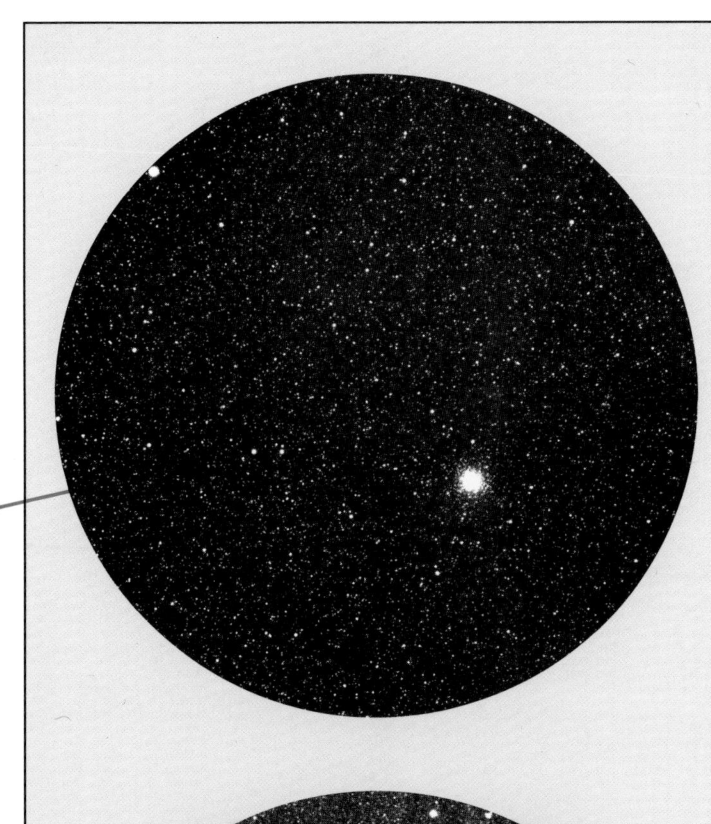

NGC 6752
Globular cluster;
RA 19^h10^m9;
Dec. $-59°59'$;
5.4^m.
You should be able to find the relatively bright globular cluster NGC 6752 in the field of stars with the naked eye (apparent magnitude around 5).

Observation Tips for NGC 6752

The star on the left top edge of the photo is λ Pavonis. It lies in the middle of the line between the star pair β and γ Arae and the star α Pavonis. Have λ Pavonis in the visual field of your binoculars so that you can also see NGC 6752. This globular cluster, NGC 6752, has a brightness of 5th magnitude, and its apparent diameter is half of the moon.

The concentration of stars is not even in a globular cluster. But as a general rule, the stars are standing so densely in the center that, even in the large telescope, you cannot resolve the individual stars there. On the other hand, you can see without any difficulties the individual stars at the edge with a small telescope. The observation of globular clusters on a clear, moonless night is again and again an experience for the amateur astronomer, especially in an area where there are few stars or clouds of the Milky Way.

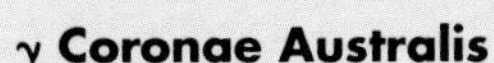

γ Coronae Australis
Double star;
RA 19^h06^m4;
Dec. $-37°04'$;
4.8^m and/or 5.1^m; $1.8''$.
The constellation Corona Australis can be already easily observed around the latitude of Florida. In the midsummer months, it stands in the hours before midnight below the constellation Sagittarius above the horizon.

Observation Tips for γ Coronae Australis

The double star γ Coronae Australis is in the center of the photo (arrow). In actuality, it consists of a pair of nearly identical yellow stars, each having an apparent magnitude of 5. The distance between them is 1.8 arc-seconds. They orbit every 120 years to form a tight double star. They are closest together in the 1990s; therefore, a small telescope with a 100 mm aperture is needed to split the two stars. But after the year 2000, the viewing will be easier. In the area to the left of γ Coronae Australis is NGC 6723, another globular cluster that belongs to the constellation Sagittarius.

The Bright Stars

α Pavonis (Peacock); RA 20^h25^m6; Dec. $-56°44'$; 2.0^m.

α Coronae Australis; RA 19^h06^m1; Dec. $-37°59'$; 4.1^m.

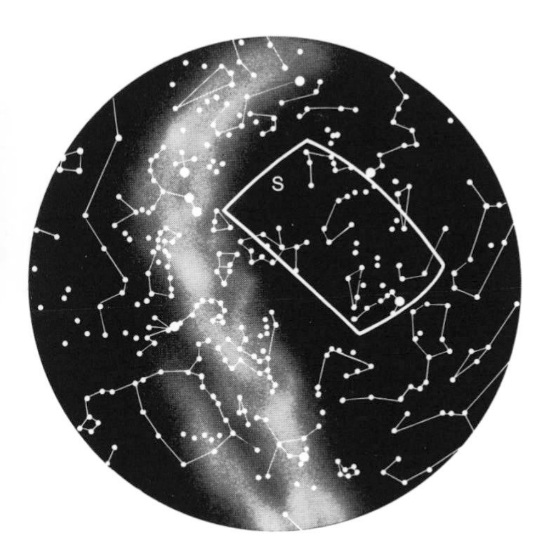

27 Indus, Grus, Octans, and Tucana

Globular clusters are "ancestors of the universe," because they consist of the oldest stars (10 billions years or older). The globular cluster 47 Tucanae, near the Magellanic Cloud, is one of the most beautiful ones.

Visibility At midnight over the northern horizon (mean local time): September. The stars in the map section stand each month about 2 hours earlier over the northern horizon: October 22^h, November 20^h.

Indus (The Indian) This narrow constellation is between Pavo (The Peacock) and Tucana.

Grus (The Crane) This constellation with the main stars of the 2nd magnitude borders on the constellation Piscis Austrinus towards the south. The star β Gruis lies on the line between the stars Fomalhaut (Piscis Austrinus) and α Tucanae.

Octans (The Octant) This constellation, covering the polar region, is low on brighter stars. There is no constellation that is comparable to the constellation Ursa Minor in the southern celestial pole.

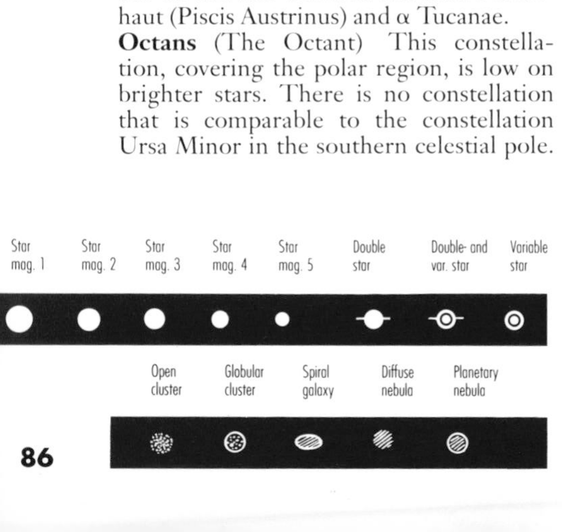

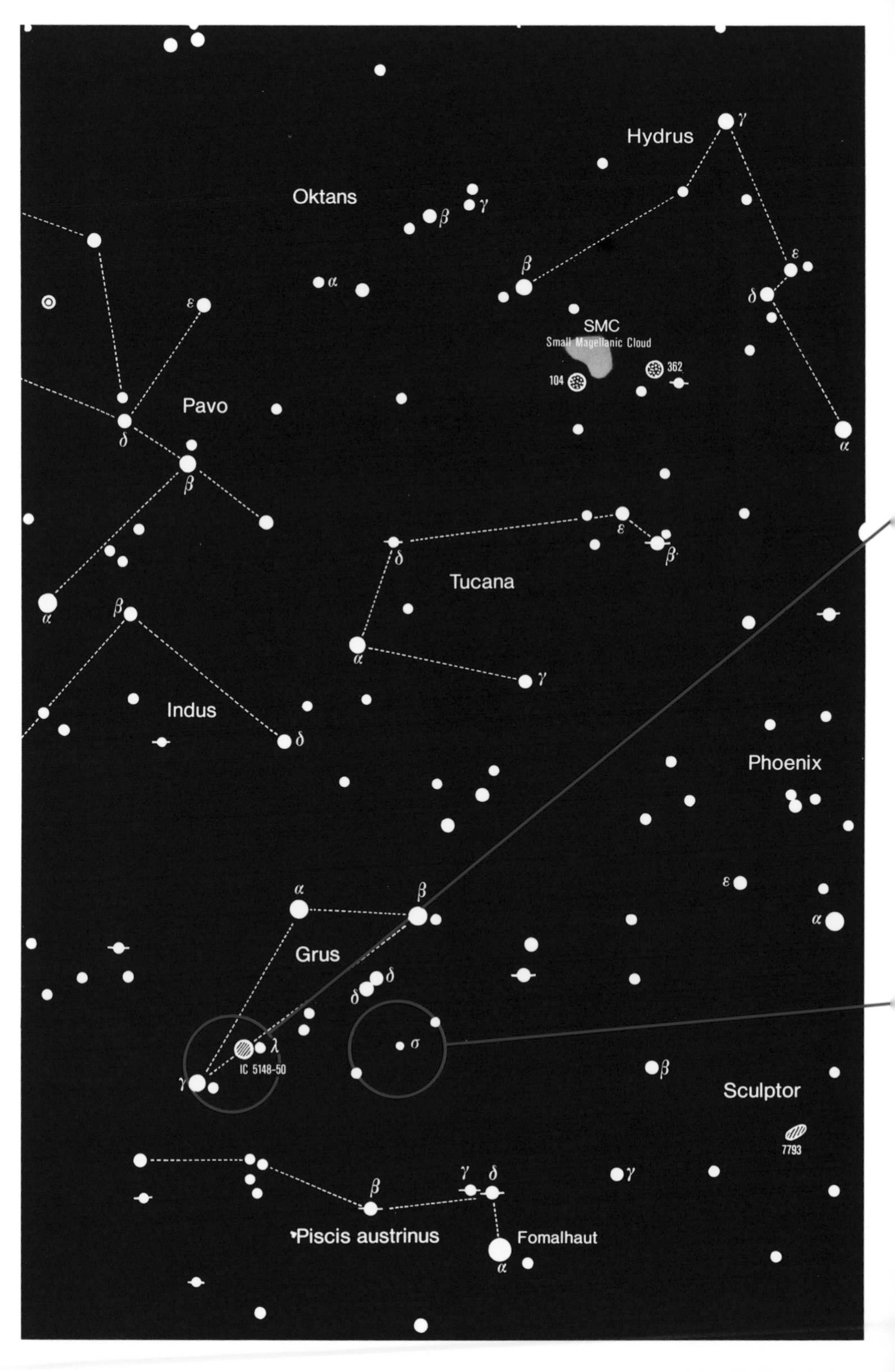

IC 5148-50
Planetary nebula;
RA 21^h59^m5;
Dec. $-39°23'$.
The field is, overall, low on objects. Therefore, there are not many observation suggestions here.

σ Gruis
Optical double star;
RA 22^h37^m0;
Dec. $-40°35'$.
In the center of the photo are two stars of approximately equally brightness: σ^1 Gruis and σ^2 Gruis. They form an optical double star like δ Gruis.

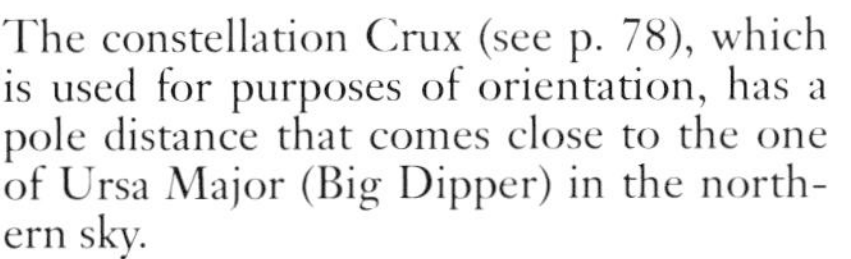

The constellation Crux (see p. 78), which is used for purposes of orientation, has a pole distance that comes close to the one of Ursa Major (Big Dipper) in the northern sky.
Tucana (The Toucan) This constellation has the Small Magellanic Cloud (see p. 71).

Observation Tips for IC 5148-50

To find it, you need not only very good visibility conditions and an optical instrument that is strong in light (binoculars 14 × 100 or astronomical telescope with a 5-inch aperture), but also enough perseverance in order to focus on it star by star. IC 5148-50 has an apparent brightness of 13^m and a diameter of 120 arc-seconds.

Begin at γ Gruis, a star of 3^m; in the photo, it is the brightest star on the bottom left. From there, draw a path to λ Gruis, the brightest star on the right, above the center of the picture. It is only 2° away from the planetary nebula (see arrow), which is close to the center of the photo. It is of some relief that there is no Milky Way in this area and that the density of stars is low. Otherwise, this object would be difficult to locate.

Observation Tips for σ Gruis

As its description, optical double star, suggests, here we have a star pair, whose closeness in the sky is a perspective deception from our Earth. In reality, they are two stars that have nothing to do with each other in outer space. In contrast, actual double stars follow the gravitational laws of Newton (see p. 73) and go on their orbit around each other.

σ Gruis forms a right triangle with the bright stars α and β Gruis (right angle at β Gruis).

The right star of our optical double star, σ^2 Gruis, is a real double star with a companion of only 10.4^m (main star 5.8^m). Their difference in brightness and their distance of 2.6 arc-seconds make the object difficult to resolve with small telescopes. You must have a 6-inch telescope.

The Bright Stars

α Gruis (Alnair);
RA 22^h08^m2;
Dec. $-46°58'$;
1.7^m.

β Gruis;
RA 22^h42^m7;
Dec. $-46°53'$;
2.1^m.

α Tucanae;
RA 22^h18^m5;
Dec. $-60°16'$;
2.9^m.

Objects Within the Solar System

The Star of the Day: the Sun

The sun is a common fixed star. At night, we see with the naked eye thousands of them. But in our life, this fixed star, the sun, is of special importance. As part of the solar system of the sun, the Earth is influenced in numerous ways: the length of the day, the seasons, the energy household (the basis for the existence of life), and other parameters. The sun is a gas ball with an enormous diameter of 1,392,000 km (863,000 miles). Nuclear fusion takes place inside: hydrogen transforms into helium. Through this, energy is produced, which is sent into outer space as warmth and visible light as well as X-rays, ultraviolet radiation, and radiowaves to reach the Earth.

A depiction of the planets and solar system by the artist Shigemi Numazawa that is off scale. Between the Earth (bottom) and sun (center) are the planets Venus and Mercury; above the sun are (from left) Saturn, Mars, Jupiter, and Uranus.

Sun's Energy for the Earth

The sun's energy supplies life for Earth. For the future, it is becoming a more and more urgent task to transform the sun's radiation into energy, which can be used by humans. Sunlight is indispensable for our personal well-being. It stimulates our blood circulation and strengthens the immune system. But too much can make you sick—for example, it can cause skin cancer.

The spacecraft Ulysses, which flew in May of 1994 over the South Pole of the sun and in May of 1995 over the North Pole, supplied the latest information about the sun. The spacecraft was about double the distance from the sun than the distance of the sun to Earth. Measurements done when flying over the solar south pole showed that the little parts of the so-called solar wind come much faster from the pole than from the solar equator. Their temperature is lower than until now assumed, and the chemical composition deviates from the hitherto existing model. The magnetic field-strength is clearly lower. The solar wind that was explored by Ulysses is the magnetic gas, which streams away from the sun's atmosphere into outer space. Surprisingly, astrophysicists are more familiar with the inside of the sun than with the sun's atmosphere, where the "solar weather" takes place with magnetic fields, spot for-

Data about the Sun

Diameter:	1,392,530 km (Earth: 12,756 km)
Surface area (in Earth-surfaces, Earth: 510 million square-km):	11,957
Mass (in Earth-mass, Earth: 5973 trillion tons):	332,950
Rotation periods in days (at the sun-equator):	
Sidereal:	25.03
Seen from the Earth:	26.8
Most frequent chemical elements of the photosphere in mass-percent:	
Hydrogen:	78.4
Helium:	19.8
Oxygen:	0.863
Carbon:	0.395
Iron:	0.140
Mean density:	1410 kg/m^3 (Earth: 5515 kg/m^3)
Temperature at the surface (photosphere):	5770°K
Temperature at the core:	15,000,000°K
Estimated age:	4.6 billion years
Sun–Earth distance:	1.496 million km

Setting sun, photo by Peter Stolzen, Volks Observatory Remscheid, prize winner of the competition "Youth Explores."

to explore them. They have a rest mass of 1.67×10^{-24} grams. A moving electron has a mass that equals its speed. And it also has a mass that does not move called a rest mass. Once we know something definite about the sun's neutrons, we will get important information about the side reactions, which occur parallel to the nuclear fusion inside of the sun.

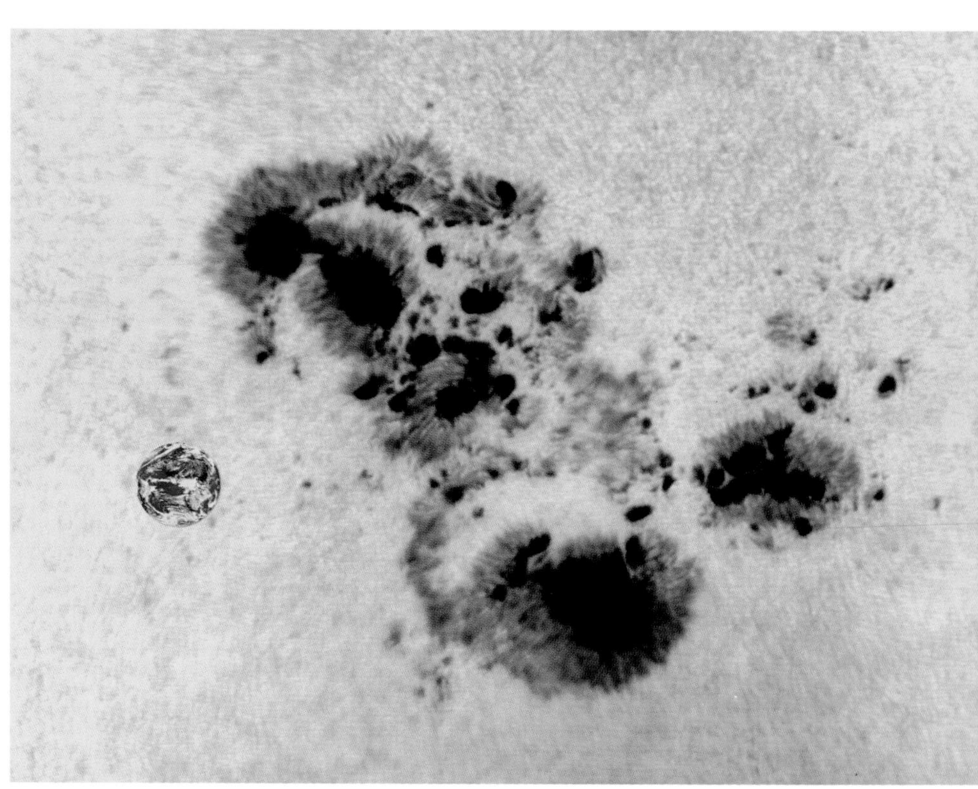

Large group of sunspots shot with a medium-large amateur telescope. The picture of the Earth was copied into the picture for a size comparison.

mation, hot corona, solar flares, and solar wind. Computers made the simulation of star formations possible as well as the time development of the sun with its nuclear reactions. The supply of nuclear energy will last for approximately another 10 billion years.

Mysterious Sun Neutrons

There are observations that support the results of computer models, such as proving that the sun conducts vibrations in a 5-minute rhythm. These vibrations continue inside the sun and thus make other observations possible, for example, the temperature dependency on the depth. This is comparable to earthquake waves, which give us information about the condition inside the Earth.

During the fusion of hydrogen to helium, neutrons, natural elementary particles, are created in the center of the sun. They come without any resistance into the atmosphere of the sun and continue to the Earth. They cover the distance from the sun to the Earth in about 8 minutes. The Earth is also no obstacle for them as they enter everywhere and go through everything. Since neutrons hardly react with other parts of the substance, it is difficult

The Sun Through the Telescope

What is there to see on the sun? With the naked eye, protected by a filtering glass, we see a round disk, which is less bright at the edge than in the middle. When observing the edge of the sun disk, we see the thinner layers than when observing the center of the disk. The temperature quickly decreases in the photosphere towards the higher located layers of the sun's atmosphere. The result is a decrease in brightness towards the edge of the sun disk, which is called limb darkening. This effect is clearly visible when its image is projected onto a white piece of cardboard. In the outer layers of the sun's atmosphere, the temperatures rise again and reach about 1–2 million degrees C in the corona.

The gas shell of the sun is also an atmosphere. It is about 6000°K in the layer (the visible radiation comes the photosphere) and it has a different composition than the Earth's atmosphere. Now and then you will see dark spots on the sun with the naked eye. Those are extremely large sunspots. But it is better to use the binoculars to see the sunspots more frequently.

What kind of spots are those? They have astonishing diameters of between some thousands to 50,000 km (31,000 miles) and they occur mostly in groups. Sunspots mark areas of special movement in the gas-shaped surface of the sun. They are dark, because their temperature is 1000°–2000° lower than that of the photosphere—the surrounding layer from which the visible sun radiation comes. At a closer look, the telescope reveals, besides the core area of the spots (umbra), the surrounding, somewhat lighter, half shadow (penumbra). The photo on page 90 makes that impressively clear. The activity of the sunspots follows an 11-year rhythm. The last spot maximum was at the end of 1989; the last minimum was in the summer of 1986. Therefore, the sun's surface will be full of spots again around the year 2000 (provided that the nuclear reactor sun plays along, because deviations of some years from the indicated rhythm can occur). The fast increase of number of sunspot after the minimum is striking. The maximum is reached 3–4 years later. The decrease lasts longer (about 6–7 years). The sunspots are produced by the eruptions that break out from inside the sun. There is no simple physical explanation up to now for the differences in the 11-year spot cycle, just as there is little proof for the relations between sunspots, planetary constellations, and catastrophes on Earth.

From one day to the next, changes occur in sunspots. Light bridges divide the umbra and give the spot a new look over and over again. The handmade illustrations above show the development of a sunspot from observations with a refractor of 100 mm opening at 100-fold enlargement. Light protection with a foil filter.
Top: August 17, 1991 10h05 MESZ
Center: August 18, 1991, 9h40 MESZ
Bottom: August 19, 1991, 10h00 MESZ

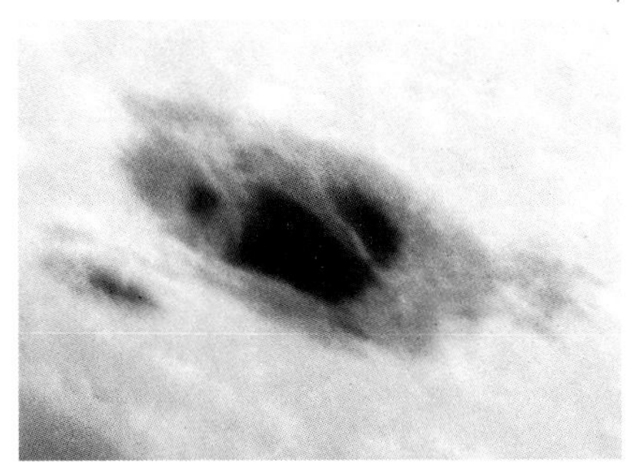

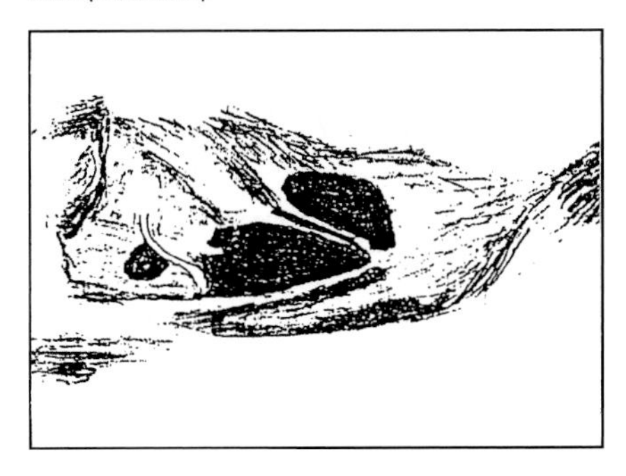

Double light bridge, on March 14, 1993, documented by observers with different types of equipment.

Top: Photographic observation with an 8-inch refractor at 10h02 MEZ (B. Veenhoff).

Below: Illustration of visual observations with a 7-inch reflector telescope at 112-fold enlargement at 13h45 MEZ (G.D. Roth).

Be Careful Observing the Sun!

Never look into the sun without a filtering glass (pane of glass blackened with soot, welder's glasses, smoky filter). The dazzling sunlight can easily damage the eyes badly. When a filtering glass is used together with a telescope, keep in mind that the filtering glass heats up a lot under the influence of the sun's heat and it can burst.

A more indirect method is safer. Use a telescope to project an image of the sun onto a screen (such as a white piece of cardboard), which is attached behind the eyepiece. Do not use any glued eyepieces!

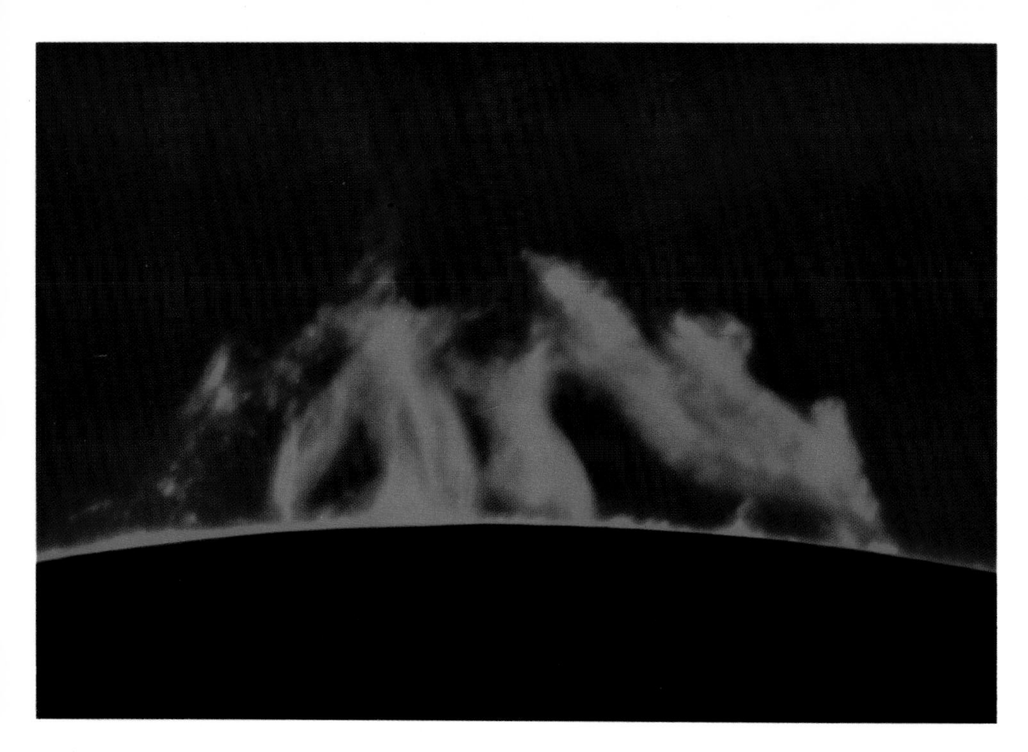

Photo of the prominences with a special telescope. Prominences are glowing gas masses, which can be observed during solar eclipses and in the coronograph. Height up to 500,000 km (310,000 miles).

In addition to the umbra and penumbra, we have "light bridges" in sunspots: bright stripes that are interspersed and optically divided by individual spots. They often change quickly and they are a task for the amateur astronomer to observe.

The Atmosphere of the Sun

Three layers form the atmosphere of the sun: The photosphere is the lowest one, about 400 km (248 miles) thick. In the telescope, it has a granulated structure. This is also where the sunspots are formed. The next-higher layer is the chromosphere, about 20,000 km (12,400 miles) thick. Here, eruptions occur in areas of disturbance (flares, spicules), which are visible at the edge of the sun during a solar eclipse or with a special telescope as prominences. The outermost layer is the corona. It reaches far into the interplanetary space. The corona is a very thin and hot gas (several million kelvins).

During a total solar eclipse, we can observe the events of the sun in a very impressive way. Then, when the moon covers the largest part of the sun disk, the corona, chromosphere (thin, reddish bright layer), and prominences (reddish shiny gas eruptions) become recognizable.

Total solar eclipes on November 3, 1994, shot in Brazil with a Schmidt-Cassegrain by Meade. Aperture 254 mm, focal length 1400 mm, exposure 1/30s on Kodak Ektar 1000. The almost colorless corona is clearly visible.

Our Satellite: The Moon

Even though the moon has already been reached by spaceships from Earth, this Earth satellite remains a popular object for observation. Anyone who sees the moon for the first time through a telescope is thrilled by the diversified play of light and shadow. There are the eternal increase and decrease of the moon:

- New moon
- First quarter
- Full moon
- Last quarter

Right: Moon crescent, 2 days old with Earthshine ("ashen light," see p. 98). Photo taken by amateur photography. Optical lens system Vixen-refractor 102/1500. 1 minute exposure on Fujichrome film.

Far right: Moon on January 16, 1989. Photo in the focal point of a 4-inch refractor (APQ 100/1000). Kodak Ektar 25, 1/15s. Negative diameter 9 mm.

And then new moon again. The light of the sun provides us the visible abundance of details of the moon. Sunlight wanders over the moon, lighting up, according to the position of the orbit, a narrow crescent or, during full moon, the entire surface that is turned towards the Earth (see p. 97).

Most interesting is the zone between the day and the night sides; experts call it the "terminator" (the dividing line between the illuminated part and the dark part of the moon). There, the sun rises above and sets behind the lunar mountains. First, only the very highest tips shine in the sunlight. Gradually, the entire lunar mountain follows and becomes clear in its shape. In a few hours, the dividing line between light and dark moves. Again and again, new craters shine in the reflected sunlight and sink back into the black of the moon's night. Binoculars give an overwhelming the three-dimensional impression of the lunar surface at the terminator. In summary, it can be said that we see three typical things on the lunar surfaces:

- the lunar oceans (maria),
- the terrae with craters and mountains,
- the bright rays.

The lunar orbit is inclined 5° to the ecliptic and therefore the position of the moon in regards to the ecliptic changes constantly. Each of the four pictures at the right show two possible lunar orbits and/or positions of the moon. The different positions of the ecliptic in the course of a year (see p. 20) and the different positions of the sun cause completely different positions, relating to space, of the moon crescent in the sky (e.g., "boat position" in the spring).

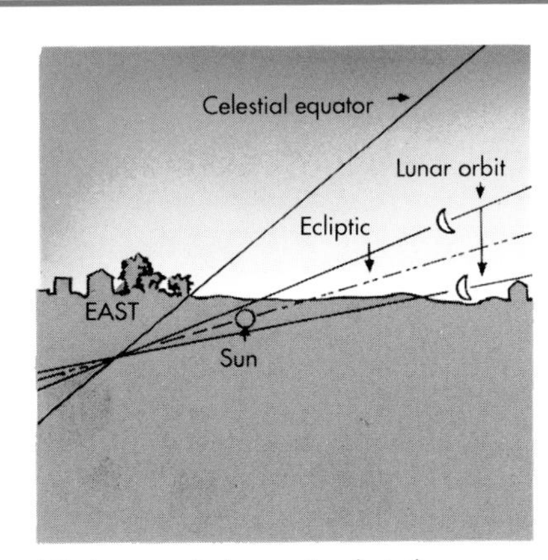

Waning moon in the morning sky in the spring.

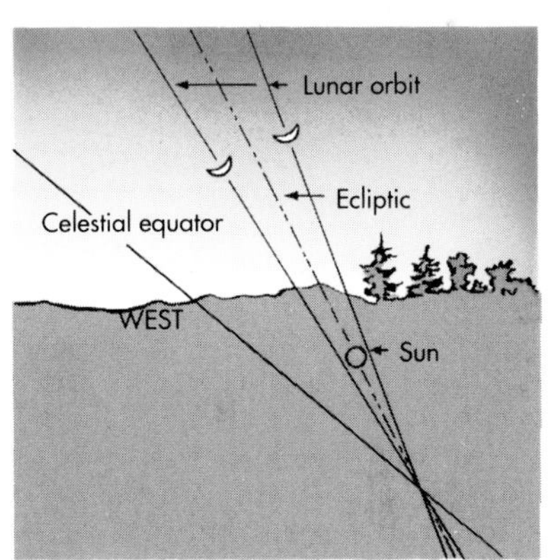

Waxing moon in the evening sky in the spring.

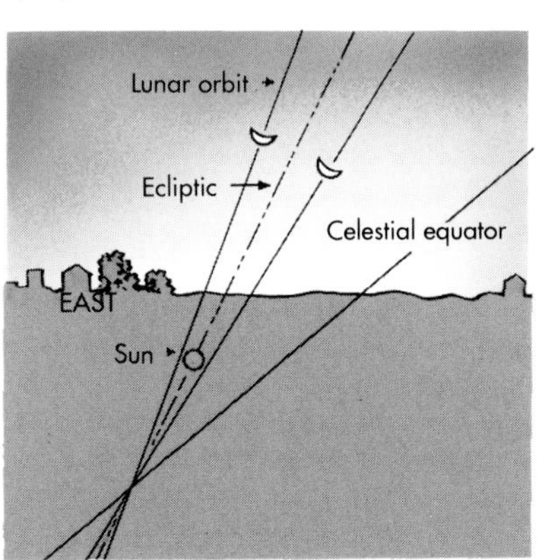

Waning moon in the morning sky in the fall.

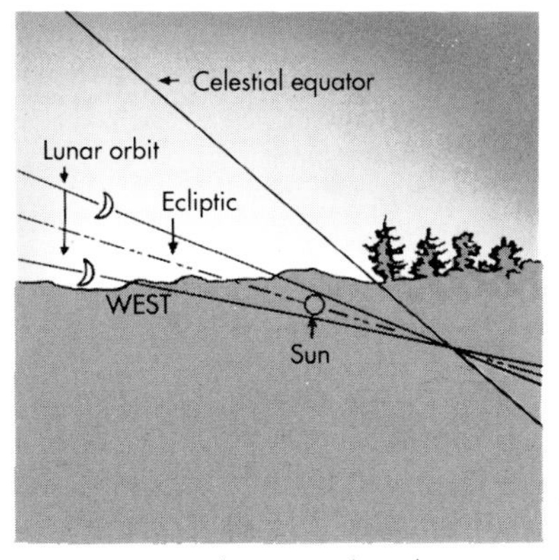

Waxing moon in the evening sky in the fall.

The changing sight of the moon offers impressive crater landscapes, especially those close to the light boundary (terminator). The picture on the far left shows a 5-day-old moon (5 days after a new moon); the picture on the left shows the moon around the first quarter (7 days after a new moon). The lunar map below gives some of the names of the visible surface structures.

Names of Structures on the Moon

1 de la Rue
2 Endymion
3 Atlas
4 Messala
5 Gauss
6 Geminus
7 Cleomedes
8 Proclus
9 Tarantius
10 Langrenus
11 Vendelinus
12 Petavius
13 Furnerius
14 Metius
15 Janssen
16 Rosenberg
17 Vlacq
18 Hommel
19 Manzinus
20 Morteus
21 Clavius
22 Blancanus
23 Scheiner
24 Bailly
25 Schiller
26 Schickard
27 Longomontanus
28 Wilhelm
29 Tycho
30 Maginus
31 Orontius
32 Cuvier
33 Maurolycus
34 Stöffler
35 Gemma Frisus
36 Rabbi Levi
37 Zagut
38 Piccolomini
39 Sacrobosco
40 Catharina
41 Cyrillus
42 Theophilus
43 Walter
44 Regiomontanus
45 Purbach
46 Aliacensis
47 Arzachel
48 Alphonsus
49 Ptolemaeus
50 Mösting
51 Hipparchus
52 Albategnius
53 Parrot
54 Pitatus
55 Fra Mauro
56 Piazzi
57 Lagrange
58 Darwin
59 Mersenius
60 Gassendi
61 Letronne
62 Grimaldi
63 Riccioli
64 Hevelius
65 Otto Struve
66 Kepler
67 Copernicus
68 Erathosthenes
69 Archimedes
70 Autolycus
71 Aristillus
72 Plato
73 Pythagoras
74 W. Herschel
75 Fontenelle
76 Epigenes
77 W.C. Bond
78 Aristoteles
79 Eudoxus
80 Callipus
81 Gärtner
82 Posidonius
83 Le Monnier
84 Fracastor
85 Pitatus
86 Riphäen
87 Carpathian Mts.
88 Apenninen
89 Caucasus
90 Alps
91 Harbringer Mts.
92 Hyginus
93 Ariadaeus
94 Altai Scarp

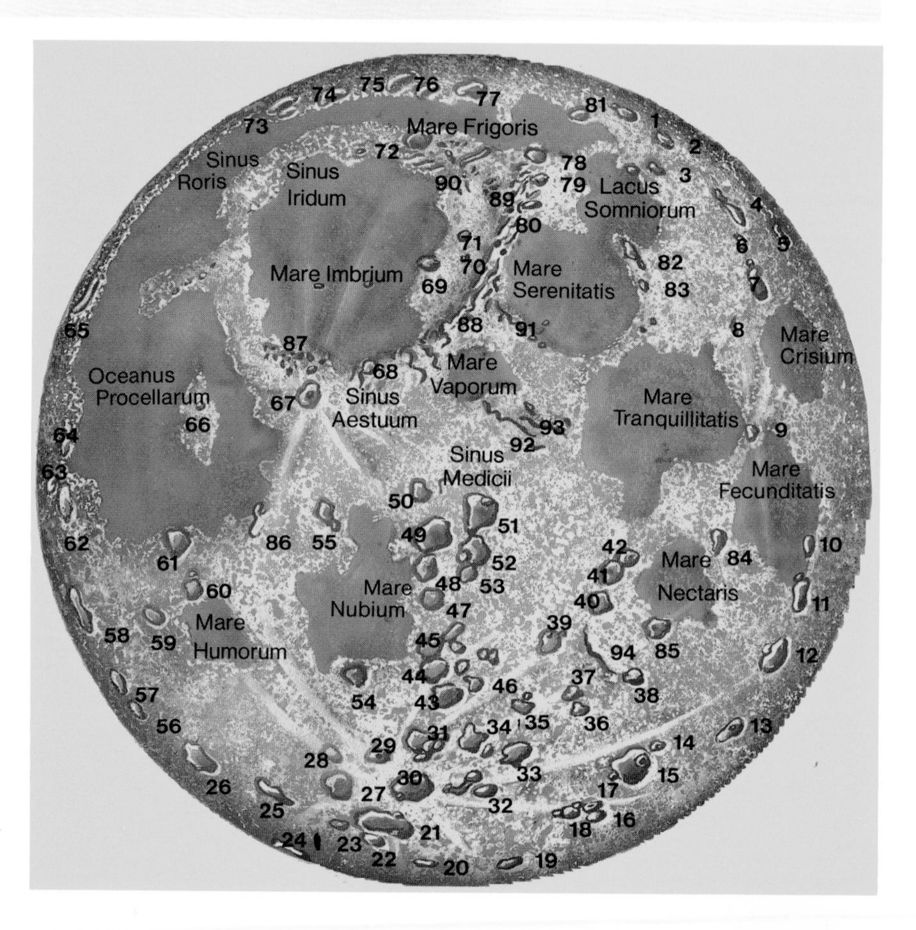

The names of the individual maria appear on the lunar maps mostly in Latin:

Mare Australe
Mare Crisium
Mare Fecunditatis
Mare Frigoris
Mare Humorum
Mare Imbrium
Mare Nectaris
Mare Nubium
Mare Serenitatis
Mare Tranquilitatis
Mare Vaporum
Oceanus Procellarum
Sinus Medii

These large dark surfaces, or "oceans" (*maria* means "seas"), have nothing to do with water. But centuries ago, the astronomers believed in a similarity with earthly conditions and, thus, gave it the misleading name. The full moon makes the large-scaled structures of the surface especially beautiful to see. Besides the dark, low mare landscapes, there are the light, seemingly high-lying surfaces of the highlands. Then there are also the famous "lunar rays," which are eye-catching and start at the craters Copernicus, Tycho and Kepler (see lunar map). Another crater that is worth mentioning is the brilliantly shining Aristarchus, which is the brightest visible point on the moon's surface. In view of all these details, the astronomer, who observes with binoculars, should try to be in the position of an aerial geologist, who takes pains to get a first overview of the geology of an unexplored part of the Earth from an airplane.

The Moon in the Telescope

But not only the surface is remarkable during full moon. The moon rotates on its axis in the same amount of time it takes to circle Earth, which means one side of the moon always faces us. Due to a slight oscillation of the moon, known as libration, about 59% of the moon's surface can be seen. A careful observer can detect that the objects at the edge of the moon are not always equally visible from full moon to full moon. Anyone who is interested in this problem can find a further discussion in *Astronomy: A Handbook*. When making observations with binoculars, it is important that the full moon fills out the visual field without losing its "globe effect," which is seen amazingly through binoculars or a telescope. Binoculars, which have greater enlargement power (10× and more), in combination with a tripod are the right equipment to use. Anyone who owns binoculars with picture stabilization (e.g., 12 × 36 IS by Canon) can easily observe without a tripod.

The brightness of the full moon occasionally causes difficulties. Using gray or yellow filters, which are commonly in the trade stuck on binoculars, make it more comfortable to observe.

Large lunar maps contain about 33,000 craters on the front side of the moon alone. Expanded craters, also called walled plains (chains of lunar craters), reach a diameter of 100 km (62 miles). They are named after astronomers and nature researchers. The most important circular forms of the terrain on the moon's frontside are shown with the names on the map on page 94.

To see more of the ragged craters, use a small astronononomical telescope and enlargements between 50- and 100-fold. For astronomical telescopes, there are also binocular attachments (e.g., Baader, Munich, and Pentax), which make it possible to observe with both eyes. Lunar mountains of 6000, 7000, and 8000 meters (19,800, 23,000, and 26,400 feet, respectively) are no rarity. Thus, the lunar mountains Apennines and Caucasus (numbers 88 and 89 on the lunar map) are 5500 meters (18,150 ft) and 5900 meters (19,470 ft) higher than their environment. The highest mountain range, the Leibniz mountain range, is about 11,350 km (7037 miles) high. (Frequently, the names of well-known scientists are used for naming the lunar objects.)

Objects suitable for the 2–3-inch astronomical telescope (50–100 mm aperture):

- Crater Clavius—can be observed 1–2 days after the first quarter and 1–2 days after the last quarter.
- Crater Copernicus—can be observed 1.5 days after the first quarter and 1.5 days after the last quarter.
- Crater Theophilus—can be observed 5 days after the new moon and 5 days after the full moon.
- Alps with Alpine valley—can be observed shortly before the first and last quarter.
- Ariadaeus rill—can be observed 1 day before the first quarter and 1 day before the last quarter.
- Hyginus rill—can be observed 1 day before the first quarter and 1 day before the last quarter.

These objects are shown on the lunar map on page 94. The observation times refer to the sunrise (first quarter) and/or sunset (last quarter) above the object.

Moon photo, which shows the area around the Alps valley (#90 of the lunar map, p. 94) at the east edge of the Mare Imbrium. The Alps valley is a transverse valley 130 km (80.6 miles) long with a groove in the bottom of the valley. The picture was taken on January 14, 1981, 18h10 UT with a Schmidt-Cassegrain of 35 cm aperture (C 14). The arrows mark objects at the limit of the resolution capability.

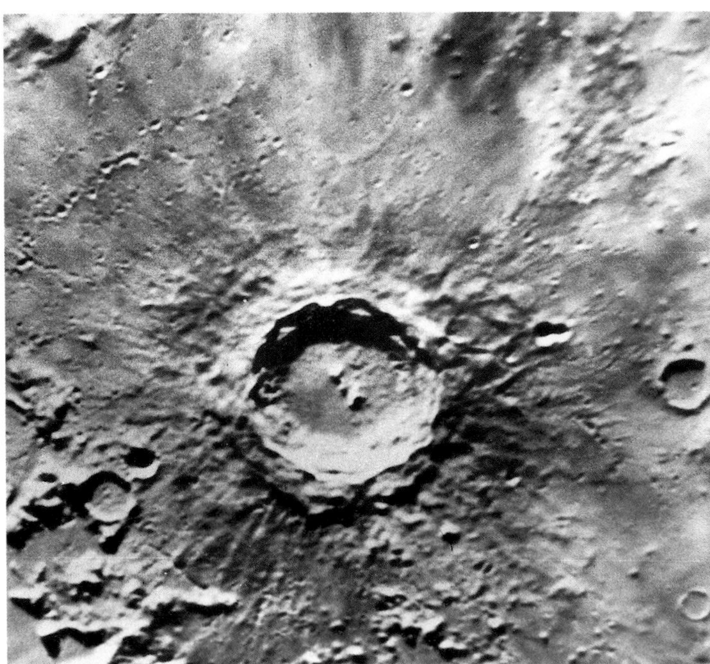

Top: The pictured rill systems are test objects for small- and medium-sized amateur telescopes. The rills Hyginus and Ariadaeus are easily recognizable. More difficult is the system of the Triesnecker rills (arrow).

Top right: Crater Copernicus, surrounded by lunar rays, deposited discharge material from the time of the formation of the crater 900 million years ago.

Below: Apollo 17. Astronaut and lunar rover on the moon in December 1972.

The Exploration of the Moon

Numerous rills on the moon resemble flowing rivulets on the Earth. But at no time was there water on the moon. The rills, which are often over 100 km (62 miles) long and several kilometers wide, were formed by lava streams that flowed out of craters. The moon is not, as often assumed, a celestial body that cooled off. On the contrary, the inside of the moon has a high temperature. Measurements, which were done during the Apollo mission, showed a rise in temperature of 1° in a depth of 2 meters (6.6 ft). During these explorations through spaceship mission (in July, 1969, Apollo 11 carried the first astronauts onto the moon), ground samples were brought to Earth. Most of the craters on the moon are impact craters. In addition to those, there are also volcano craters. The surface of the moon is covered with a layer of rubble and dust, which is partially several meters thick. Half of the moon dust consists of glass grains. The most abundant minerals are pyroxene, plagioclase, olivine, and ilmenite.

There are also moonquakes on the moon, but no atmospheric remains that can prove them. The layer of rubble on the surface was created, on the one hand, through hits of meteorites or, on the other hand, through erosion as a consequence of cosmic radiation and of the particle radiation of the solar wind.

Despite very comprehensive exploration of the moon through lunar orbiter and the Apollo missions with astronauts until 1972, there were blank spots on the lunar map until the lunar exploration flight "Clementine" in 1994. This mission gave scientists the first reliable, almost global topographical map of the moon. The weight- and densitiy-map of the outer moon-layers, which was learned from these measurements, documented how much the form of the moon had changed through hits of all sizes in the course of time. These hits caused the moon's crust to become very thin at some spots. The medium thickness of the moon's crust of 70 km (43.4 miles) decreased below the South Pole to 12 km (7.44 miles).

With the help of of radiowaves, the tide effect of the Earth and sun onto the moon was also determined. Deformations of the lunar body indicate magma areas inside of the moon. This might be the strongest indication of a liquid lunar core in the form of melted minerals.

The spaceship "Clementine" did not conclusively answer whether there is ice on the moon. The bottom of some lunar craters, which are close to the pole, lie so deep that they are never sunlit. Gases in frozen condition could exist here. Measuring the time it takes for radiowaves to bounce back indicates that ice may make up the surface of some craters.

Low Tide and High Tide

For the mechanics of sky, lunar orbit and rotation are something special. But the oscillations of the moon's rotation require a lot of mathematical calculations.

The celestial bodies, Earth, moon, and sun have a mutual effect on each other. The gravitational forces of the moon, sun and, in a weakened form, planets cause the Earth's axis to rotate. In addition, there is the movement of the Earth and moon around a mutual center of gravity.

Mainly the moon and, to a lesser extent, the sun exert gravitational and centrifugal forces at different places on and in the Earth. The consequence are the low tides and high tides of the oceans. You can also measure vibrations in the atmosphere and in the Earth's body, which are nothing else but tides. The tides cause a tide friction, which has a braking effect on the Earth's rotation. Therefore, it also has influence on the determination of time.

Formation

There are different views concerning the formation of the moon. Moonstones recovered from spaceship explorations were shown to be about 4.4 million years old. This means that the moon was at that time an already completely formed celestial body and suggests that the other planets of the solar system came into existence at the same time. There are three classical formation theories for the moon: the splitting off of the Earth, the capturing into Earth's orbit while flying by Earth, and the formation out of the original matter together with the other planets and moons. The impact theory is more recent. According to that, there was a crash between a planet as big as Mars, which was coming into existence, and the Earth, which was also in its early stages. An enormous explosion resulted and matter fell onto the Earth as well into the Earth's orbit, where gradually the moon formed.

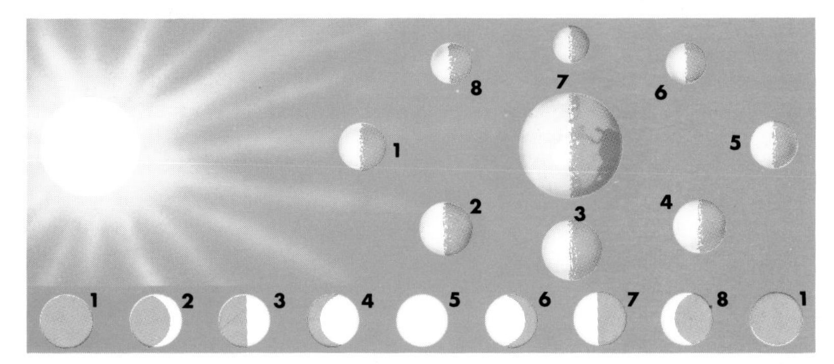

Formation of the moon's phases. 1 = new moon, 2 = waxing moon, 3 = waxing moon (first quarter), 4 = waxing moon (almost full moon), 5 = full moon, 6 = waning moon (still almost full moon), 7 = waning moon (last quarter), 8 = waning moon, 1 = new moon again.

Lunar eclipse. The moon wanders into the Earth's shadow, which is then exactly between the sun and the moon.

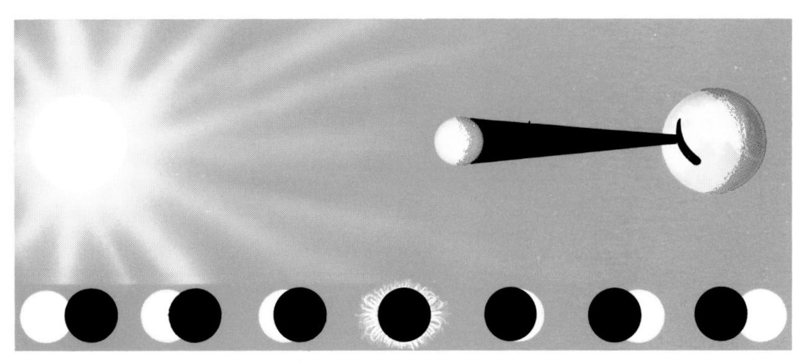

Solar eclipse. The moon, lying between the sun and Earth, throws its shadow onto the Earth. The eclipse is complete when the moon completely covers the sun. A partial eclipse comes about when the moon cannot cover the sun completely because of the position of the orbit.

Eclipses

When the crescent phase of the moon is very narrow just before or after a new moon, you will at first discover a strange phenomenon: the sky does not only have a bright shining crescent but also the remaining moon disk in a very pale, ashen gray luster (called the "ashen light of the moon"). This is especially impressive to observe with binoculars. This ashen gray light of the moon is caused by the narrow crescent shining in the sunlight. There, where the ashen-gray light glitters, is night on the moon. But now, the Earth is also lit by the sunlight, and the Earth radiates part of this light back into space, thus onto the moon. And this remainder of "Earthshine" is sufficient to produce a mat, gray reflection glimmer on the still (or again) unlit part of the moon. It is so weak that it becomes invisible when the phase of the moon becomes larger (see photo on p. 93).

On its orbit around the Earth, the moon lies now and then exactly between the sun and the Earth so that it completely or partially covers the sun from the Earth. That causes a total or partial solar eclipse. It can also happen that the Earth lies between the sun and the moon. Then we have a lunar eclipse, because the moon does not receive any sunlight. Solar eclipses only happen at new moon. Lunar eclipses only occur at full moon (see p. 97). More information about the eclipses are in the chapter "Actual Events in the Sky" (from page 165ff) and on the back jacket of the book.

Due to its inclination of 5° against the Earth's orbit, the lunar orbit changes constantly towards the ecliptic. That often causes strange positions of the moon's phases towards the horizon. Four examples for the mid-northern latitude are shown in the illustrations on page 93. In each drawing, the moon appears twice according to the possibilties that are given through the inclination of the lunar orbit against the Earth's orbit level.

An Overview of the Planets

1) For the earth, seen from the sun
2) At the surface
3) At the upper edge of the clouds
4) In the area of the reflecting layers of the atmosphere

	Mercury	Venus	Earth	Mars	Jupiter	Saturn
Mean distance from the sun (in 10 km)	57.909	108.209	149.598	227.941	778.38	1424.3
Mean distance from the sun (in AU)	0.387	0.723	1.000	1.524	5.203	9.521
Smallest distance from the sun (in AU)	0.31	0.72	0.98	1.38	4.95	9.00
Largest distance from the sun (in AU)	0.47	0.73	1.02	1.67	5.45	10.04
Smallest distance from the Earth (in AU)	0.53	0.27	—	0.38	3.95	8.00
Largest distance from the Earth (in AU)	1.47	1.73	—	2.67	6.45	11.04
Circumference of the orbit (in million km)	360	680	940	1400	4900	9000
Mean rotation-speed (in km/s)	47.9	35.0	29.8	24.1	13.1	9.6
Inclination of orbit to ecliptic (in degrees)	7.005	3.395	—	1.850	1.305	2.487
Eccentrity of the orbit (see also p. 170)	0.2056	0.0068	0.0167	0.0934	0.0482	0.0550
Diameter of the equator (in km)	4878	12104	12756.280	6786.8	142796	120000
Diameter in Earth-diameters	0.382	0.949	1.000	0.532	11.19	9.41
Flattening off (see also p. 169)	0	0	1/298	1/192	1/15	1/10
Mass in Earth-masses (without moons)	0.05527	0.8150	1	0.10745	317.8	95.16
Escape-speed, referring to the planet's equator (in km/s)	4.2	10.4	11.2	5.0	59.6	35.6
Gravity-acceleration at the surface, equator (in m/s)	3.71	8.85	9.78	3.72	24.80	10.50
Rotation period in Earth time	58.6462^d	-243.01^d	$23^h56^m4^s$	$24^h37^m23^s$	9^h56^m	10^h30^m
Day-night cycle	176^d	116.75^d	1^d	1.027^d	0.41^d	0.43^d
Axial tilt	0°	177°20′	23°26′	25°11′	3°08′	26°43′
Geometric Albedo (see also p. 168)	0.106	0.65	0.367	0.150	0.52	0.47
Largest apparent magnitude (^{m})	−1.9	−4.28	−3.86[1)]	−2.52	−2.7	−0.6
Temperature range (in °C)	−185 bis +425[2)]	+455 bis +525[2)]	−65 bis +60[2)]	−140 bis +15[2)]	−135 bis −125[3)]	−190 bis −180[3)]

Neighbors in Space: Observing the Planets

The first time you observe the planet Jupiter with a 3-inch telescope (at 100-fold enlargement), you will not easily recognize details on this tiny disk, even though the planet itself is relatively easy to see. As an amateur astronomer, you should use enlargements when watching planets through your astronomical instrument. As a comparison, imagine observing the full moon without a telescope. You will see dark and light spots, but everything else is more or less coarsely outlined and unclear. Only binoculars bring out more exact details. Nevertheless, it is worthwhile to jot down whatever you saw with the naked eye and to compare it to the lunar map on page 94. In order to get the individual planets through the telescope in at least the same seeming diameter (as the full moon offers for the naked eye), you need the following enlargements:

Mercury	280-fold (6.5″)
Venus	70-fold (25″)
Mars	70-fold (25″)
Jupiter	40-fold (48″)
Saturn	100-fold (21″)
Uranus	500-fold (4″)
Neptune	750-fold (2.5″)

The number in parentheses indicates the apparent diameter (in arc-seconds) of each planets. This number in fact is not always the same; it depends on the planet's distance from the Earth. The indicated values for Mercury and Venus are at the time of the largest elongation (angle between sun and planet); for the other planets, the values are at the time of the opposition. Although you cannot see the details of the planets through binoculars alone, it can help you observe the phases of Venus and the fascinating movement of Jupiter's moons.

For observing the planets and the moon, you can follow these optimal enlargements:

- Telescope between 5 and 10 cm aperture or diameter: 100–150-fold,
- Telescope between 10 and 15 cm aperture or diameter: 150–250-fold,
- Telescope between 15 and 20 cm aperture or diameter: 200–300-fold.

As an amateur astronomer, you should be aware that one does not immediately see well with the aid of a telescope. Anyone using a telescope for the first time has to train his or her eyes by observing with and without the telescope. By the way, the exercise of observing the full moon with the naked eye is a good start. Practice using the telescope on as many evenings as you can to become efficient at it. Once you are, the observing becomes a real pleasure.

Uranus	Neptune	Pluto
2866.6	4492.3	5887.3
19.162	30.029	39.354
18.27	29.71	29.7
20.06	30.34	49.1
17.29	28.71	28.7
21.07	31.31	50.1
18000	28000	37000
6.8	5.4	4.7
0.772	1.771	17.147
0.047	0.010	0.246
50800	48600	≈2400
3.98	3.81	0.19
1/33	1/39	?
14.50	17.20	0.0023
21.4	23.7	1.2
9.00	11.60	0.6
-17^h15^m	16^h07^m	-153^h17^m
97°51′	29°34′	117°34′
0.51	0.41	0.52
+5.5	+7.5	+14.3
−210[4)]	−220[4)]	−235[2)]

Mercury

His patron name is the Roman god of merchants, which corresponds to the messenger of the gods, Hermes, of the ancient Greeks. Mercury moves on a very eccentric orbit around the Sun. With 7°, it has the largest inclination of orbit to the ecliptic after the planet Pluto (see table). The planet Mercury, closest to the sun, moves in only 88 days around it. Its orbital speed is 48 km/sec—the fastest among the planets.

Mercury is a difficult object to observe, and there are certainly many people who have not seen it. It stands at the equator comparatively low over the horizon. In medium latitudes, on the other hand, the closeness to the horizon, with its abun-

Sensational shot of Mercury taken by the spacecraft Mariner 10 in 1974. It shows a surface, fissured by craters, which strongly resembles the Earth's moon.

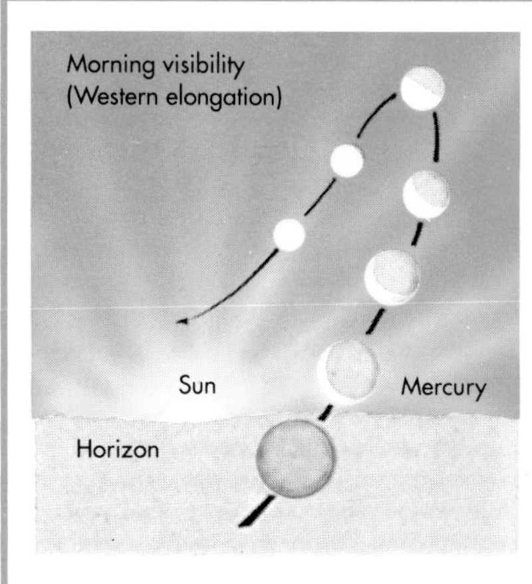

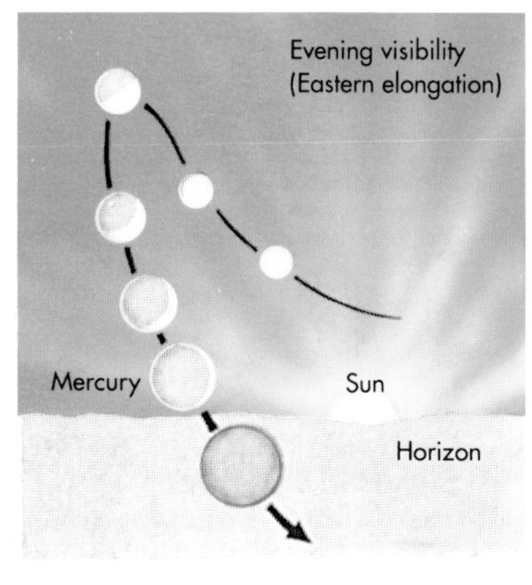

Morning and evening visibility of the planet Mercury. The distance to the Earth increases with increasing phase, and the apparent diameter decreases. Elongation = angle between the Sun and a planet as viewed from Earth.

dantly scattered light, makes it an exercise in patience to find. We can watch Mercury almost only in the twilight. Yet, the planet reaches, at the time of its highest luster, an apparent magnitude of almost –2, which makes it brighter than Sirius in the constellation Canis Major, the brightest of all fixed stars. Mecury moves within the Earth's orbit and shows phases like Venus and the moon—an "increasing Mercury" in the morning sky (Western elongation) and a "decreasing Mercury" in the evening sky (Eastern elongation). Due to the position of the ecliptic in relation to the horizon and taking into consideration the inclination of the orbit of Mercury towards the ecliptic, these are the best visibility times:

- Northern hemisphere of the Earth: in the spring evening and the fall morning;
- Southern hemisphere of the Earth: in the fall evening and the spring morning.

The planet reaches its brightest at the time of the superior conjunction. It is possible to find Mercury at only a small elongation, i.e., close to the sun. Thereby, it is better when the sun is covered by clouds. Observations in bright daylight are also successful when the planet has a distance of 5°–10° from the sun.

It is important to remember that the sun's surrounding has to be low on scattered light and you have to avoid any direct observation into the bright sunlight. Therefore, search for Mercury in the daytime only in the shadow of a building.

It is best to use binoculars when looking for Mercury in the morning or evening sky. During the day, using a 2–3-inch telescope with 40-fold enlargement is better. Astronomical year books give information about the respective locations of the planets. *Astronomy: A Handbook* and the *Handbook for Planet Observers* also contain observation tips with individual details.

In March 1974, the spaceship Mariner 10 sent the first pictures of Mercury to Earth. The pictures showed many craters, similar to those on the Earth's moon that were also created by impacts of cosmic bodies (see photo on p. 99). Even before the mission, astronomers assumed that there was a similarity between Mercury's surface and the moon's surface. Their assumptions were based on photometric and polarimetric observations done from Earth. Also on Mercury, there are large basins with diameters of 1000 km (620 miles) and small craters of 100-m (330-ft.) diameter. The small craters correspond to the limitation of resolution of the photos that were taken from Mariner 10. The discovered craters have names of famous artists. The largest crater was given the name Beethoven.

The similarity of the surface of Mercury and the Earth's moon allows the assumption that meteorites existed everywhere in the inner solar system at the time of formation of the planetary system. The bombardment of Mercury and the moon probably took place at the same time. It released volcanic activities. Lava filled craters and basins that had been created in the early epoch. New crater formations marked the genesis of this planet.

Mercury has an extremely thin atmosphere. It partially consists of solar wind, that comes from the sun. The major component of its atmosphere is helium. Mer-

Phase formation of the inner planets Mercury and Venus. When it is farthest away from the Sun, Mercury has an elongation of 27°, Venus has an elongation of 47°. The best observation opportunity of the two planets is the time around the elongation.

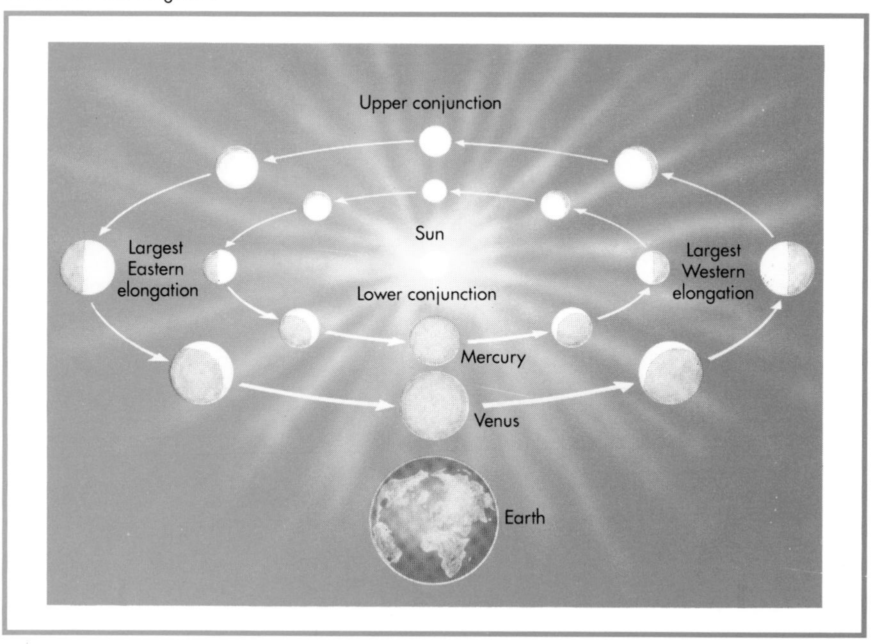

cury's atmosphere has no oxygen, carbon dioxide, or nitrogen.

For a long time, astronomers assumed that the rotation of Mercury lasted 88 days. The study of radar signals in 1965 came in with the surprising result of 58.6 days. The time of rotation is influenced by the tidal forces of the sun, which affect daily Mercury's body. The strongly elliptic rotation orbit is another factor.

The sun shines during the long Mercury day onto the surface so that the noon temperatures reach higher than 430° C (700 Kelvin). At night, on the other hand, the temperatures sink to under −180° C (100 Kelvin).

Venus: the "Morning and Evening Star"

You may have already seen Venus, the radiant, bright "evening star." Many people think that Venus can only be seen at night. But it can be seen, though not as brightly in the morning sky like other constellations or other bright planets (Mercury, Mars, and Jupiter) that dominate the evening and the morning sky. None comes close to Venus' brightest luster, (whose apparent magnitude is –4.3).

Venus moves within the Earth's orbit around the sun; it is like Mercury, an "inner" planet. And just like Mercury, it comes on its orbit between the Earth and sun, and it shows, as seen through binoculars, the increasing and decreasing crescent forms similar to the moon's that we observe for months with the naked eye. The distances between the Earth and Venus during a synodic period are not constant (a synodic period is the measure of time it takes to complete a full cycle of orbit positions with respect to the sun and moon i.e., the time period between two superior conjunctions with the sun, seen from our planet, Earth: 583.92 days. The sidereal rotation time, on the other hand, is the time period of an entire rotation of the planet around the sun: 224.7 days). The apparent diameter of Venus is different at the individual phases. It is largest in the weeks before and after the inferior conjunction and smallest at the time of the superior conjunction.

At the time of the superior conjunction, Venus is, like Mercury, at its longest distance from Earth; at the time of the lower conjunction, it is closest to the Earth (see picture on p. 100). What is a conjunction? It is when the planet is between the Earth and sun, and it becomes lost in the sun's glare. Another important aspect is when it is the opposite: when the planet is on a straight line with the Earth and sun, but opposite the sun so that it cannot be seen (see illustration on p. 105). Neither Mercury nor Venus can ever stand in opposition. Only the outer planets, such as Mars and Jupiter can do this.

We observe that Mercury and Venus are never in a very large angle from the sun (elongation). The largest angle for Venus is 48°, for Mercury only about 28°. That means that their visibility is always restricted either to the evening or morning sky. But the elongation, which for Venus is essentially larger, makes it possible to still see the planet several hours after sunset or before sunrise. In addition, there is the enormous apparent magnitude. Venus can therefore become a dominating and striking celestial object.

Venus through the Telescope

When observing the planets, it is important to recognize the two different movement conditions:

1. the movement of the planet around the sun,
2. the movement of the planet in regards to the Earth, which itself rotates around the sun.

Aspects and phases result from these two conditions. The bottom table shows it for Venus. In order to observe the different phases, you need binoculars or a small telescope. Venus passes by the sun several degrees to the north or south in inferior conjunction. It is then worthwhile to try to recognize the narrow crescent beyond the conjunction. Numerous observers have done this in the past. Because of the large apparent magnitude of Venus, you can also watch it during the day with the naked eye. All you have to know is where the planet stands approximately in the sky: on the arc of the sign of the zodiac before or after the

Venus as an evening star taken May 5, 1967, 21h30 UT. Photographing the change of the phases is possible with amateur telescopes.

Aspects and Phases at Venus

Aspect	Phase
Inferior conjunction	New Venus
Largest Western elongation (Morning sky)	First Quarter (waning phase)
Superior conjunction	Full Venus
Largest Eastern elongation (Evening sky)	Last quarter (waning phase)
Inferior conjunction	New Venus

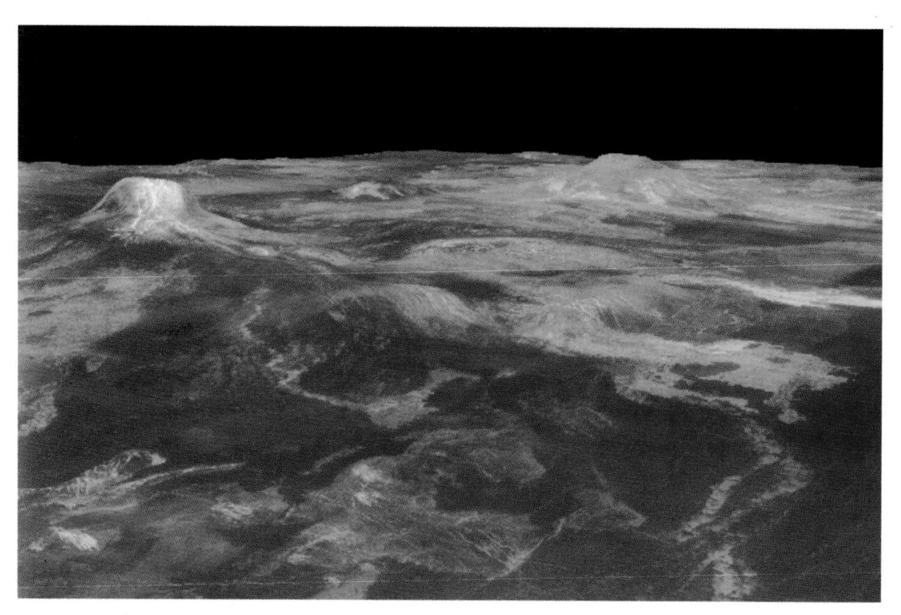

The radar enterings on the map, which began in 1990, of the surface of Venus seen from the spacecraft Magellan, show a surface, that is strongly characterized by volcanoes.

sun (elongation). Then search the sky while standing in the shadow of a building. Werner Sandner writes in his book *Planets: Siblings of the Earth*, which is full of experiences about Venus: "When it was accidentally seen in the sky by people who did not know what it was, it caused a sensation, which led to superstitions. Thus, in May, 1609, when Venus was seen at many locations in France during the day, it was interpreted as an omen for the murder of King Heinrich IV. Also in 1798, Venus' visibility in Paris so attracted the attention of the people that no less than the Emperor Napoleon I declared it to be his star of destiny . . ."

It is not possible to recognize details on the surface of Venus with a telescope. Spacecraft sent to Venus have confirmed that the planet is enveloped by a dense atmosphere, which makes it very difficult to observe. It is ninety times denser than Earth's atmosphere. The measurements taken from the "Venera" spacecraft hinted at a high content of carbon dioxide, a high air-pressure at the surface, and ground temperatures between 400° and 500°C. If you see bright and dark shades through a 4-inch telescope (at 100- and 150-fold enlargement), they are the upper cloud-layers of Venus' atmosphere (up to 70 km/43.4 miles) that do not allow any viewing of the firm details on the surface. The surface of Venus has an atmosphere similar to Earth's on a gray, rainy day, except it is much hotter. The texture of the surface is comparable to the desert with mountains, craters, and volcanoes.

The cloud cover has several layers of sulfur. The largest particles, in greatest density, are in the lowest layer. The atmosphere is clear up to a height of 30 km (18.6 miles), because the dominating heat evaporates all the particles. The cloud cover on Venus reaches 40 km (24.8 miles), compared to Earth's, which is 6 km (3.72 miles) thick.

The surface

Wind movement is very slow on Venus' surface. Its speed increases with height and reaches in the upper atmosphere-layers up to 360 km/hr (223 miles/hr). The winds blow from east to west because Venus rotates "backwards" from east to west (called retrograde motion). That means, while stormy winds predominate in the high atmosphere (comparable to the jet streams in the Earth's atmosphere), it is almost calm on the surface and, because of the greenhouse effect, it is several hundred Celsius degrees. Liquid water would evaporate immediately under these circumstances. Nevertheless, it cannot be excluded that a lot of water existed at one time on Venus.

The American space probe Magellan, which circled around the planet from August 1990 to May 1993, is to date the climax of the space flight program. Radar

Venus, shot from 59,000 km (36,580 miles) from Pioneer's Venus Orbiter on May 28, 1979. The cloud whirls are clearly recognizable at the upper layer of the atmosphere.

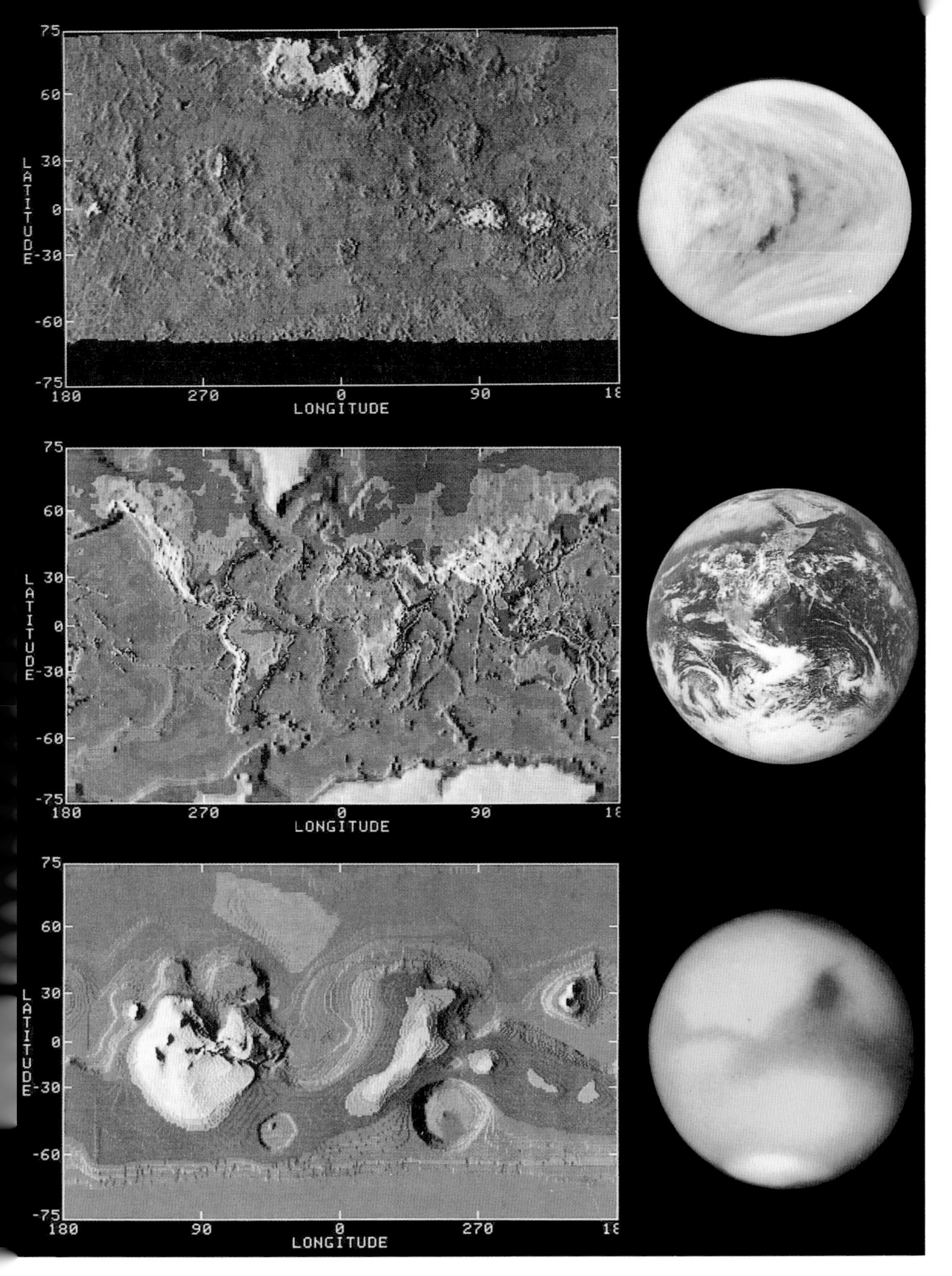

pictures, taken by the probe, penetrate Venus' clouds to give us a comprehensive topography of its surface. In general, it is level. Over 80% is less than 1000 m (3300 ft.) away from the planet's radius (6052 km/3752 miles). Lava, which streamed out, formed these levels. In between, there are volcanoes of different sizes and mountains, which rise up to 11,000 m (36,300 ft.) above the surrounding areas. The mountains dominate high plateaus, some of which are extensive. The two largest ones, Aphrodite Terra and Ischtar Terra, have the expanse of Africa and/or the United States. All important terrain formations on Venus have the names of women. The surface rocks consist of basaltic lava.

The measurement data of the Magellan probe also brought in information that Venus rotates more slowly than all the other planets. It rotates in 243 Earth days once around its axis. Thus, the rotation takes longer than the movement around the sun, which requires only 225 Earth days. And another special thing about Venus: The sun rises in the west on Venus, not in the east.

Venus is one of the most-observed celestial objects, and was so even long before the invention of the telescope. Ancient peoples were interested in the planet's movements for very obvious reasons: they served as a way of calculating time, as a date calendar for festivities, and for sowing and harvest. The Mayans in South America even had a proper Venusian calendar and the Venusian year of 584 days. This astronomical tradition goes back to an epoch which began several millenniums before our time calculation. That's why astronomy is considered the oldest among the sciences.

People realized early on that the planets, on the one hand, take part in the daily rotation of the celestial vault from east to west; on the other hand, they also execute highly independent and unequal movements in the sky. All possible theories have been developed in order to interpret the latter. The illustration on page 106 shows how the movements actually come about.

Computer produced radar maps of Venus, Earth, and Mars.
Top: Topographical map of Venus in Mercator projection. Each color corresponds to a height interval of 500 m (1650 ft.); the resolution at the equator is about 100 km (62 miles). The data for this map was taken through the Radar Mapper experiment of Pioneer's Venus Orbiter.
Center: Topographical map of the Earth in Mercator projection with the same height interval as on the Venus map. The resolution at the equator is about 111 km (68.8 miles). The data for this map was compiled by the Rand Corporation, Scripps Institute of Oceanography and by the U.S. Defense Mapping Agency.
Bottom: Topographical map of Mars in Mercator projects with the same height interval as on the Venus map. The resolution at the equator is about 600 km (372 miles). The data for this map was compiled by the U.S. Geological Survey through different space missions and earthbound radar experiments.

Mars: the Red Planet

The people of antiquity thought this planet represented the god of war, because of its reddish color. The Babylonians, ancient Greeks, and ancient Romans called their war gods Nirgal, Ares, and Mars. When the telescope was invented, the astronomers discovered a number of characteristic features that had comparable similarities to Earth: atmospheric objects, polar caps, and seasons. A Martian day is 24 hours, 37 minutes, and 22.6 seconds, which is similar to the Earth day. The Martian year—the rotation time of the planet around the sun—is 687 Earth days—twice as long as the year on Earth.

The first observer to determine the length of the Martian day was Christian Huygens. In 1669, he discovered a dark triangular formation on the Martian surface, which is today called the Syrtis Major, and he derived the time of rotation from the meridian transit observations. Observers after Huygen detected changes of the light and dark areas on Mars together with the changes of the seasons. Gradually, it became clear that currents in Mars' atmosphere played a role. Winds blow away dust on the surface, collecting it in one place to uncover Mars ground in another place.

The Mariner probes helped dispel many old theories about Mars by supplying us with photos. One such theory was the existence of "canals." For almost an entire century, generations of planetary observers had occupied themselves with these canals, which were "discovered" by the Italian astronomer Giovanni Schiaparelli in 1877. The suggestion of canals fueled speculation of an artificial origin and, thus, led some to believe a higher form of intelligence on Mars had created them. Then, on July 1, 1965, the Mariner probes supplied photos showing craters that were similar to those on the moon. It helped cast aside many old models and theories about Mars. In 1971, the Mariner probes also brought information about two moons on Mars. The photos showed the moons to be surprisingly potato-shaped (see page 108).

Left: Map of planet Mars with numerous "canals" according to observations by Giovanni Schiaparelli in 1886 at the observatory Nizza.

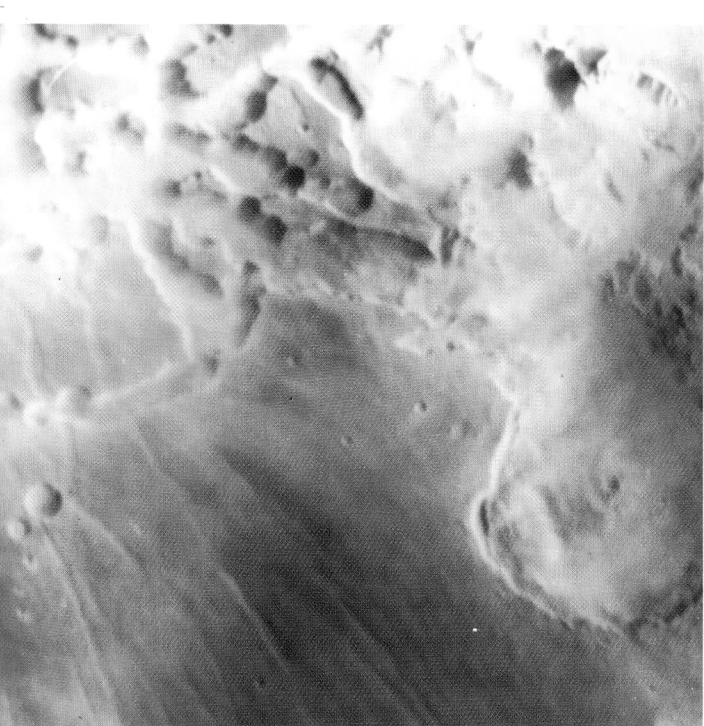

Bottom left: Noctis Labyrinthus, part of a large canyon system on Mars. The picture was taken on October 12, 1976 from Viking 1, when it orbited Mars while the sun was rising over the Martian landscape. You can see light clouds made from water ice in the valleys.

Bottom: Mars, photographed in February 1995 from the Hubble Space Telescope. The distance between Mars and the Earth is 103 million km (64 million miles).

Viking I and II landed on Mars in the middle of 1976. The results of the space missions showed a planet that has some resemblance to the Earth's moon, such as craters that were the consequence of impacts of cosmic bodies. Unlike those of the moon, the large volcanoes have considerable dimensions. The volcano Olympus Mons, for example, has a diameter of 600 km (372 miles) and reaches a height of 27 km (16.74 miles) above the medium level of Mars. It is the largest known volcano in the solar system. As a comparison, Olympus Mons is almost three times higher than the Loa volcano in Hawaii measured from the bottom of the ocean. Furthermore, spectacular valley landscapes were discovered on Mars, which are best compared with the Grand Canyon in Arizona. Valles Marineris would stretch all the way across the United States!

The surface

As mentioned before, the "canali" of Schiaparelli did not only spur on the observers but also the fantasies of many people, such as H.G. Wells' *War of the Worlds* and Ray Bradbury's *The Martian Chronicles.* Since then, Mars has been the most observed and noticed planet. It also played a role in antiquity. Because of its strikingly reddish color, it was interpreted as a "fire star" and called the god of fire. It eventually became the god of war.

There is no doubt that rivers once flowed on Mars' surface. The photos from the space probes showed dried out riverbeds and stream valleys. It is assumed that masses of ice melted on the surface of Mars and, thus, rivers, carrying the water, were formed. It is most likely that even today there is still ice on Mars' surface. The ground layer, which is frozen throughout the entire year, similar to the tundra, is estimated to be around 1000 m (3300 ft.) thick. Other ice caps are at the poles. During the Martian winter, "dry" ice (a combination of water ice and frozen carbon dioxide from the atmosphere), forms at the North Pole. In the summer, the dry ice melts into the atmosphere. Then, a layer, which consists of water ice, becomes accessible for observation. The conditions at the South Pole are similar.

Mars' ground, rocks, and dust accumulations, which were photographed, belong to a layer of volcanic rocks that was weathered by chemical and mechanical influences.

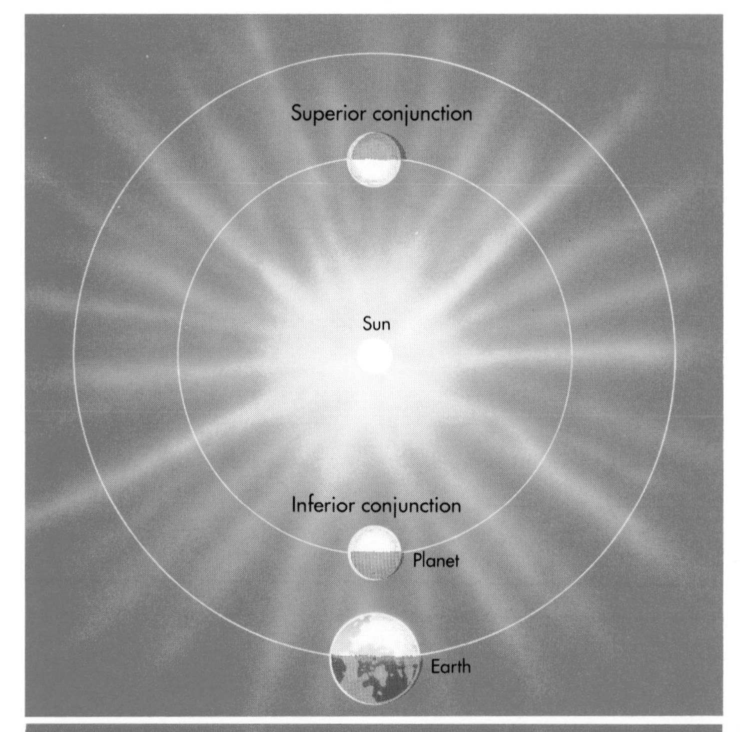

The two planets Mercury and Venus circle within the Earth's orbit around the sun. When they are on their orbit between Earth and sun, it is called inferior conjunction. When they are behind the sun, i.e., the sun then stands between the planet and Earth, it is called superior conjunction.

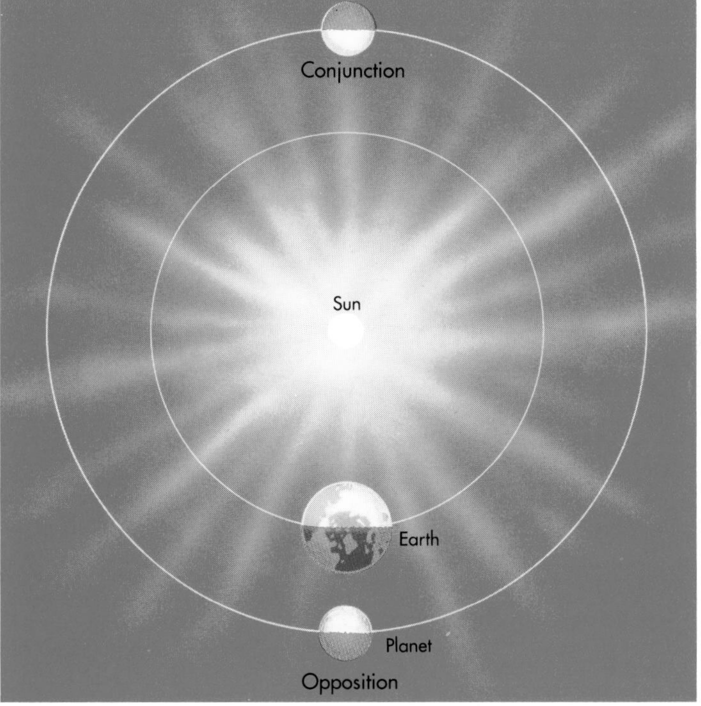

Mars, Jupiter, Saturn, Uranus, Neptune, and Pluto circle around the sun on orbits that are outside the Earth's orbit. When the Earth is between the planet and sun, the planet stands in opposition and it can be well seen during the night. When the sun stands between Earth and planet, the planet is in conjunction and it is invisible.

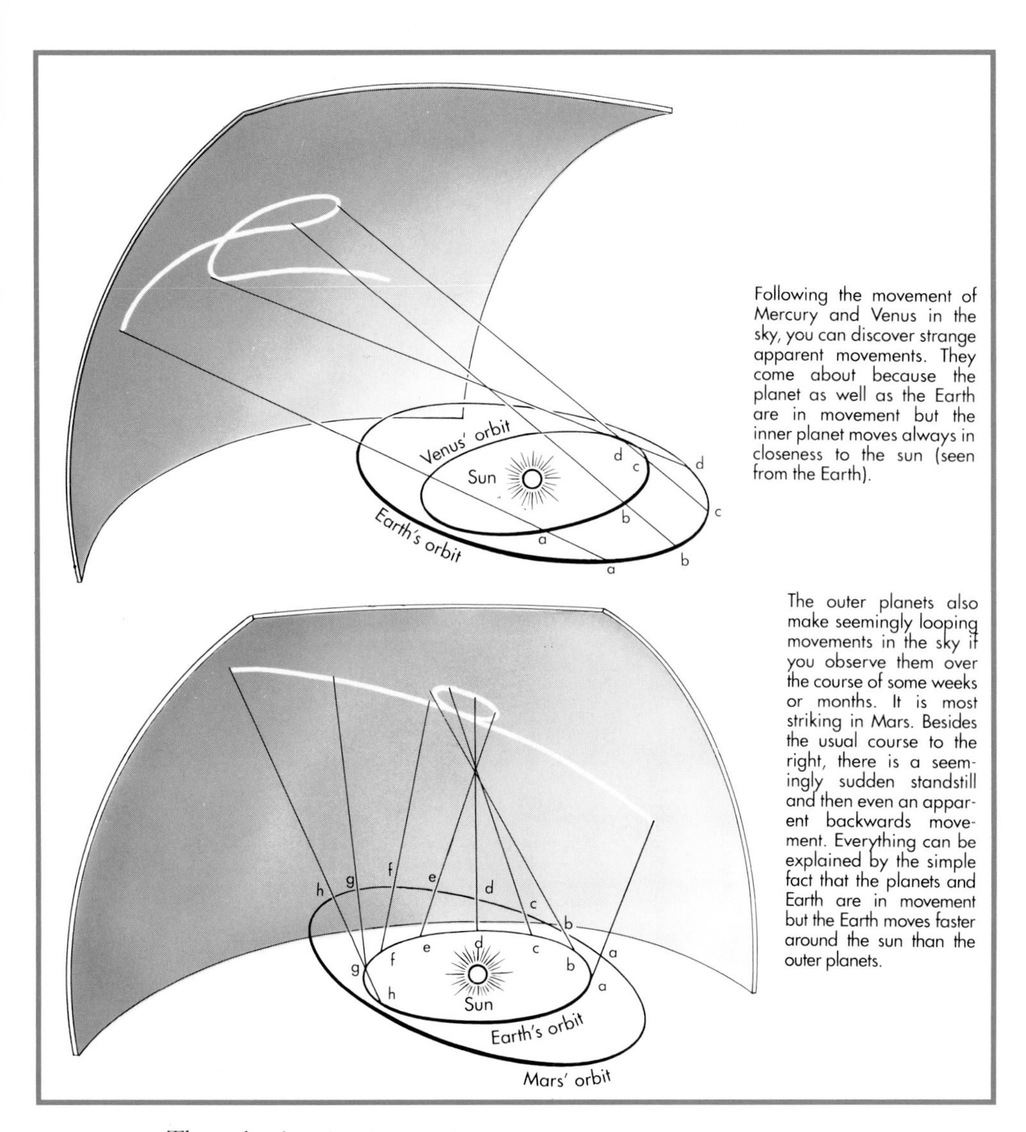

Following the movement of Mercury and Venus in the sky, you can discover strange apparent movements. They come about because the planet as well as the Earth are in movement but the inner planet moves always in closeness to the sun (seen from the Earth).

The outer planets also make seemingly looping movements in the sky if you observe them over the course of some weeks or months. It is most striking in Mars. Besides the usual course to the right, there is a seemingly sudden standstill and then even an apparent backwards movement. Everything can be explained by the simple fact that the planets and Earth are in movement but the Earth moves faster around the sun than the outer planets.

The red coloration is caused by mineral iron-combinations, which make wide parts of Mars' ground look like a rust-brown to orange-colored desert. The dust accumulations are often like shifting sand dunes, which slowly move over the surface. Strong sandstorms occasionally rage on Mars and envelop the planet in a yellow veil.

Orbit Loops in the Sky

Mars is the second closest planet to the Earth. Its orbital position results from the deviations of the distance between Mars and the Earth and the combined changes of the apparent disk diameter, which are larger than those of Venus. Venus shows us, when it is standing closest to the Earth, its unlit side for the most part. In the case of Mars, the conditions are completely different. It gets into the opposition position (see illustration on p. 105) when it stands opposite the sun and, at the same time, close to the Earth. From Earth, we directly face Mars and its fully lit disk. For this reason, Mars has, among all the planets, the strongest brightness deviations. The apparent brightness varies between -2.5^{m} and 2.0^{m}. Mars thus can become brighter than the brightest fixed star, Sirius in the constellation Canis Major (Great Dog).

Mars is an outer planet, because its orbit takes it outside of the Earth's orbit around the sun. At the time of opposition, it stands closest to the Earth and can be observed throughout the entire night. Thus, its viewing conditions are considerably more favorable than of the two inner planets, Mercury and Venus. Mars' orbit is characterized by a relatively large eccentricity (a perfect circle has an eccentricity of 0; the most elongated ellipse possible would have an eccentricity of just under 1.) What does this mean? The distance between Mars and the sun is between 207 million km/128 million miles (the point of the orbit closest to the sun is also called Perihel) and 249 million km/154 million miles (point of the orbit farthest away from the sun is also called Aphel). As a result, the distance between Mars and the Earth is also not the same at each opposition. Their deviations are between 56 million km (35 million miles) and 101 million km (63 million miles). Because of that, the apparent diameter of Mars' disk can grow from 14 arc-seconds (diameter in the Aphel oppositions, which are favorable for observation) to 25.5 arc-seconds (diameter in Perihel oppositions). Perihel oppositions can be especially well observed in the southern hemisphere of the Earth—to the regret of the astronomers in northern latitudes—because then Mars is in the constellations Capricornus and Aquarius and has reached a low height above the horizon. Visibility conditions are much worse closer to the horizon because of air disturbances, such as the long way for light to travel through the atmosphere and air pollution that is concentrated there.

The orbit of the planet points to another characteristic feature. As a general rule, the Earth, at the time of opposition, overtakes an outer planet. When this happens, it seems to the observer that the planet is changing its usual west-east direction of movement. The expert refers to this as the planet becoming retrograde. This phenomenon is most striking in Mars because of the small distance to the Earth. That means that the part of the orbit, which is traveled in reverse in the sky, is considerably long. In addition to that, there are movements vertical to the ecliptic. The result is that the astonished observer will detect orbital loops of considerable size, which are the more impressive the more bright stars there are close by at a certain point in time. Then you can see the change well from week to week in comparison to the fixed position of the stars.

The loops in the sky, which the outer planets make and which are so strongly pronounced in Mars, caught the attention of the astronomers of the Hellenistic world. They pondered the question of how such unrhythmed movements could be brought into line with the desired harmony of nature. After all, they came with the idea that the stars were gods, which had only the most perfect movement: the movement in circles. As a result, complicated theories about the planets were formed, the best known of which was that of the Egyptian astronomer, geographer, and mathematician, Claudius Ptolemy (A.D. 85–160). His theory was officially accepted for 1500 years until Nicolaus Copernicus switched the view of life from that of a geocentrical to a heliocentrical existence, i.e., from the Earth as the center to the sun as the center. German astronomer Johannes Kepler got on the track of the planets' loops by replacing the circular path with the ellipse. Observing Mars especially helped him to do this. Kepler reported to his Emperor, Rudolf II: "On command, your honored majesty, I will present for public view the very noble prisoner [Mars], whom I have captured some time ago in a burdensome and laborious war. Through mother nature, he conveyed to me the admission of my victory. After he had asked for freedom within the voluntary handcuffs, he soon changed merrily and cheerfully into my camp, accompanied by arithmetics and geometry." Kepler formed three laws to explain the movements of Mars:

1st law: The orbit of each planet is an ellipse with the sun at one focus of the ellipse.

2nd law: Each planet revolves around the sun so that the line connecting planets and sun, the radius vector, sweeps out equal areas in equal times.

3rd law: The squares of the rotation times of the planets are proportional to the cubes of their mean distances from the sun.

Although these laws are obvious today, they were not back then. When observing Mars, or any of the other planets, anyone has the opportunity to move on Kepler's tracks. Many astronomy books contain overview maps with the planetary movements in the sky (see "Actual Events in the Sky" on page 165ff).

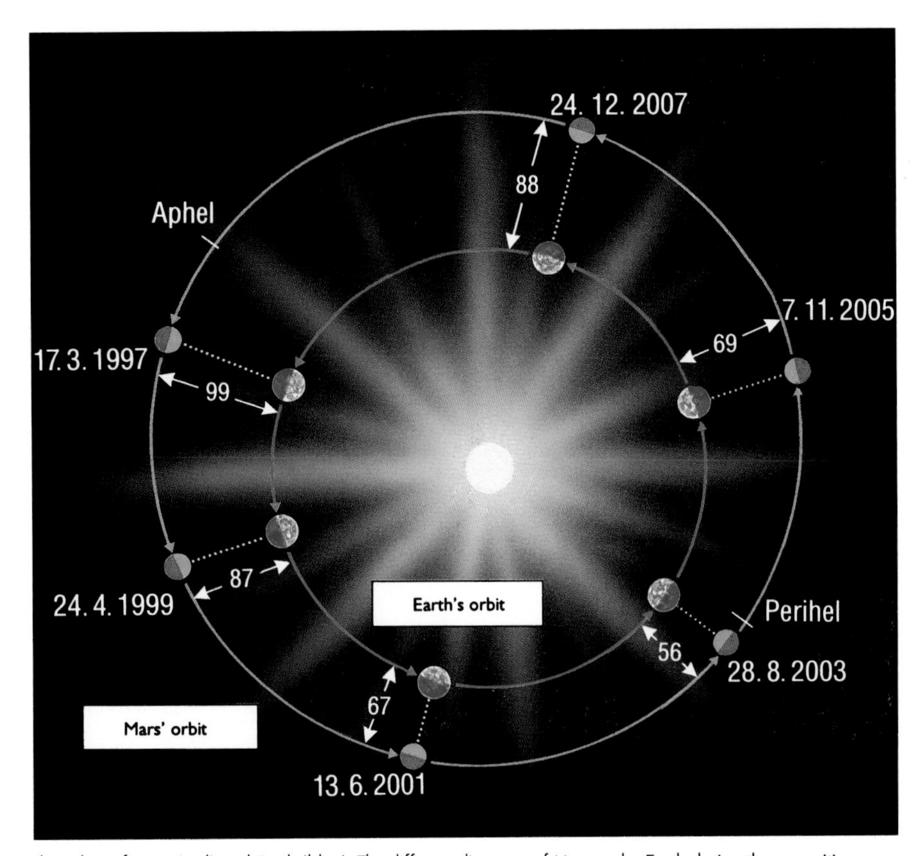

The orbits of Mars (red) and Earth (blue). The different distances of Mars to the Earth during the oppositions as well as the varying distances from the sun becomes clear. There are oppositions farther away from the sun (Aphel oppositions) and oppositions closer to the sun (Perihel oppositions). Distance between the Earth and Mars are in millions of kilometers.

Mars through the Telescope

What does the astronomer see through his or her telescope when observing Mars? The planet has undeniable advantages, which you can see. Even with a small telescope (2–3 inches), you can observe details on it that belong to its firm (as opposed to gaseous) surface. This is different from Jupiter and Saturn (see pages 110 and 115 for details on these planets). The disk shows the typical yellow-red color. Besides the so-called dark areas, some bright spot-like formations stand out on Mars as well as the polar caps with their seasonal changes. In the Martian fall and winter, the polar caps increase in size and can cover up to 10% of the planet's surface. In the Martian spring and summer, they become smaller again. That offered a lot of information for astronomers to speculate about snow and ice and the melting process on Mars when it becomes warmer again because of the seasons. Since the firm surface can be observed, maps of Mars were already made early on. On the whole, the formations turned out to be constant. When they become in one opposition less visible, or even partially invisible, then it is generally because of atmospheric conditions, such as cloudiness or dust storms,

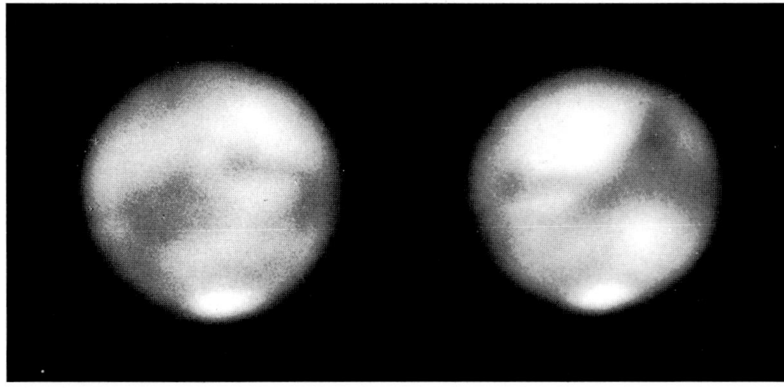

The color photo (left) makes it immediately obvious why Mars is called the "red planet." Its yellowish-red color draws the attention of any observer without the need of a telescope. The photo, taken with a large telescope, shows on the surface bright and dark details, which partially have the typical color as well.

Compared to the color photo, the two black and white photos (above) also show astonishingly many details. And consider that these two photos were taken with a small telescope (20 cm opening) by an amateur photographer.

on Mars. Such phenomena can be observed with a 4–5-inch amateur telescope. Weather exists on Mars, but be careful not to compare it too closely with the Earth's. For example, the enormous oceans and large land masses which substantially influence the weather conditions on Earth are missing in Mars. The ocean atmosphere is also positively dry. Mariner probe measurements hint at a lot of carbon dioxide, a little bit of water vapor, carbon monoxide, and some oxygen. The discovery of nitrogen through direct measurements of the Viking spacecraft was important. The atmospheric ground pressure on Mars reaches not even one hundredth that of the Earth's. But let's not forget that Mars is quite a bit farther away from the sun than is the Earth. This translates into the measured mean temperature of −40° C; the top readings are between 30°C and 100° C.

Was there or is there life on Mars? This is maybe the most moving question for everyone investigating this planet. The Viking spacecraft did not find any clues of microorganisms, but the speculation that there was formerly life on the planet still remains. Fossil evidence of that epoch before the cooling off of Mars is currently being debated (some have believed that they found evidence of fossils of microorganisms on a Martian meteorite).

Deimos and Phobos

The idea that Mars had two moons already existed in the centuries before their discovery. Johannes Kepler did not exclude the possibility. English satirist, Jonathan Swift described in his book *Gulliver's Travels* two Martian moons, whose rotational orbits came astonishingly close to reality. The satellites were actually discovered in 1877 by Asaph Hall at the U.S. Naval Observatory in Washington.

They received the names of the companions of the god of war in Homer's *Iliad:* Deimos and Phobos—fright and fear. Again, it was the space probes that delivered more precise information about these strange formations, whose force of gravity is much too small to be a real globe and to hold an atmosphere. The potato-shaped celestial bodies were hit with many mete-

One of the two moons of the planet Mars: Deimos. Its surface is characterized by grooves made by craters. It is believed that the Martian moons were captured from the Asteroid Belt between Mars and Jupiter.

orites during the formation of the planetary system. Their surface, covered with craters and grooves, shows this. Phobos is strongly governed by the tidal forces of Mars and approaches the surface of the planet in spiral-shaped movements. In about 100 years, it will hit. Phobos travels around Mars in 7 hours and 38 minutes, as compared to Deimos, which takes 30 hours and 18 minutes.

Asteroids, or minor planets

Strange celestial bodies move between the planets Mars and Jupiter. They are strange in regards to their size as well as their movement conditions in the solar system. Because of their smallness, they were discovered relatively late. But astronomers had realized, by calculation, that a gap existed between Mars and Jupiter long before the discovery of Ceres, the first of these asteroids (planetoids, or minor planets, as they are also sometimes called) in 1801. The discovery of Uranus in 1781 confirmed Bode's law, which expressed an apparent numerical relation between the average distances of the planets from the sun. This law also predicted the existence of a planet between Mars and Jupiter. In the 17th and 18th century, much was discussed about this possibly still undiscovered planet, encouraging observation. At the beginning of the 19th century, Ceres was discovered, and the first of these asteroids fitted perfectly into the mysterious gap. But this minor planet was not the last one to be discovered. Quickly, discoveries increased so that today we know over 4000 asteroids, whose orbits have been secured. By far, most of them were discovered photographically, which is not surprising: Only about two dozen asteroids reach or surpass at a Perihel opposition the 10th magnitude. The asteroid Vesta reaches the largest apparent brightness of 6^m.

Astronomers estimate that there are billions of asteroids with a diameter larger than 1 km (half a mile). Of those, about two hundred have an expanse of over 100 km (62 miles). About 90% of the asteroids in the Asteroid Belt are over 100 km.

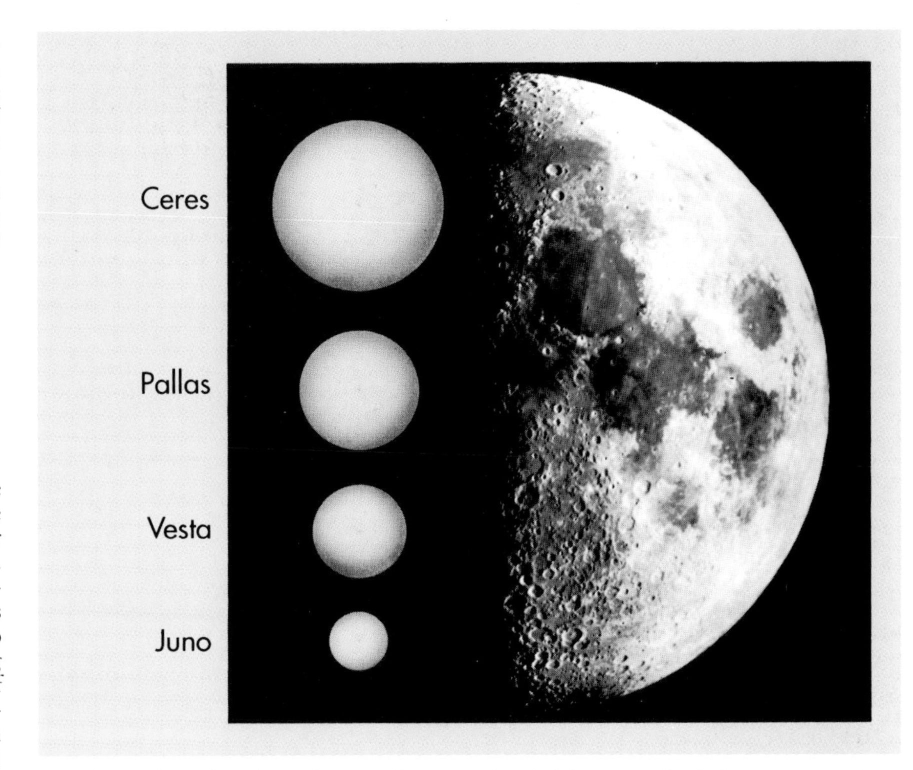

Size of asteroids in relationship to the moon. These pictured on the left correspond to the four largest ones. From top to bottom: Ceres (940 km/583 miles diameter), Pallas (588 km/365 miles diameter), Vesta (516 km/32 miles diameter), and Juno (248 km/154 miles diameter).

Rocks in outer space

Magnitude measurements tell us something about the rotation and shape of the asteroids. The fastest among them rotate in a few hours around their axis. The observed irregular light-curves point to irregular-shaped asteroids that look like a potato or cigar. The spacecraft Galileo, on its way to Jupiter, in October 1991, came so close to the asteroid Gaspra that photos of the surface were taken (see photo on page 110). The many craters immediately stand out. The body looks like a piece of rock in outer space, which confirmed the irregular look of asteroids.

In the case of Ceres, we deal with a celestial body of almost 914 km (567 miles) diameter; the others are much smaller. Asteroids are regarded as splinters of planets rather than actual planets. This idea is not far-fetched. An existing theory traces the formation of the asteroid swarm back to the explosion of a larger planet and/or to its development out of protoplanets in the early stages of the planetary system.

Maybe it was the influence of the planet Jupiter, and its force of gravity, which forced the small protoplanets onto strongly elliptic orbits. There, collisions happened again and again. The small protoplanets broke apart and formed the Asteroid Belt between the planets Mars and Jupiter. All asteroids revolve, just like the large planets, in the direction to the right around the sun. In the main, asteroids orbit the sun in an elliptical path. But there are a number of asteroids with extreme deviations from this norm. Some have highly eccentric orbits, with the result that their distances from the sun, Earth, and other large planets are noticeably changing. The orbits of the asteroids Eros and Icarus, for instance, go beyond the

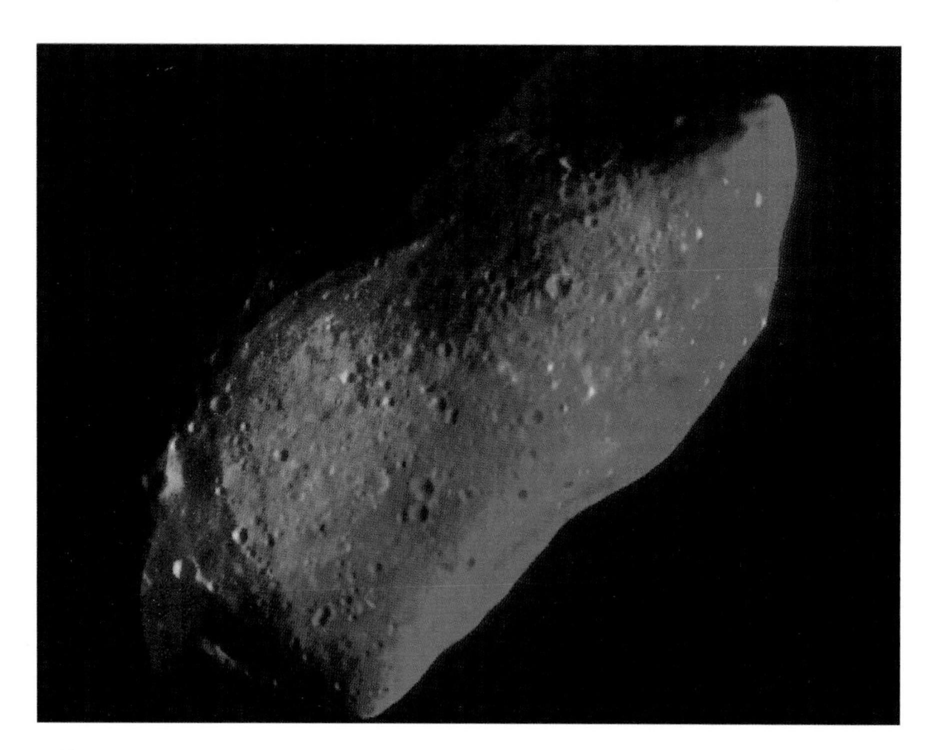

The Gaspra asteroid (1951) as it was seen from the Jupiter probe Galileo on October 29, 1991 from 5300 km (3286 miles) away. The sun is shining from the right. The total length of the illuminated part is 18 km (11 miles). The formation of craters on this minor planet puzzles science.

orbit of Mars. Eros approaches, at the time of its opposition, relatively close to the Earth; Icarus intersects the orbit of Mercury close to the Perihel. An extreme counterpart is the asteroid Hidalgo, whose strongly eccentric orbit beyond Jupiter's orbit brings it very close to the orbit of Saturn. Then, there are a number of asteroids, collectively called Trojan asteroids, whose distance from the sun is equal to the distance of the planet Jupiter. There is justified suspicion that the special case of the Trojans is caused by orbital disturbances, which emanate from Jupiter and which entrap asteroids into the area of Jupiter's orbit. In addition to that, broken matter, in the form of meteoric particles, can have an effect between the planets, which favor the diversion of the orbits of asteroids, low in mass, into the orbit of the giant planet Jupiter, which is so much more enormous in mass. In other words, a lot is going on in outer space and in regions where one assumes everything to be in order. Scientists consider asteroids an interplanetary substance, along with certain meteor streams, comet groups, and the zodiacal light. The relationships that have been discovered between meteorites and minor planets are very informative. The analysis of meteorites, which fell on Earth, suggests that they could be the fragments of a planet.

For the visual observation of the brighter asteroids, you need binoculars or a 2- or 3-inch astrotelescope. You can also try experimenting with photography. Besides estimating magnitude, you can, with a little patience, determine their position. *Astronomy: A Handbook* and *Handbook for Planet Observers* give advice for that.

Even people who are not astronomy buffs are particularly interested in asteroids for two reasons. For one thing, there is always the question about a threat to the Earth through the impact of asteroids. After all, about 1300 asteroids with a diameter of 1000+ meters (at least 3300 ft.) intersect the Earth's orbit. Therefore, there is always the possibility of collisions. If one of these small bodies hits the Earth, it creates a crater that will be much larger than the object itself. A number of craters discovered on Earth were created by asteroids, meteorites, and comets. The largest craters are caused by asteroids. The most recent and best-preserved crater is the Nordlinger Ries in Germany. It is about 14.5 million years old and measures 25 km (15.5 miles) in diameter. The probability of a larger asteroid (with a diameter of 10 km/6 miles and more) hitting Earth is small. Such a natural catastrophe happens only once in 100 million years. The second reason for the interest in asteroids concerns the so-called metal asteroids. They are asteroids with a large amount of metal, from iron to nickel and copper. A small metal asteroid, with only a 1000. m (3300 ft.) diameter, has several billions tons of metal in it. Spacecraft can bring this asteroid to Earth. Since the metals are at the surface of the asteroid, they can be easily extracted.

Jupiter: the Giant among the Planets

Jupiter is the largest of the nine large planets. It is an outer planet, like Mars. It also becomes in opposition to the sun so that it can be seen throughout the entire night. In contrast to Mars, which is only visible every 2 years for a couple of months, Jupiter can be observed every year for about 10 months. Despite its greater distance from Earth, compared to Mars, the apparent diameter of the planet's disk grows to over 40 arcseconds at the time of the opposition. Therefore, also from this point of view, there are other favorable conditions for you to observe it (see table on p. 98–99). In fact, Jupiter produces, with the slightest enlargement, the view which the full moon offers to the naked eye.

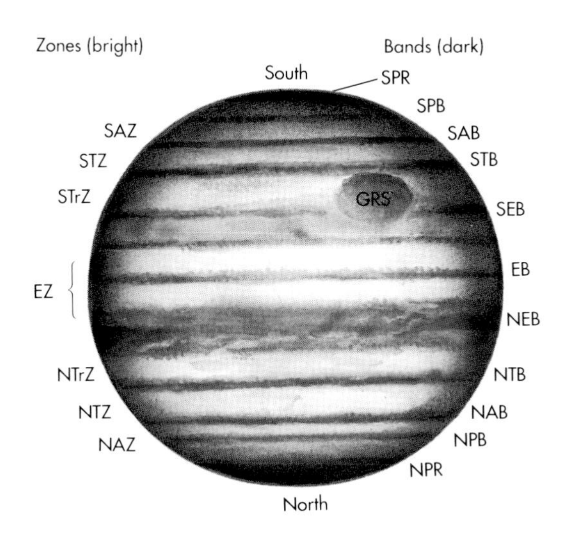

The bright zones and dark bands quickly stand out when we observe Jupiter. Our picture shows the view in reverse through an astronomical telescope. Thereby, south is on top and north on the bottom of the picture. The names of the zones and bands are abbreviations that stand for the following: SAZ = South Arctic Zone; STZ = South Temperate Zone, STrZ = South Tropical Zone, EZ = Equatorial Zone, SPR = South Polar Region, SPB = South Polar Belt, SAB = South Arctic Belt, STB = South Temperate Belt, SEB = South Equatorial Belt, EB = Equatorial Belt.
The analogous assignment of the respective zones and bands on the northern hemisphere of the planet are marked with an N (for North). On the upper right side of Jupiter is the Great Red Spot abbreviated as GRS. It is overall one of the most interesting objects on a planet of our solar system.

The giant planet Jupiter from the view of the Hubble Space Telescope (HST), photographed on May 18, 1994. The distance from Jupiter to the Earth is 670 million km (415 million miles). The black spot is the shadow of the moon Io, which is to the left of it, seen as a yellow disk. Cloud bands of different colors form the upper atmosphere. These are storm clouds, which are drawn out into long streaks by Jupiter's fast rotation. Thunderstorms with enormous flashes of lightning dominate the atmosphere without rain.
Since December 1995, the space probe Galileo has been sending new data about the planet, which confirm tornado-like storms in the dense atmosphere. But Jupiter's atmosphere is supposed to be drier, and there are supposedly fewer thunderstorms than originally assumed.

Jupiter clearly presents itself as a disk to binocular observers. You can then also follow the constantly changing movement of Jupiter's four large moons: At one point, all four stand on one side of the planet, then two stand on the right and two on the left, showing that everything in the cosmos is in movement. You need at least a 2-inch telescope with at least a 40-fold enlargement to see the zebra stripes on Jupiter's disk. The almost parallel, alternating bright and dark stripes look completely different, for example, from Mars.

We encounter a new type of planet with Jupiter. Its equatorial diameter of 142,984 km (88,650 miles) is more than twenty times that of Mars and more than ten times that of Earth and Venus. Jupiter's mass is 318 times greater than that of the Earth. Mercury, Venus, Earth, and Mars are similar in regards to their size, shape, mass and density. Therefore, they are considered terrestrial planets (*terra* is Latin for "earth").

The close-ups of Jupiter and some of its moons, taken by the American Voyager spacecraft in 1979, showed clearly the completely different cosmic landscape in contrast to the terrestrial planets. The photos show an atmosphere that is rich in currents and turbulences. It has a current speed of a maximum of almost 500 km/hr (310 mph). Because of its enormous mass it has a large supply of energy. Astronomers speak of "Jupiter-like planet", because Saturn, Uranus, and Neptune, which are farther away from the sun, have similar characteristic features.

The striped bands that we see across Jupiter are cloud formations of the planet's dense atmosphere. If you have a larger telescope on hand (with enlargements of between 100- and 200-fold) and good viewing conditions, you can look closer into the cloud and see within the stripes a variety of light and dark spots, round, oval, and longish. The experienced observer sees, especially through a mirror telescope or apochromatic refractor, different shades of color: orange, brown, yellow, cream, light blue, and reddish.

Larger swirls frequently suck up the smaller ones and thus create a new push of energy. Jupiter's fast rotation also contributes to the consolidation of larger spots, which brings new movement into the atmosphere. Jupiter's lack of a firm core causes atmospheric objects to immerse into deeper layers of the planet's atmosphere.

As a consequence of the fast rotation, the clouds of the upper atmosphere, those visible to us, are drawn out. In the light zones, the gas rises from bottom to top; in the dark bands, the gas sinks down into the deeper atmospheric layers.

The Great Red Spot

The so-called Great Red Spot (abbreviated GRS) on the southern hemisphere of Jupiter, in about −25° latitude, is a very special, well-known, and famous cloudy and colorful object. This cloud formation has an astonishingly long life. The GRS was discovered over 300 years ago. Since 1831, it has been regularly seen, drawn, and, with the advent of photography, photographed by astronomers. American scientist Gerard Peter Kuiper developed a model, which explained the condition of the GRS and, at the same time, allowed an insight into the procedures that occurred within Jupiter's atmosphere.

Thus, it can be assumed that the GRS is a matter of one of the overall highest cloud-columns in the atmosphere. Between 1831 and 1970, the GRS rotated three times around the planet with irregular speed. The energy that is necessary to give the GRS this stability is provided through the constant possibility in the atmosphere for thunderstorms of unforeseen extent. It can be compared to the conditions on Earth on both sides of the equator except that thunderstorm clouds on Earth are much less constant. The thunderstorm

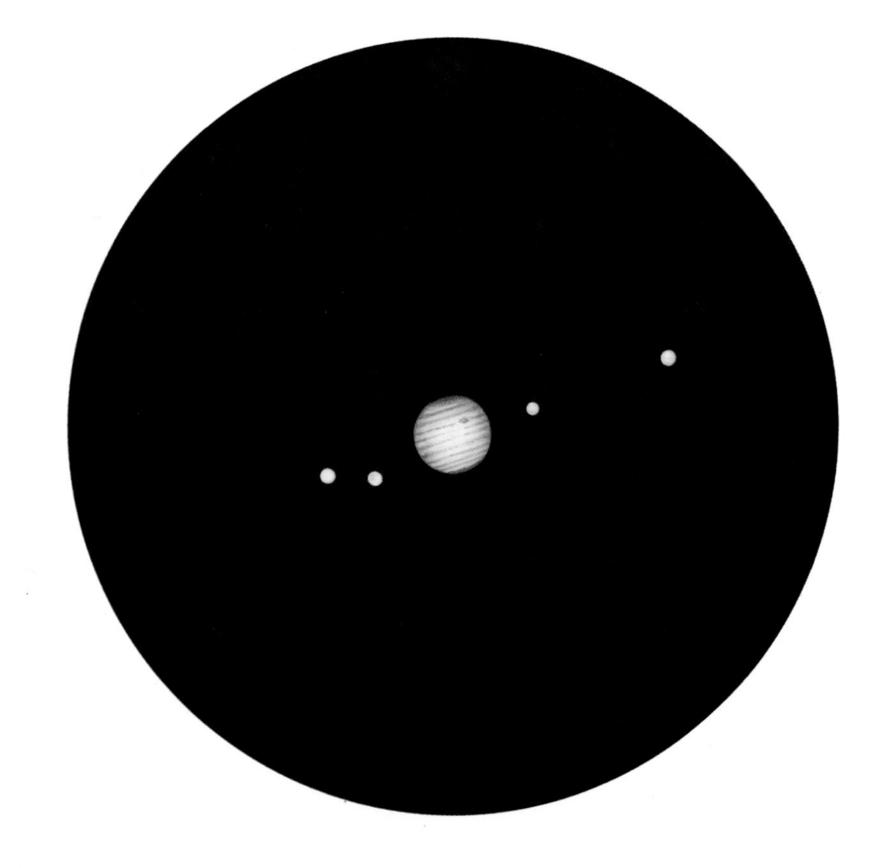

A view of Jupiter and its four largest and brightest moons through a small telescope. Actually, you should be able to see these four moons with the naked eye, but Jupiter outshines them. Through binoculars, the movements of the moons and their different positions to each other and Jupiter can be well observed.

clouds lose the water, which is quickly condensed into the ocean again. Tropical thunderstorms and storms are held in action only with the help of a slow evaporation of the transport of humid air. It is different on Jupiter: The condensation falls into the deeper, hot atmosphere and immediately evaporates. Kuiper assumed that this is the start for a recurring circular course in which the GRS is one enormous never-ending storm. In addition, Jupiter's lower atmosphere is not cooled at night by an ocean as Earth's is. The column of GRS reaches into a gaseous, hot subsoil. Jupiter's permanent condition of reciprocal action restricts the movement of the GRS, with the exception of that slow movement into length.

The other cloud formations have much shorter lives, but individual objects can exist for months, even years.

This photo, taken by Voyager 1 in February 1979, from 9.8 million km (6,076,000 miles) away, shows the enormous dynamics of Jupiter's atmosphere with its many clouds and streams. Cloud formations of several 100 km (62 miles) diameter can be seen. The famous Great Red Spot (GRS) is very large in the picture.

Jupiter through the Telescope

By systematically observing one of the light or dark spots of Jupiter, you will find after several hours that the spot clearly changes its position. In reverse through an astronomical telescope, the spot, along with its other objects that are close by, wander in the latitude from right to left; this is a sign of the rotation of the planet—and a surprisingly fast one. The giant planet rotates in somewhat less than 10 hours once around its axis. One point at the equator of Jupiter, thereby, covers 12.5 km (7.75 miles) in one second (as comparison the value on Earth: 0.46 km/0.29 miles). This large rotation speed of the planet with its 142,984 km (88,650 miles) diameter is also why hardly any cloud movements can be observed from the equator towards the poles. The light and dark cloud stripes almost always surround the entire parallel circle and are interrupted only by manifold currents and disturbances, which suggests turbulence in the enormous atmosphere.

There are explanations for the magnificent colors of the clouds. There are chemical processes taking place there. They are dominated by hydrogen. At closer examination, one again and again comes across hydrogen compounds in the atmosphere: methane, ammonia, H_2O, and hydrogen sulfide. Under the influence of the ultraviolet radiation of the sun, chemical transformations take place. Ammonia and hydrogen sulfide, for example, combine and polymerize to ammonia polysulfides, which can be colored yellow and orange. When the temperature drops, white colorations occur. When the polysulfide row is long enough, it tends to have a reddish color. Kuiper assumed, among other things, that the more prolonged influence of ultraviolet radiation onto the ammonia polysulphurs results in the formation of elementary sulphur structures and, thus, to yellowish shades.

Jupiter is, like Mars, a much-observed planet. Amateur astronomers have especially observed it over many decades and, thereby, have provided science with much valuable data by taking photographs, which have become more and more important, and making drawings. The photos by G. Nemec on page 114 show what amateur astronomers can achieve. Besides determining the systematic positions of Jupiter, which is the main goal, it is also a fascinating experience to follow an entire rotation during the time of the opposition in which the planet stands in the sky the entire night. That does not take longer than from the evening until the morning, because Jupiter rotates in 9^h55^m. It is relatively easy to follow at the light and dark spots how far this giant planet has turned below the other planets from one hour to the next. On this astronomy night, there are many other interesting things to observe in the starry sky. You can experience the apparent rotation of the sky sphere, which is caused by the rotation of the Earth.

The Inside of Jupiter

Jupiter's top atmosphere layer that we see with a telescope is just 1000 km (620 miles) thick. All of its layers added together are estimated to be about 16,000 km (9920 miles). Scientists assume that below the cloud tops the atmosphere begins to turn to liquid hydrogen. Closer towards the inside of the planet, the pressure increases and metallic hydrogen develops. Metallic hydrogen is a very good electricity and

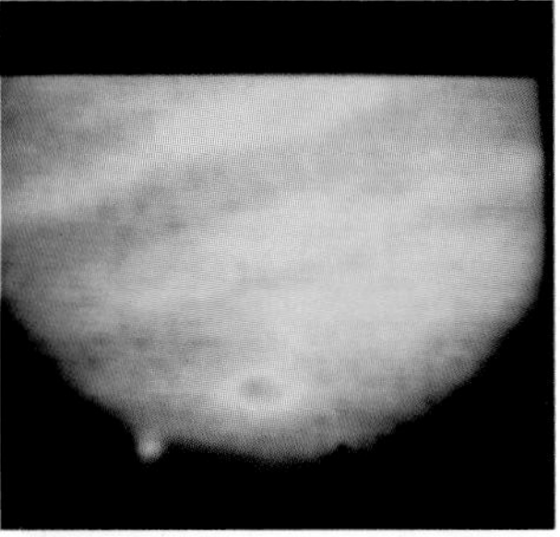
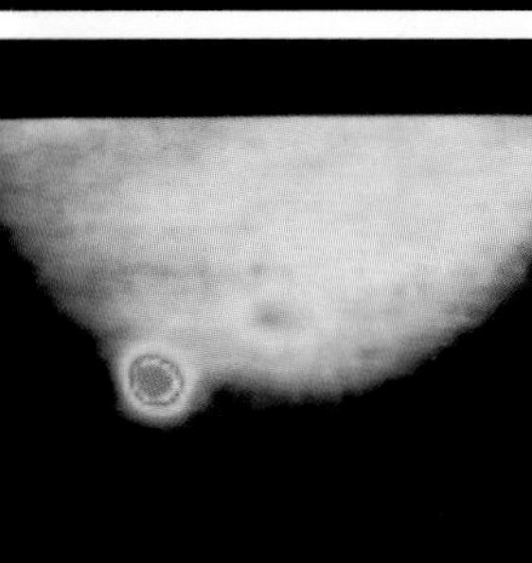
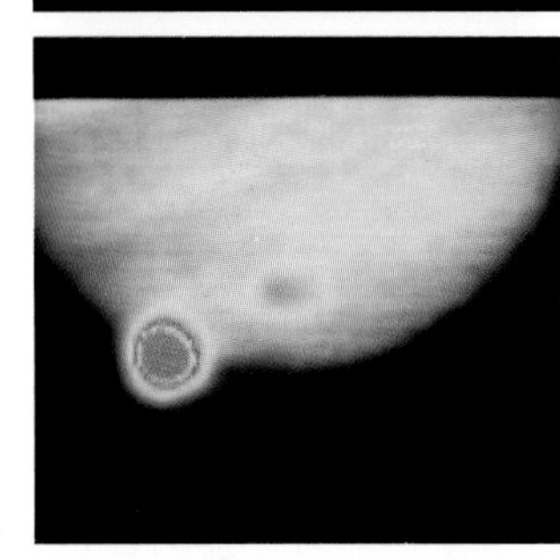
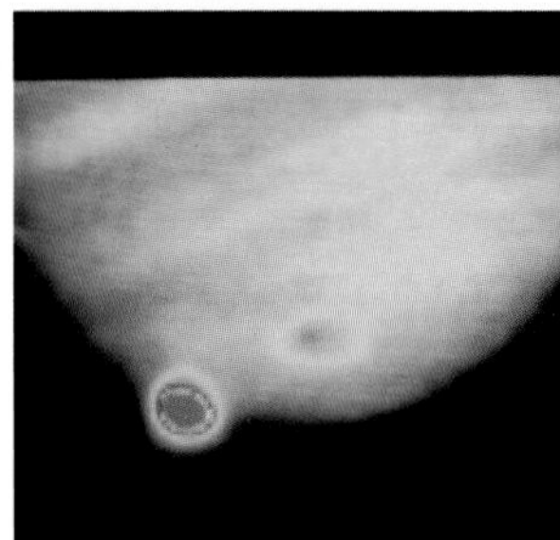
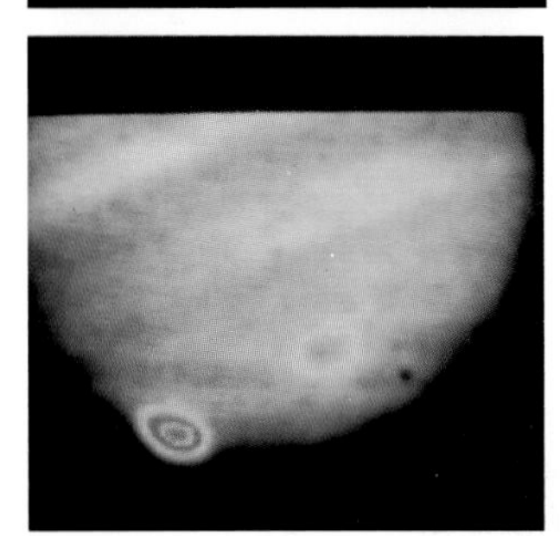

Impact of a part (fragment H) of the shattered comet Shoemaker-Levy 9 in Jupiter's atmosphere. The pictures were taken with the 3.6-meter telescope from the European Southern Observatory (ESO) on July 18, 1994. The development of the huge cloud explosion (from top to bottom): 19h36UT, 19h39UT, 19h42UT, 19h48UT, 20h15UT. The maximum brightness of this cloud explosion reached the 50-fold brightness of Jupiter's surface; temperatures reached 300 Kelvin. The other circular phenomenon on the photos comes from the impact of the fragment G from Shoemaker-Levy 9, which happened about 12 hours earlier.

heat conductor. The firm core has a diameter of 20,000 km (12,400 miles). It is composed of iron and silicon compounds.

Jupiter is a heat generator. Measurements of the infrared spectral field show that Jupiter gives off more heat than it receives from the sun. This means that the planet still holds inner heat from the time of its origin. This heat escapes more slowly from Jupiter than from the smaller planets. The metallic hydrogen is partly responsible for the strong radio radiation and the magnetic field. The necessary electric currents are produced by the inner heat and the high rotation-speed. Measurements, taken by the spacecraft Pioneer and Voyager, came up with a much stronger magnetic field than that on Earth: at Jupiter's equator it is twelve times stronger, and at the poles it is twenty-five times stronger than that of the Earth.

The strong magnetic field causes a magnetosphere that reaches far into outer space, whose outer parts are then exposed to the influences of the solar wind. Ions and electrons become trapped into the inner sphere of the magnetosphere. Jupiter's innermost moon, Io, is subject to the constant influence of the highly energetic particles of the magnetosphere. Thus, sodium atoms are loosened out of the surface of this satellite to produce a shining cloud, which can be seen with larger telescopes from Earth in the optical spectral-field (589 nanometers).

Jupiter's Moons and Ring

We spoke briefly before about Jupiter's four large moons. The planet has altogether sixteen moons. It is certainly possible that some of them are captured asteroids (see p. 109). Only the four bright moons can be seen with the telescopic instrument of an amateur astronomer. The discovery of these

Jupiter with two of its moons, I and II. Moon II throws its shadow onto the surface of the planet. Photos taken on October 26, 1965; top 0h35UT; below 1h45UT.

Jupiter's moon Io. You can see a volcano eruption in the photo. The blue-white spot on the bottom is the volcano. Io has approximately the same density and size as the Earth's moon. Instead of impact craters, there are active volcanoes.

four moons was made in 1610 simultaneously and independently by Galileo Galilei in Italy and Simon Marius in Germany. Even though the first telescopes were rather primitive constructions (hardly on a par with the optical force of today's telescope), they still found the four bright moons, which have the names Io, Europa, Ganymede and Callisto. They have diameters between 3000 km (1860 miles) and a little bit over 5000 km (3100 miles); in comparison, the Earth's moon is 3480 km (2157 miles).

The spacecraft Voyager 1 and 2 revealed the real look of these four moons, which are also known as the Galilean satellites. Sensational pictures gave us a volcano eruption on Io, an ice shell around Europa, crater impacts and mountain ranges on Ganymede, a crater landscape on a bed of ice on the fourth moon, Callisto. According to everything that we know today, Io is the most volcanically active celestial body in the entire solar system. With medium-sized amateur telescopes and good visibility conditions, you can recognize the four Galilean moons as little disks. Observers with instruments with a 30-cm aperture and larger can even see surface shades on Ganymede, the largest moon.

The other twelve moons are much smaller: Moon V and moon VI have a diameter of 160 km (99 miles); the measurements of all the other moons are between 30 km (18.6 miles) and 60 km (37.2 miles).

In March 1979, Voyager 1 confirmed an old assumption that Jupiter is surrounded by a thin ring of fine dust. The ring has a width of 600 km (372 miles), but it is only 30 km (18.6 miles) thick. A halo expands up to 5000 km (3100 miles) above and below the ring level. It is caused by the collision of tiny ring-particles (some thousandth millimeters) with micrometeorites.

Saturn: the Planet with Rings

Somewhat farther away from the sun is the second-largest planet in the solar system, Saturn. It is the farthest planet that can still be easily visible with the naked eye. It also completes the number of the planets that have been known since ancient times to the people of all cultures: Mercury, Venus, Mars, Jupiter, and Saturn. The outer ring (the A-ring) reaches an opposition magnitude between -0.3^m and $+0.9^m$, which makes it a striking object to see in the sky.

With the naked eye and/or through binoculars, nothing unusual stands out about Saturn. It was first recognized for its motion and unblinking, constant light, which made it an object distinct from the stars before the earliest planetary records. In 1610, Galileo Galilei watched Saturn for the first time through his telescope and found it unusual. He discovered that Saturn was not quite round and it had appendages on both sides of the planet, for which he had no explanation. Surprisingly, these "handles" disappeared again after a couple of years. Only decades later, Christian Huygens discovered, with a more strongly enlarging telescope, the rings around the planet. In 1675, Gian Domenico Cassini noticed a dark gap in the rings, which later was named after him: Cassini's division.

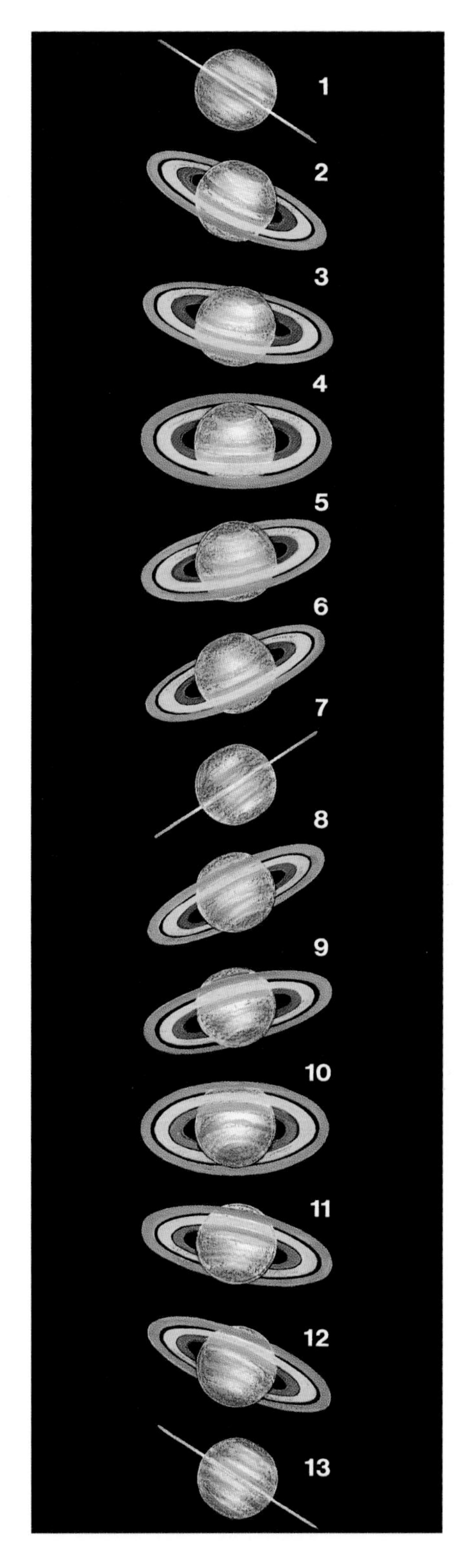

Alternating Sight of Saturn's Rings
From top: 1 = ring system from the edge (Earth goes from north to south through the orbital level of the rings, sight 1995); 2 and 3 = in-between positions, view to the south side; 4 = largest ring-opening, south side (sight 2002); 5 and 6 = in-between positions, south side; 7 = ring system from the edge (Earth goes from south to north through the orbital level of the rings); 8 and 9 = in-between positions, north side; 10 = largest ring opening, north side; 11 and 12 = in-between positions, north side; 13 = ring system from the edge (Earth goes from north to south through the orbital level of the rings). The northern and/or southern visibility of the rings lasts about 15 years each. Therefore, in about 30 years, all possible kinds of sightings of Saturn's rings occur once, according to the rotation time of the planet around the sun (in 29 years and 167 days).

Saturn. Photo taken on October 29, 1970 by an amateur photographer. Cassini's division of the ring can be seen.

Photo of Saturn taken from the space probe Voyager 1. The impressive rings, which surround the planet, are clearly visible. You can also see the cloud bands in Saturn's atmosphere.

Saturn's rings from a distance of 8.9 million km (5.5 million miles). Contrasting and enhancing colors were added to the photo to clearly show the large number of rings. This photo was taken from the space probe Voyager 2.

The rings become clear through a small astrotelescope at enlargements from 50-fold on. It is important to note when observing the rings that, because of the angle, their appearance changes. It is easier to observe the bottom or the top of the rings to get their entire view. It is more difficult when we view them from the edge, because then you only see at most thin, bright lines. The level of the rings lies exactly at the equatorial level of the planet. They are, therefore, tilted on the axis about 27° against the orbital level of Saturn. The illustrations of p. 115 show the changes, along with the possible appearances of the rings.

There are several rings around Saturn. You need a sufficiently large, 3-inch telescope (with 100-fold enlargement) to observe the dark gap, Cassini's division, which divides the outer ring (A-ring) and middle ring (B-ring). Through the A-ring is actually a narrower gap called Encke's division, which is extremely hard to detect. The three main rings of Saturn are the A-ring, B-ring, and C-ring (a semitransparent inner ring that is hard to see).

In November of 1980, Voyager 1's flight past Saturn brought the big sensation. The analysis of the pictures showed that the three main rings were actually made up of hundreds of ringlets and narrow divisions. Horst W. Köhler said about this: "The comparison of Saturn's system of rings with a cosmic record is by all means to the point. At first sight, the rings look like grooves of a record. But the structures are by no means uniform. Matter accumulations form spokes and other irregularities. Close examination shows that the firm particles of the rings range from the size of a dust particle to the size of rock pieces. Dynamic reciprocal actions between these particles and Saturn's moons still have to be explored."

That was, even more than the discovery of the many individual rings, the real sensation of the pictures taken by the Voyager probes. Even in those gaps, which, when observed from the Earth, had no details (e.g., in Cassini's division), ring-shaped arranged matter-accumulations were found. Observing the ring system today with available instruments, the astronomer will see only three rings and five divisions, even with a ring that is widely opened. The Voyager photos (see photos on p. 116) demonstrate the entire diversity of the expanded ring system

The thickness of the rings is probably less than 100 meters (330 feet). The total mass of the rings reaches only 1/25,000 of the entire mass of Saturn.

In the solar system, the rings around Saturn are not unique anymore. Jupiter and Uranus also have ring systems. There are reasons to assume that the ring systems around these planets were formed by the smashing of moons and asteroids. According to a law proposed by Edouard Roche, a satellite that orbits within a certain distance of a planet (2.45 the radius of the planet) would be torn apart by tidal forces. This "danger zone" is referred to as the Roche limit. This distance of Saturn's ring from the planet corresponds with the Roche limit. In the already quoted book *Planets: Siblings of the Earth*, Werner Sandner points out the research of astronomy professor Bucerius from the Munich observatory: "Bucerius discussed at length this question and came to the conclusion that Saturn's rings owe their existence to the tidal forces of the planet onto an original moon, which existed before this one. At a low temperature, which is assumed to be −180°C (−292°F), the material was brittle i.e., the molecular cohesion was so low that this original moon, through the tidal forces which came from Saturn, had to fall apart against its own gravity into a heap of disconnected particles."

Everything points to the fact that the particles of the rings consist of ice or that they are rock particles with an ice cover. The formation and long-term existence of rings probably can only be explained by the very complex gravitational forces of Saturn and its moons. Rings around planets can be remainders of moons, which came into a certain gravitational area (Roche limit) and which were ground down in collisions. There is also the theory that moons never formed the rings, but that original matter-particles formed to become rings after the planet's formation.

In connection with the description of the planet Jupiter, we pointed out the physical-chemical relationship of the planets Jupiter, Saturn, Uranus, and Neptune. Saturn is not much smaller than Jupiter: Its diameter is 120,800 km (74,896 miles) at the equator. The rotation time is somewhat longer: between 10 and 11 hours. The atmosphere is dense, and the oblateness is even greater. At 1/10, it has the greatest overall oblateness of all planets. Saturn's disk does not appear circular in the telescope, rather it is clearly elliptic; anyone can see that with a small telescope by looking at Saturn and Jupiter. The oblateness is the result of the quick rotation of both planets. As on Jupiter, light and dark stripes can be observed on Saturn. Only, it is more difficult in this case, because Saturn has a considerably smaller apparent diameter: maximally 21 arc-seconds at the time of the opposition. In order to see the light and dark spots within the stripes, you need at least medium optic instruments, 5–10-inch telescopes with enlargements between 200- and 300-fold.

Also as with Jupiter, the quick rotation stretches the clouds to bands and zones, which are parallel to the equator. Close to the equator, the wind speeds reach extreme values of more than 1800 km/h (1116 mph)—almost quadruple that of Jupiter!

The interior of Saturn is similar to that of Jupiter. Saturn also has an inner heat-source, which is the result of remaining heat from the time of its formation that is still actively producing energy. In the case of Jupiter, it gives off more heat, just through its internal heat source, than it receives from the sun. The smaller Saturn, which has less mass, needs additional energy in order to be able to give off double the amount it receives from the sun. This energy is obtained similarly when raindrops are heated as they fall onto Earth. Thereby, kinetic energy, which is obtained from the gravitation, is transformed into heat. Helium and hydrogen form the bulk of Saturn's upper atmosphere. Thereby, helium drops fall into the direction of the planet's center.

Despite the similarity in their rotation rates, Saturn's magnetic field is twenty times smaller than Jupiter's. Probably a smaller electric current is produced in the less massive layer of the metallic hydrogen. But as in the case of Jupiter, the coming together of the magnetosphere and the solar wind produces a magnetic front, which crosses the orbit of the largest of Saturn's moons, Titan. Thereby, hydrogen comes from the atmosphere of Titan into an orbit around Saturn. This results in the formation of an expanded hydrogen ring.

The bright Saturn moon can be seen with larger amateur telescopes, as can five of the overall twenty-one moons currently known: Titan, Rhea, Iapetus, Tethys, and Dione. Titan is, with a diameter of almost

Photo of Saturn taken by the Hubble Space Telescope on May 21, 1995. On that day, Earth traversed the level of Saturn's rings. You can look directly onto the rings' edge. The bright dots on the left in the picture are Saturn's moons Tethys and Dione.

5150 km (3193 miles), slightly smaller than Jupiter's main moon, Ganymede. But since Titan is so much farther away, even this largest of Saturn's moons reaches only the apparent magnitude of 8.3. Still, you can look for it with a 2–3-inch telescope. Refer to astronomical yearbooks for the precise date regarding visibility.

Saturn's moon Enceladus, photographed by Voyager 2. The icy surface reflects almost 100% of the sunlight that falls onto it.

Like Jupiter's moons, the larger moons of Saturn were photographed by the Voyager probes. The existence of volcanoes on some of Jupiter's moons was cause for surprise. Saturn's moon Titan was to be one of the only moons in the solar system with an atmosphere. Titan's atmosphere consists mostly of molecular nitrogen. The small moons are characterized by an irregular shape (e.g., Saturn's moon Hyperion is almost disk shaped) and has a surface that is covered by craters.

Uranus, Neptune, Pluto: Planets at the Edge of the Solar System

Saturn is the farthest planet that is still effortlessly visible to the naked eye. The assumption that there is a planet on an orbit between Mars and Jupiter encouraged the astronomers in the 18th century to look for a still undiscovered planet. Even though they did not find a planet between Mars and Jupiter, they found instead a new planet beyond Saturn's orbit. On March 13, 1781, Sir William Herschel discovered the planet Uranus, which at first he thought was a comet. It is actually surprising that this discovery was not made earlier. Uranus reaches in the opposition an apparent magnitude of about 5.5 and is, therefore, barely visible to the naked eye. Interestingly, Uranus had been observed twice, in the 17th and early 18th century, but it was thought to be a fixed star.

Uranus

Uranus is, with a 51,118 km (31,693 miles) diameter, a large planet, but the distance makes it difficult to observe. The apparent diameter of the planet's disk reaches, at best, barely 4 arc-seconds. Uranus' characteristics, such as its oblateness, density, and quick rotation, resemble those of Jupiter and Saturn. But it distinguishes itself essentially from these two and all other planets by the fact that its axis lies almost in its orbital plane. In other words, its axis lies nearly in the plane of its orbit, so it looks like it's on its side. The theory is that a collision with another celestial body caused this unusual position.

The position of the axis of Uranus is very strange. Its equator stands almost vertically on its plane of orbit (i.e., the axis lies almost on the plane of orbit). Therefore, its appearance, seen from the Earth, changes in the course of the years (bottom row of pictures).

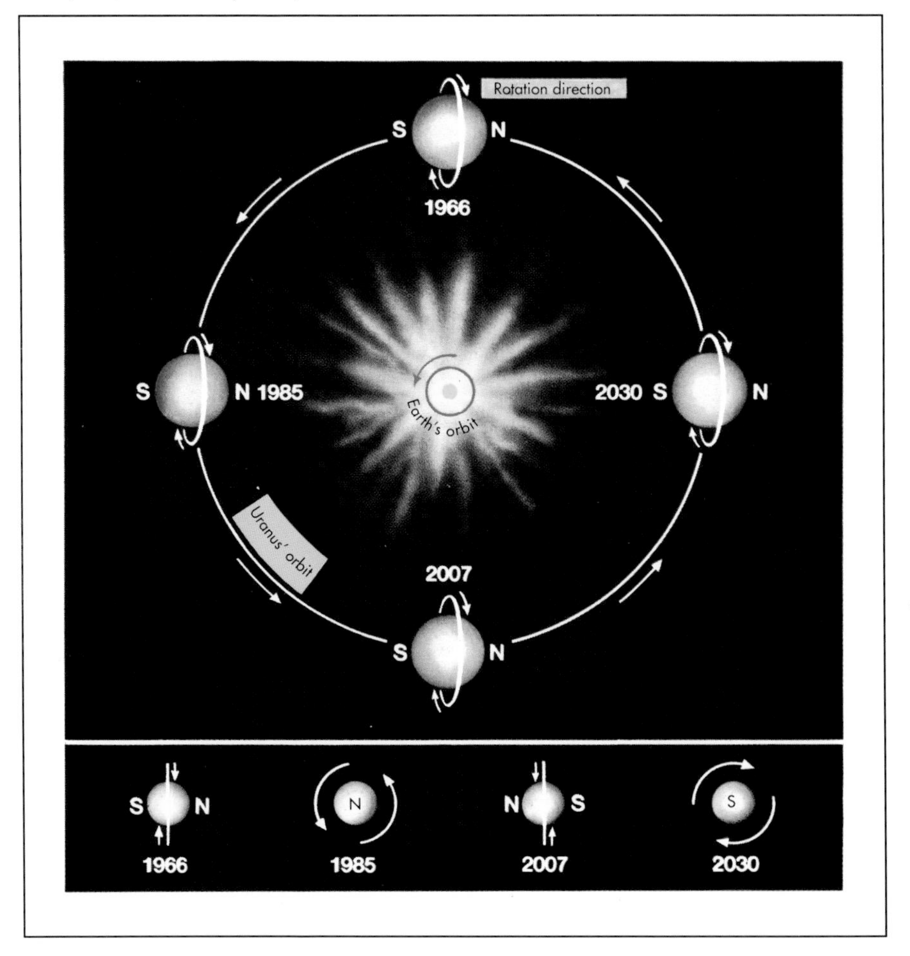

On January 24, 1986, Voyager 2 flew past Uranus at 93,000 km (57,660 miles) distance. The pictures, taken by the probe, show a structureless atmosphere that is caused by a strong haze surrounding the planet. Filtered photos show clouds in deeper layers of the atmosphere, which move with a speed of 100 m/s. Jet streams on Earth at a height of 9 km (5.6 miles) have the same speed. The west winds on Saturn reach, in latitudes near the equator, up to 500 m/s.

At the surface of the atmosphere is a cloud layer that consists of frozen methane. The methane gas absorbs the red part of the sunlight and reflects a bluish-green light, which the observer on Earth can see with a small telescope. The clouds that are deeper in the atmosphere consist of ammonia and water. Above the visible cloud layer, the planet is surrounded by a hydrogen corona, which reaches far into outer space.

As with the other planets, the apparent magnitude of Uranus is not constant, since the varying sun-planet and Earth-planet distances cause changes. But at Uranus, there are other circumstances as well. First, there is the changing position of its axis. Second, there are obviously periodical changes of the reflection capability of the atmosphere ("pulsation appearances"). Since the individual brightness changes superimpose on each other, an overall light-change in magnitude 1.3 can occur.

Uranus has very thin rings that are very difficult to observe from Earth. The rings (less than 10 km/6.2 miles wide) are separated by large interspaces. The dark or colorless ring particles reflect barely 2% of the sunlight. They were accidentally discovered from Earth in 1977, when Uranus occulted a star, causing the star to "wink" (or blink) repeatedly, indicating that rings were blocking the star.

Uranus has fifteen moons. The two brightest moons (about 14^m) Titania and Oberon were discovered by William Herschel in 1787. William Lassell, found the moons Ariel and Umbriel with a 61-cm mirror-telescope on Malta in 1851. They have a magnitude of 16. Gerard Peter Kuiper discovered the 17^m bright moon Miranda in 1948. Finally, ten new moons were discovered from pictures taken by Voyager 2 in January 1986.

The surfaces of Oberon and Umbriel show craters and different traces of ice. Photographs from Voyager 2 were able to capture mountainous structures with elevations and valleys. The well-equipped astronomer can photograph the two outer moons Titania and Oberon, but without the mentioned structures.

This photo, which was taken by Voyager 2, shows the large number of the rings that also surround the planet Uranus. The opposite light shot (in direction to the sun) shows the dust between the rings as being strikingly bright.

Neptune

Uranus owes its discovery to previous theoretical considerations, which is also the case for the planets Neptune and Pluto. Mathematicians helped in their discovery. Uranus' orbits were calculated in the 19th century more and more precisely, partially with newly developed calculation procedures. As a result, an unknown planet was predicted when there were irregularities (called disturbances) in Uranus' expected orbit. One calculation suggested that the planet was located beyond Uranus' orbit. English astronomer John Couch Adams and French astronomer Urbain Le Venir both independently predicted the location of the unknown planet. On September 23, 1846, the astronomers Johann Galle and Heinrich D'Arrest discovered the new planet, which was named Neptune.

This planet can become 7.7^m at the time of the opposition. Six of its eight moons were discovered only in August 1989 when the probe Voyager flew past it. Voyager also showed the existence of enormous storm

The Voyager 2 photos, taken in August 1989, found three objects on Neptune: "Great Dark Spot," surrounded by bright white clouds (center right of the picture); to the south of it is a light cloud formation ("Scooter"), which is adjacent to an object with a light core called "Dark Spot 2." North is on bottom, south is on top.

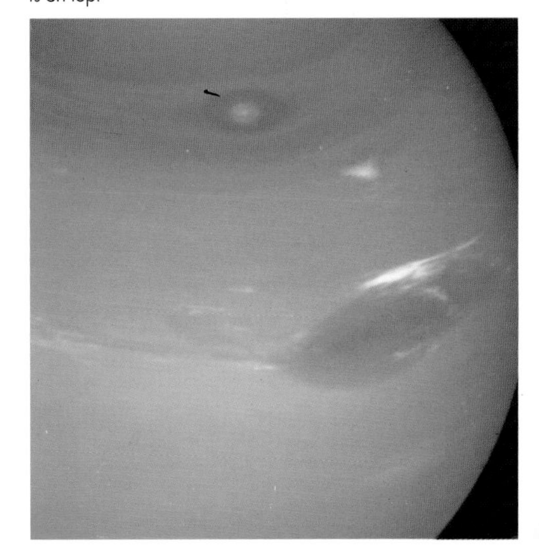

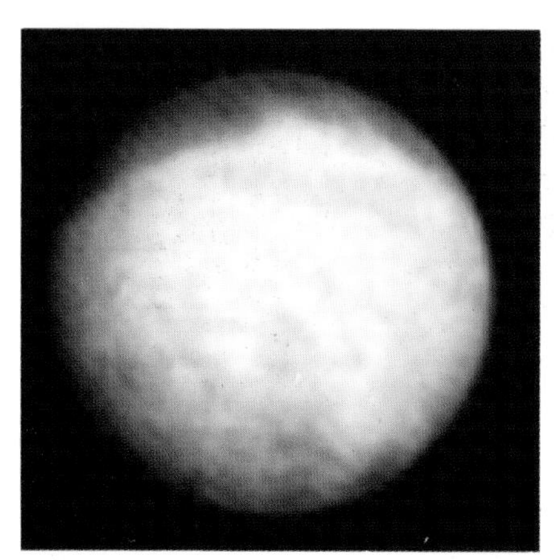

Neptune's moon Triton, photographed on August 20, 1989 by the space probe Voyager 2. Individual objects up to 100 km (62 miles) in diameter can be seen. The pale pink color is possibly due to a reddish matter, which consists of methane and ice. The dark areas on the top of the picture seem to belong to a belt of dark surface structures.

swirls in the atmosphere of the planet and of a Great Dark Spot, comparable to the GRS on Jupiter.

Neptune's moon Triton has a surface temperature of 37 Kelvin. It has, thus, the lowest measured temperature in the solar system. The surface shows lava, volcanoes, and craters of ice. Triton was probably at some point in time caught by Neptune's gravitational pull. Its retrograde orbit around Neptune indicates that. The six moons discovered by Voyager are dark, misshapen formations. They are so small that their force of gravity is not sufficient to form globes.

Neptune's rings were observed for the first time in 1984 by R. Häfner and J. Manfroid from the European South Observatory in Chile. Voyager 2 confirmed four rings. Two of these rings are very thin. The other two are much wider. One is within the two thin rings and it may even reach up into the upper layer of Neptune's atmosphere. Like the rings around Uranus, Neptune's rings are very dark. The rings contain concentrations of small particles called ring arcs. Analysis of Voyager data showed that a radial distortion with an amplitude of about 30 km (18.6 miles) travels through the ring arcs.

Pluto

The discovery of this planet is also due to orbital disturbances that led to a position calculated by American astronomer Percival Lowell, who believed that another planet, besides Neptune, had to have an influence on Uranus' orbit. In January 1930, Clyde Tombaugh discovered the searched-for planet on two photo-plates as an object with a brightness of 15^m. In fact, Pluto can never be brighter than 14.3^m, and most of the astronomers will have to be satisfied to know that this planet really exists. In 1978, a moon of Pluto was discovered, which was given the name Charon.

Whether another planet is beyond Pluto is something always preying on the minds of astronomers and mathematicians. They want to make the family of the planets even larger, and, thus, there are numerous calculations for a "trans-Pluto" for whose discovery the astronomical world is still waiting.

Many things point to the fact that objects beyond Pluto exist, which are similar to our planets and/or which existed in earlier epochs. The orbital positions of Uranus or of Neptune's moon Triton could have been caused by the collision with objects from the outer solar system. In the case of Triton, it could be the matter of such an object that Neptune forced into its system.

Comets: Wanderers in Space

Comets once caused fear and terror among people on Earth. Their striking and relatively rare appearances in the sky were often interpreted as signs of impending danger. The comet's often impressive tail also contributed its share of comet scare to

The first infrared photo of the comet Hale-Bopp taken on August 5, 1995 with a 2.2-meter telescope from the European Southern Observatory La Silla in Chile. Comet–Earth distance is 860 million km (533 million miles); comet–sun distance is 990 million km (614 million miles). For this large distance, the comet is astonishingly bright. The prediction that in spring 1997, this comet would be brighter than 0^m came true. It was a striking phenomenon in the sky.

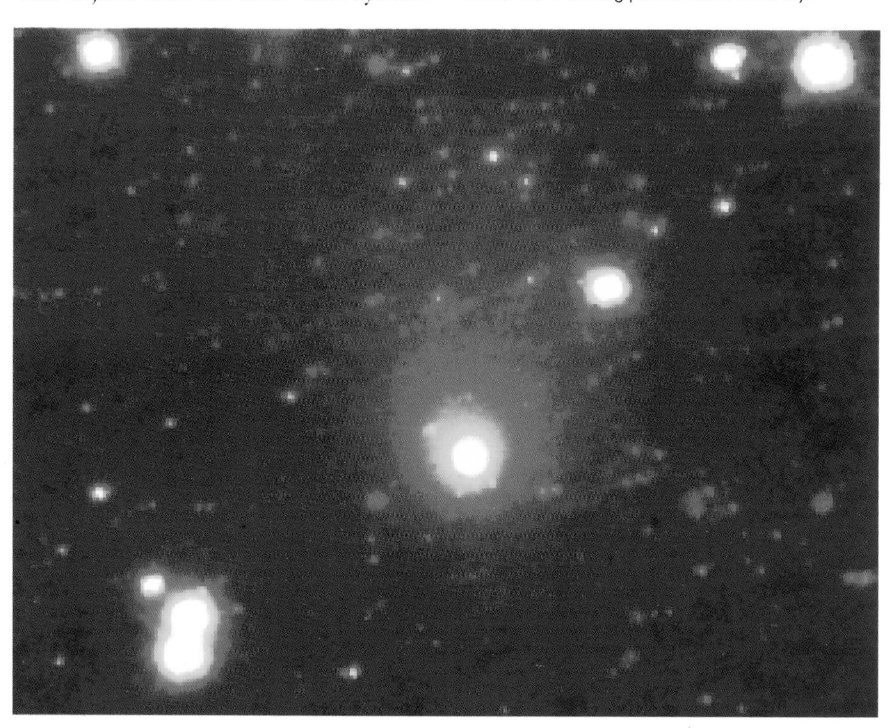

people. Striking comets, on the other hand, seem to have become rare in our century. Since the appearance of Halley's famous periodic comet in 1910, not much has happened in this respect. The recurrence of Halley's comet in 1986 was not the large spectacle that had been expected. Especially in the northern hemisphere of Earth, the comet has remained a little eye-catching object in the sky. Although it only made a number of medium-sized appearances, the comet West stirred the most comments when it appeared in 1976. But every year, six to nine comets are found, of which a third are rediscovered periodic comets. But mostly, they are objects of little brightness, which can be seen only in large telescopes and are also mostly discovered photographically. Nevertheless, the hunt for comets is a pursuit that amateur astronomers have tried with success, and there are many amateur astronomers all over the world that systematically comb the sky for comets. Binoculars with high intensity of light and a large visual field are an ideally suited instrument for that. The common enlargements between 10- and 20-fold are sufficient. Rudolf Brandt recommends to the observer: "The western horizon after sunset and the eastern horizon before sunrise offer the best chances to find a comet. But one thing is important: One must know the small nebulae and clusters in order to be able to distinguish them from a comet, because in the beginning comets often resemble certain spots of fog a lot. If you believe, as you are doing such close inspections, that you have found an object that could be a comet, then there is an infallible sign to see whether it is a comet: the movement of the object among the stars."

That means, you have to be familiar with the sky as well as with your star map. Otherwise, embarrassing surprises can occur, as when (according to D.W. Heintz) "the assumed comet is actually a hazy-stripe in the floodlight of the airport!"

The new, light-intensive refractors ("comet searchers") with a focal ratio of 1:8 to 1:5 (e.g., 5-inch Starfire and 4-inch Genesis) and wide-angle binoculars are suitable for the search.

Comets are named after their discoverers. Also, the comets of one year are at first organized with the date of the year

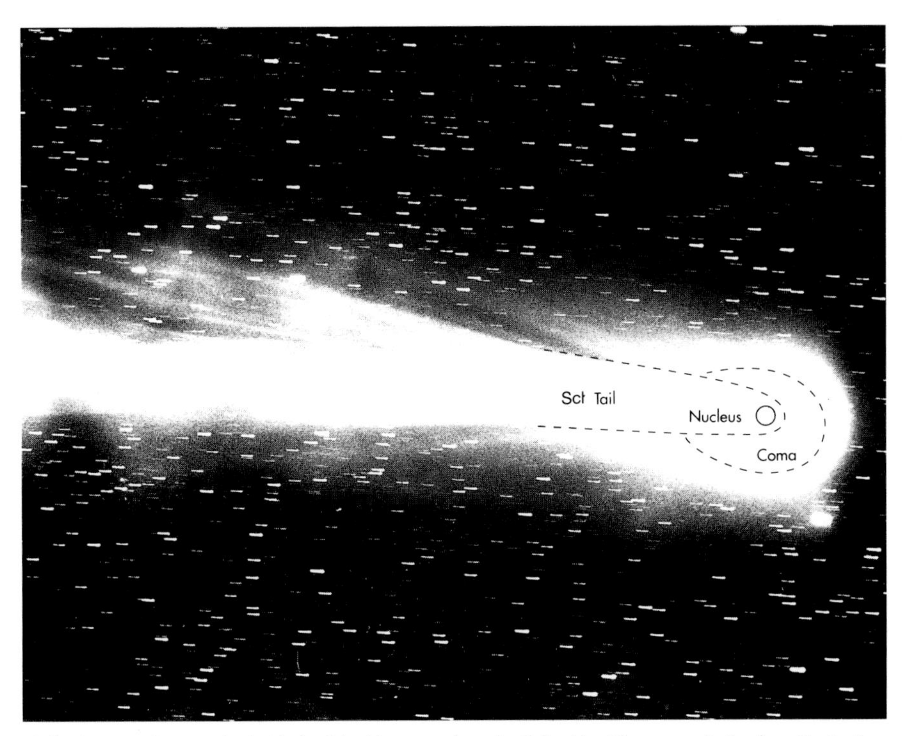

Halley's comet photographed with the Schmidt camera from the Calar Alto-Observatory in Southern Spain. Top picture: January 9, 1986; Bottom picture: January 10, 1986. The fast movement (compare with the bright star at the comet head below) and the change of the comet's tail within one day are striking. In the picture above, the most important components of a comet are marked (see also p. 122).

and a letter (comet 1972a for the comet that was the first one to be discovered that year). When the orbit is definitely known, the comets of that year are marked in the sequence of their Perihel transits (orbital point closest to the sun) with Roman letters (comet 1972 I).

If an observer has discovered several comets, then there is another number after his or her name, which indicates which of the discovered comets of this observer we are dealing with.

The resemblance of comets to spots of fog at a certain point in time can be confusing, when the comet actually makes itself known through its tail. To alleviate the confusion, here are some remarks about the orbit of comets. Like the planets, comets move in the gravitation field of the sun. They are small, compact bodies of the solar system with diameters between 1 km (.62 miles) and 100 km (62 miles). The nucleus of the comet can only be observed from Earth when it is close enough to reflect sufficient sunlight. Short-period comets move in elliptical orbits; long-period comets move in more parabolic orbits. Comets come out of the space outside of the solar system and move back there. That means, the comet comes from the most outer distance of the planetary system close to the sun. Only from about Saturn's distance on, the nucleus becomes bright enough to be registered from Earth. The comet's tail development has not even begun at this point in time. But with its increasing approach to the sun, the nucleus warms up more and more, through which changes in the structure are triggered. Matter at the surface of the nucleus begins to evaporate and produce a gas shell around the nucleus (called the coma). Interestingly, the coma shines in its own light. F. Gondolatsch from the Astronomical Mathematical Institute in Heidelberg, Germany, wrote about this in the monthly magazine *Stars and Outer Space*: "The spectroscopic examination of this self-shine is the main source of our knowledge of the construction of the comets and of the physical processes in them."

As the comet comes closer and closer to the sun, gases in the nucleus begin to thaw and evaporate. The tail develops. It always points away from the sun. Before the Perihel transit of the comets, it stays back behind the nucleus; after that, it precedes the nucleus. The tail's length differs. There have been comets whose tails stretched over the entire visible section of the sky. Tails are usually in straight lines, but they can also be crooked. Sometimes, there are deviations in length as well as in visibility, probably because of occurrences in the nucleus and in the coma under the influence of the sun's radiation. The matter in the tail is very thin, as the mass of the nucleus is extremely small—one millionth of the mass of our Earth.

The nucleus is a mixture of ice (frozen water, hardened gases, ammonia ice, methane ice, and others) and dust made of elements like iron, calcium, magnesium, manganese, silicon, nickel, aluminum, and sulphur. During evaporation, some of the dust from the dust tail, which follows the same orbit of the comet, remains. When a planet gets into such a dust tail, a "collision" happens, and we experience it on Earth in the form of a swarm of shooting stars (see p. 124). Whether organic molecules were transported through outer space and came to Earth that way is a scientific speculation. For the entire nucleus to be used up through evaporation takes quite some time. Thus, the periodic comets execute dozens to hundreds of rotations around the sun.

There are about one thousand known comets that have been discovered from Earth and have been observed in the course of the centuries. Only barely one hundred of them were repeatedly observed. A comet discovered by the infrared satellite IRAS in 1983 came closest to Earth so far. Its orbit led it past the Earth at a distance of 5 million km (3 million miles).

As stated before, since the most frequent orbital shape of comets is the elliptic, the path forces the comets to stay in the solar system. This causes their reappearance near the sun, which provides us the possibility of repeatedly observing them. The reappearance of Halley's comet is the best known example for a periodic comet.

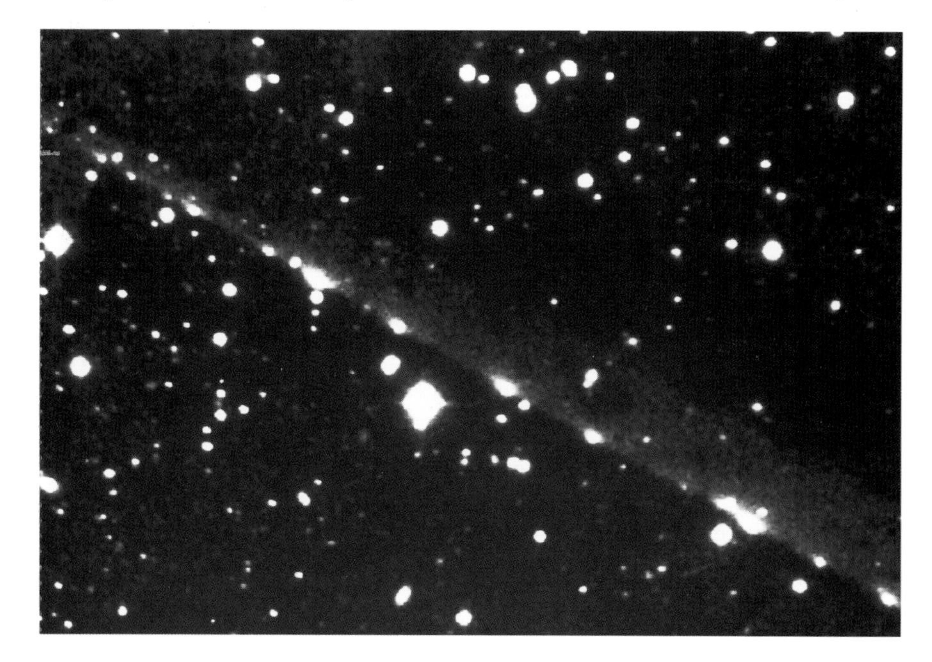

Comet Shoemaker-Levy 9 on July 1, 1994, about 2 weeks before its fall into the atmosphere of the planet Jupiter. Under the influence of the gravitational forces of the planet, the comet has already been torn up into several pieces (see also photos on p. 114).

The dark core of Halley's comet, which is rich in carbon. The bright gas and dust eruptions are also recognizable. The comet is 16 km (9.9 miles) long and 8 km (5 miles) wide. This photo was taken by the space probe Giotto in March 1986 from 600 km (372 miles) away.

Its fame is based on the fact that it was the first comet to reappear at its predicted point of time. At the end of the 17th century, Edmond Halley calculated the orbit of this comet for 1682 and predicted its reappearance in 1758. Back calculations showed that Halley's comet had already been repeatedly observed in the Middle Ages and ancient times. During its last appearance in 1986, Halley's comet was observed very intensively. On March 14, 1986, the European space probe Giotto flew past the comet from a distance of 600 km (372 miles) to take photos and measurements. American, Japanese, and Russian probes were also sent into the vicinity of the comet.

The force of gravity of a planet, especially Jupiter's, which is massive, repeatedly forces comets from their long-stretched orbit onto a shorter orbit. Thus, the orbits of the short-period comets come about with rotation times of less than 20 years.

The end of the comet 1993e Shoemaker-Levy 9 was sensational. It was catapulted by Jupiter onto a short rotation-orbit, which brought it directly into the gravitational field of Jupiter. There, it was torn into numerous pieces, which fell onto Jupiter at the end of July 1994.

The overall orbits of the comets hint at a common comet birthplace, which reaches far into the interstellar space at the edge of the solar system. Even though far away from the sun, its force of gravitation is still sufficient to keep all comets on elliptic orbits, which will bring them at some point in time into the inside of the solar system. A comet needs several million years from this distance for a rotation. Its travel speed averages 144 km/h. The border of our solar system is marked by the comet nursery called the Oort Cloud.

Comets are interesting for research. Since they have barely changed during their existence, they are documents from the time the solar system originated. Because of their coldness, some wonder about whether they could transport deep-frozen viruses from outer space.

Did comets bring life to Earth?

Shooting Stars: Matter from Outer Space

On November 13, 1833, a celestial fireworks erupted. In surprising frequency, shooting stars sparkled from the constellation Leo—up to twenty stars a second! People on Earth took pictures of this natural event to pass on to future generations.

Shooting stars do not always fall in such large numbers, but there seem to be certain seasons in which one can count on a larger frequency. Warm summer nights are good for staying up and looking at the starry sky. Shooting stars can be seen frequently, especially in the first half of August, which is the time of the Perseids shower. The name comes from the constellation Perseus, from which the flashing shooting stars seem to come. Of course, the shooting stars have nothing at all to do with the constellation itself and/or with its stars.

Two shooting stars during the fall of the Perseids on August 12, 1993. The Perseids belong to the regularly reappearing meteor showers.

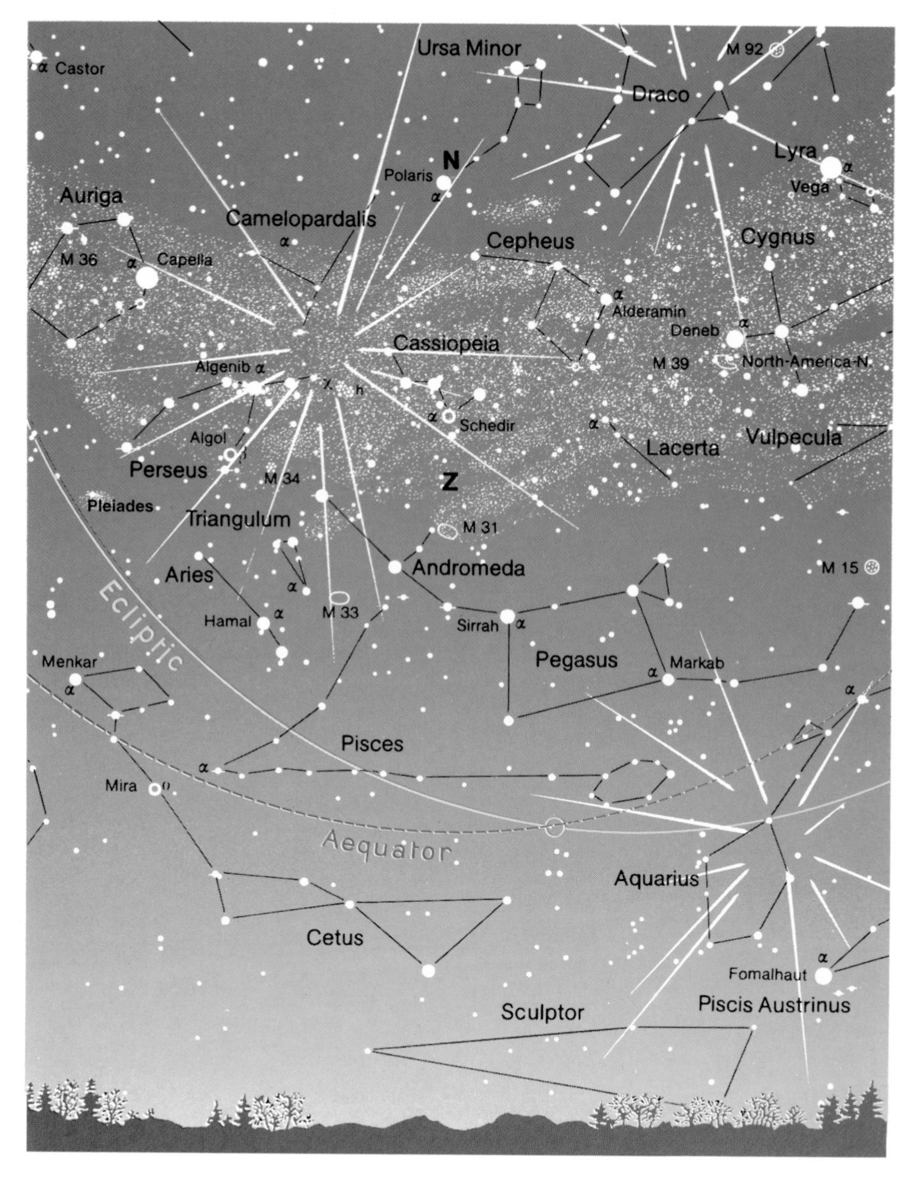

Map with radiants (the seeming point of radiation of shooting stars in the sky) and three meteor showers: Delta Aquarids, Perseids and Glacobinids (see also table below).

But the fact is that this point of radiation is the easiest one to classify in the sky when bringing it in connection with a constellation. This point is called radiant.

Besides the Perseids, the Quadrantids (which appears at the beginning of January in the northern part of the constellation Boötes), the Giacobinids (appears near the dragon's jaws in the constellation Draco on October 10), and the Geminids (appears near Castor and Pollux in the constellation Gemini at the beginning of until mid-December) are showers which are especially rich in shooting stars.

Generally, the orbit that a meteor describes in the Earth's atmosphere is almost a straight line. When they light up, shooting stars are still over 100 km (62 miles) high; most of them are extinguished when they are 70–80 km (43–50 miles) high. A tip for observers: "During the movement of our planet around the sun, a front side and a back side can be distinguished on the Earth. The first one is the hemisphere for which the aiming point of the movement lies above the horizon. The aiming point lies on the ecliptic 90° to the west of the sun—i.e., it will rise above the horizon on average about 6 hours before the sun so that each location in the first half of the night belongs to the back side, and in the second half to the front side of

Meteor Showers

Name	Time	Radiant (α, δ)		Productiveness	Comet
Quadrantids	1/1 to 1/6	232	+50	large	no comet known
Lyrids	4/19 to 4/24	272	+32	little	comet 1861 I
η-Aquarids	5/1 to 5/8	340	0	medium	Halley's comet?
δ-Aquarids	7/15 to 8/15	344	−15	medium	no comet known
Perseids	7/25 to 8/18	40	+55	large	comet 1862 III
Giacobinids	maximum 10/8	262	+54	large	comet 1900 III
Orionids	10/16 to 10/26	97	+15	little	Halley's comet?
Taurids	10/10 to 11/30	57	+14	little	Enckeschian comet?
Leonids	11/15 to 11/19	152	+22	little	comet Temple 1866 I
Geminids	12/7 to 12/15	113	+32	large	no comet known
Ursids	12/17 to 12/24	217	+78	medium	comet Tuttle 1939 k?

Meteor Showers

In the course of the year, different meteor showers can be observed whose orbits, extended backwards, seem to run together into one point in the sky. This point is called radiant. These meteors describe in space parallel orbits, and they are members of a meteor shower. Their relatively regular appearance points to the fact that swarms of meteoroids traverse the space between the planets (interplanetary matter in interplanetary space). Here, a connection exists between comets and meteor showers. The orbit of a certain comet coincides with the orbit of the meteor shower. In the first half of August, the best-known meteor shower, Perseids, occurs. The most important meteor showers are summarized in the table on the left. Some seeming locations of their occurrence in the sky (radiant) is indicated on the map on top left.

The almost-circular Barringer Meteor Crater near Flagstaff, Arizona. It has a diameter of 1200 m (3960 ft.) and a depth of 175 m (577.5 ft.). A mound, 37 m (122 ft.) high, rises above the surrounding desert.

the Earth. Now it is easy to understand that during the movement of the Earth through shooting stars, which are distributed in space and which in general do not have a directional preference of movement, the front side has to catch more meteoroids than the back side, from which results the well-know daily deviation of the frequency of the shooting stars" (C. Hoffmeister).

Shooting stars are actually meteoroids (tiny grains of dust), which enter the Earth's upper atmosphere and burn up because of friction with the air. They are heated up to a white heat and then momentarily light up. The mass of a shooting star has burnt out before it reaches the Earth's surface. Of course, there are sometimes huge lumps, which impact onto Earth with often impressive brightness and even thunderous noises. Once a meteoroid enters Earth's atmosphere it is called a meteor. If it survives to land on Earth, it is called a meteorite. For these meteorites, we have the name fireballs (bolides). These remainders are highly demanded discovery pieces for science, because until they had the moon-mineral tests, these fragments of meteorites were the only cosmic building blocks that researchers could test in laboratories. Beautiful meteor pieces are exhibited in natural science museums all over the world. The Hayden Planetarium in New York has the largest meteorite that has been found up to now in the U.S.: It weighs more than 30 tons.

Meteorites can consist of metal (nickel, iron) as well as stone. The masses of extremely large meteorites can produce real craters in the Earth's surface. Near Flagstaff, Arizona, is the Barringer Crater, which has a diameter of 1200 m (3960 ft.) and a depth of 175 m (577.5 ft.). Of course, not all craters stand out as much as this one. Often, nature covers up some of them so that scientists can only assume what caused the strange forms of the landscape. This is the case, for example, for the Nordlinger Ries, a flat, fertile basin between the Swabian and Franconian Alps (in Germany).

Shooting stars are small bodies in the solar system that have a tight connection to comets. Dust from comets is very roughly comparable to the dust along a road. Through these loose accumulations of matter, the Earth describes its orbit around the sun, so a "collision" with this matter takes place from time to time, and we experience a shower, like the one of the Perseids in August. But that does not mean that every shower and every shooting star necessarily have to have a connection to a comet's orbit. There are good reasons to assume that meteors have also come into existence out of fragments of small planets (asteroids). A shooting star can make an appearance anywhere in the sky, especially as a lone star, because tiny as well as larger particles are constantly on the "road" in the solar system. There are "micrometeorites," which are so tiny that they land undamaged on the surface without burning out or floating in the Earth's atmosphere. It is estimated that 2 million tons of nickel-iron dust and rock dust land on Earth each year. It is a fluid transition for interplanetary matter.

Small piece of a meteorite, which is assumed to have come from the asteroid Vesta.

The meteorites can be divided into three groups by their chemical property:

- Stony meteorites: about 95% of all meteorites,
- Iron meteorites: about 4% of all meteorites,
- Stony-iron meteorites: about 1% of all meteorites.

The iron meteorites have the largest density ($7.8 g/cm^3$); the stony meteorites have the lowest density ($3.4 g/cm^3$). In comparison with Earth rocks, meteorites are denser. The most frequent rock on Earth, silicate, has a density of about 3.2 g/m^3. But there are exceptions, such as the coal-like chondrites, which are one of the original substances in the solar system. They have many things in common with the chemical composition of the sun. They have a density of only 2.2–2.9 g/cm^3. In contrast to the normal chondrites, a type of rock meteorite, the coal-like chondrites are rare and very fragile.

Exotics among the meteorites come from the moon and Mars. During meteor impacts onto these celestial bodies, pieces are knocked off of the surface and hurled onto Earth. The age of these oddities is lower than that of most of the asteroids. Chemical analysis show them to be moon rocks and volcanic Mars rocks. These splinters that were catapulted to Earth are less than 1 billion years old. Meteorites that stem from the Asteroid Belt are 4.5 billions years old—old enough to witness the coming into existence of the solar system.

The zodiacal light above the western horizon on May 13, 1983, taken from the tip of Mauna Kea in Hawaii. At the tip of the zodiacal light cone is the planet Venus, about 43° away from the sun. Directly above the horizon is the moon, 19° above the already set sun.

Zodiacal Light

This section could have also been called "Particles of the Solar System." We have already talked about small and tiny particle matter in the solar system in another connection (see p. 123). The outer space that is between the large celestial bodies is by no means so "empty," as some have in the past assumed. Astronomers today speak of interstellar and interplanetary matter. The zodiacal light is considered a part of interplanetary matter.

A cloud of dust particles rotates around the sun, which scatters the sunlight so that a visible light cone forms in the evening or morning sky. Because this dust cloud is rather close to the ecliptic level, the zodiacal light appears as a band alongside the ecliptic. As a result, alternating visibilities occur with the seasons. In *Handbook for Star Friends* (3rd edition, 1981), Werner Sandner describes as follows: "As it can be easily shown with the help of a celestial globe or a rotating star map, the ecliptic rises for intermediate northern latitudes steeply above the horizon in spring after sunset in the western sky and in the fall before sunrise. Therefore, the phenomenon in the fall morning-sky is analogous to the evening sky in the spring, but the phases take place in reverse succession. In the other seasons, the ecliptic runs more or less flat along the horizon, and the pale shimmer of the zodiacal light disappears in the dust. In intermediate southern latitudes, the observation conditions are in general the same, but then the best visibility is, in February/March, in the morning sky in the east and, in October, in the evening sky in the west."

For a long time, there was a bitter dispute of theories as to whether the zodiacal light could be attributed to a particle belt around Earth or to interplanetary matter. Very recently, scientific research points to a transition of the sun's corona to the zodiacal light. In total solar eclipses, observations showed that the zodiacal light comes from sunlight reflecting off interplanetary dust that lies in the plane of the ecliptic.

Between the planets, around the sun, and between the fixed stars, is matter in dust and gas form. Also in the solar system are numerous small bodies that form the interplanetary matter. The comets are important sources, which gradually fall apart and form meteor showers. Meteoroids are ground down from constant collisions into interplanetary dust. Some meteoroids stem from asteroids that were smashed to pieces. The phenomenon of zodiacal light is, thus, a part of the interplanetary matter. Without needing big instruments, anyone can experience the zodiacal light under the indicated visibility conditions. While the zodiacal light shines in the tropics bright like a Milky Way cloud, it is difficult to observe in intermediate latitudes and especially near a big city. The shine from a full moon as well as a bright planet and grouping of stars can disrupt the observer's viewing. Hence, this observation can be a treat for anyone who is planning a trip to the equator.

Space probes have also helped in the exploration of zodiacal light. They have brought particles that have a fantastic diameter of 0.001–0.1 mm (0.00004–0.004 inch) and strange forms back to Earth.

The Northern Lights

There are, of course, also southern lights, when the phenomenon takes place on the southern hemisphere of the Earth. The northern and southern lights are also known by their Latin names *aurora borealis* (northern lights) and *aurora australis* (southern lights). They are geophysical phenomena that have close connections to solar flares and occur, as a general rule, in the Arctic and/or Antarctic regions. One never observes auroras near Earth's equator. These lights are triggered by high-energy atomic particles from the Sun that hit the upper atmosphere of the Earth and make the atoms there glow. The solar particles cover the distance from the Sun in about one day with a speed of 2000 km (1240 miles) a second. Because the solar particles are charged, they are deflected by Earth's magnetic field to the magnetic poles.

This kind of solar radiation causes so-called magnetic storms. In high regions of Earth's atmosphere, electric currents are produced, which come into reciprocal action with the magnetic field of the Earth. Through this, currents are produced in this field, the magnetic force-lines change and they induce electric currents in telephone and telegraph lines that interrupt communications.

At the end of the last century, science pointed out the relations between the magnetic field of the Earth and the auroras. The traditional country of aurora research is Norway. The auroras stood especially in the centerpoint of research during the "International Geophysical Year 1957–1958" and the "Year of the Calm Sun 1964–1965."

Auroras can take on different colors and forms: green-yellow, but also reddish and silver rays and stripes. In intermediate latitudes, where the frequency of their visibility is very low, the light appears as an accumulation of vertical bands above the northern and/or southern horizon. Also flaming waves in different possible colors can come about. The closer the observer stands to the northern or southern pole area of the Earth, the clearer and richer in forms does this phenomenon appear. The brightness can also increase considerably. There are lights that hardly stand out close to the horizon. Other lights reach the brightness of bright parts of the Milky Way. Werner Sandner points out in the *Handbook for Star Friends*: "Yes, in high latitudes [that are] close to the brightness of the full moon, one is able to read large printed letters in the shine of the northern light." He also writes about the frequency: "While one can count on more than one hundred northern light-nights per year on average, in the zone of the largest northern light-frequency, their number is about thirty in Scotland, three in Northern Germany, one in Southern Germany per year, while but one northern light in Southern Italy in 10 years."

It is safest to observe the lights through an oval, almost circular ring at the 200-km (124-mile) distance around the geomagnetic pole. The first photos of this light oval were made in 1982 from the satellite Dynamics Explorer 1.

The light is, when it is accordingly pronounced, a fascinating spectacle of nature. The triggering force comes from the sun. That means we deal with a solar-terrestrial relationship, which can make its presence known to us in our everyday lives via disturbance of broadcast reception, burning out of fuses, disruption of news transmissions by teleprinter, telephone, or telegraph.

Artificial Satellites

In the space around the Earth, there is more traffic than many among us assume. For a long time, the spectacular events of space travels, such as the Apollo spaceship on the way to the moon, have not been the only topic of discussion. Research continues—through sending small weather satellites or the Venus probe—so that all possible kinds of device carriers can orbit around the Earth or to an adjacent celestial

The photo shows the typical phenotype of the northern light. On the bottom in the picture are the "Bright Night Clouds" (very high clouds that are still lit up by the sun after sunset).

In the center of the picture is the spiral galaxy M51. Below, the light trails of two artificial Earth-satellites cross.. Satellite traces are now found more often on sky photos—not always to the delight of astronomers. In the picture, we are dealing with two Soviet rocket carriers of the Cosmos type. The interrupted trails hint at a light change due to the rotation. On the left top of the picture is the pale light trail of a meteor.

body in order to gather information. In this, the artificial satellites of Earth play by far the largest role. Hundreds of them are on the "road," mostly with very special tasks. There are, for instance, the news satellites, which make it possible to air soccer world championships from Mexico and the Olympic Games from Greece onto the television screen in the U.S. There are the satellites "on secret command missions" with strategic tasks—space spies, you may say. There are other satellites that photograph the Earth, in search of still unused economic areas. And let's not forget the weather satellites, whose photos are shown daily on TV. Weather satellites are described in detail in the book *Collins Guide to the Weather*.

October 4, 1957 is an historical date: the first "sputnik" started its flight around the Earth. Many readers may remember "Echo 1" and "Echo 2," which after all, were at times very striking objects in the sky, because one was able to observe them with the naked eye. The slow travel of the mini Earth-satellites among the stars is indeed a special kind of experience. Anyone today who views an area of the sky with binoculars will come unexpectedly again and again across artificial satellites, which have become so numerous by now. But only the celestial photographer, who registers on his or her film and plates the light trails of the artificial satellites more frequently, will really become aware of the space traffic that takes place all around the Earth. *Atlas of Deep-Sky Splendors* by Hans Vehrenberg, a photographic presentation and description of all Messier objects as well as of two hundred more celestial objects, in which a whole chapter discusses satellite trails in sky photos, proves how far that is already going: "When taking photos of objects with large surfaces (nebulae), it can happen, because of the present spheric density of the brighter satellites,

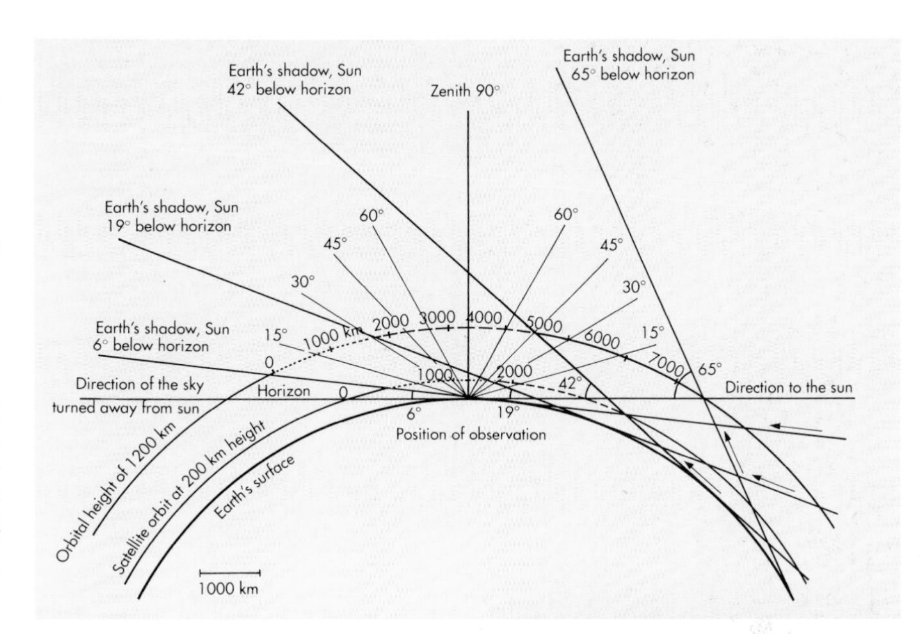

Sun's position below the horizon and Earth's shadow border: ——— visible part of the orbit when the sun stands 42° below the horizon; - - - - - additional visible part of the orbit when the sun stands 19° below the horizon; · · · · further visible parts of the orbit when the sun stands only 6° below the horizon. When the sun stands 65° deep, satellites, which move in a height below 1200 km (744 miles), are not visible in any of the parts of the orbit because their orbit runs exclusively in the Earth's shadow (from: Roland Primas, Satellite Observation in: *Stars and Outer Space 5*, 1966, p. 142).

that the trail crosses the object and, thus, thwarts a photographic analysis of the photo. Also, anyone who today wants to venture upon putting together a photographic atlas of the stars with more recent photos should be prepared for considerable difficulties. A trail-free making of *Mount Palomar Sky Atlas*, for instance, would seem to me today almost impossible, unless one could always then interrupt the exposure according to a "satellite travel-schedule," when a satellite of larger brightness came, at any given time, dangerously close to the targeted field of exposure. Luckily, all satellites, including high-flying stationary news relays, represent themselves as traces and are as such immediately recognizable."

There is growing concern about having a "clean" sky. And it is not quite unjustified, insofar as there are various plans to even increase the traffic frequency in the direct cosmos surrounding the Earth. If we then add the disturbance factors within the Earth's atmosphere—the haze formations above the cities, cloud increases as a result of fuel usage of the planes—"the question remains open, in view of this little encouraging development, how long there will be still places on our planet that will remain untouched by the pollution of civilization and on which ideal astronomical conditions are still met " (G. Klare in the magazine *Stars and Outer Space*).

Binoculars are the ideal observation tool for artificial satellites. Two procedures are especially interesting for the observer. First is the gradual decrease of the brightness of the satellite when entering Earth's shadow (evening observations) and/or the gradual increase of the brightness when coming out of Earth's shadow (morning observations). Second is the short-term brightness changes that are caused by the rotation of the satellites, which frequently deviate from the ball shape into cylindrical, elliptical, or windmill-shaped complicated shapes.

Satellites circle on very different orbits. Those that always go via a large circle, more or less, form ellipses in which the focal point coincides with the center point of the Earth. The orbits can run via the equator as well as via the Earth's poles. As a general rule, satellites are only visible because they are lit up by the sun and reflect the sun's light, just like the planets and their moons. Thus, the course of the border of the Earth's shadow is decisive for their visibility. The illustration on the top of this page should make it clearer.

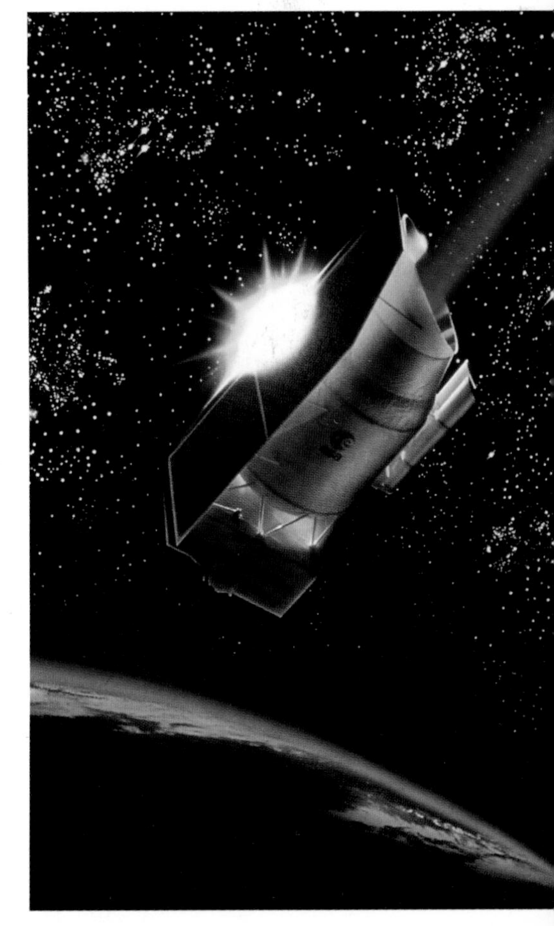

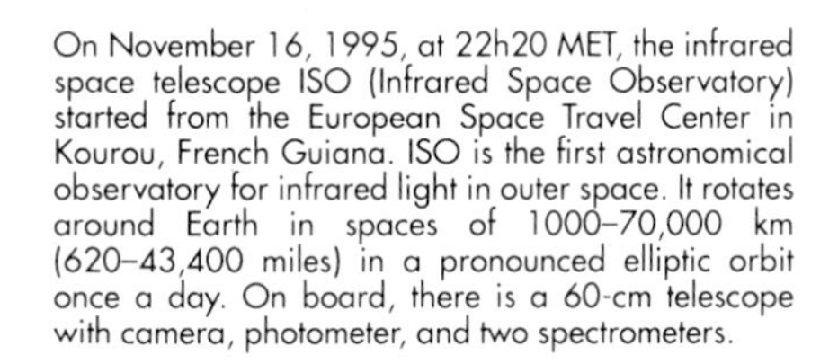

On November 16, 1995, at 22h20 MET, the infrared space telescope ISO (Infrared Space Observatory) started from the European Space Travel Center in Kourou, French Guiana. ISO is the first astronomical observatory for infrared light in outer space. It rotates around Earth in spaces of 1000–70,000 km (620–43,400 miles) in a pronounced elliptic orbit once a day. On board, there is a 60-cm telescope with camera, photometer, and two spectrometers.

The Exploration of the Star Systems

We'll now leave the solar system. For the astronomer, the fixed stars' world is characterized by large distances and objects with little apparent magnitude. But the occupation with stellar astronomy, as this special field is called, supplies basic knowledge about the construction of the universe. How did the stars come about? What is the age of our Milky Way system? What kind of radiation do the fixed stars send into outer space? How do the elements come about? These are all questions the answers to which concern the astrophysicist and which play a very large role in modern astronomical research.

In the following chapters are first the technical possibilities for the collection of data from the universe, which generally are not accessible to the amateur astronomer. Then, the description of the celestial objects follows.

Methods of Modern Astrophysics

Astrophysics can be loosely defined as the exploration of the physical (or mathematical) laws by using astronomical observations. The development in this field of work has been stormy in the last decades. Observation of outer space is no longer restricted to "optical windows" alone (see photos on p. 28–29). Since certain physical processes send out certain rays, observing the right wavelengths is the precondition for a successful astrophysical research. This feat was at first not easy to accomplish. Electromagnetic radiation passes through the atmosphere to Earth's surface only in the visible spectrum of light ("optical window") and in the spectrum of radio waves (see graph on p. 133). The remaining electromagnetic radiation can at best be seen from an airplane, but mostly only in outer space. The latter was made possible through space travel. The development of very efficient telescopes for all wavelengths went hand in hand. This technical goal will, for the most part, be reached by the year 2000. An essential step into this direction is the Hubble Space Telescope, which constantly sends information to Earth about celestial bodies. It works in the visual field as well as in various other spectral fields.

"Efficient telescopes" also applies to modern design of material for measurement tools for the recording of radiation and, of course, also data processing. The need for better material by astronomers frequently results in new product lines, which enliven business. An example is the development of the glass-ceramics Zerodur®, which makes it possible to produce large mirrors for telescopes. In the meantime, there are numerous application possibilities for glass-ceramics. One is Ceran cooktop surfaces, which can now be found frequently, that are part of glass-ceramic production.

Gamma Ray Astronomy

Gamma rays are that part of the electromagnetic spectrum which is the highest in energy. Measurements can be taken only outside Earth's atmosphere. On April 5, 1991, the NASA research satellite the Compton Gamma Ray Observatory (CGRO) was launched by the space shuttle Atlantis. This satellite, which carries four gamma-ray telescopes in an orbit around Earth, is 16 tons. This mission will take about 10 years, but it will supply astrophysicists with data about gamma-ray burst and solar flares—just to name a few.

X-Ray Astronomy

The energy for X-rays is lower than gamma-ray emissions. Here, it is also the goal to discover new radiation sources in outer space. While in the gamma-ray spec-

M83 = NGC 5236 (M for Messier Catalogue, NGC for New General Catalogue) is one of the ten largest galaxies known to us. The extragalactic star system is 8 million light years away. It has an illuminating power that is the same as the Andromeda Galaxy (see p. 30) and our Milky Way system.

ROSAT, the German X-ray satellite, was launched on June 1, 1990. The X-ray sources it discovered were recorded in an atlas "Sky in the X-Ray Light."

New Technology Telescope (NTT) of the European Southern Observatory (ESO). Modern telescopes are remote controlled from a control room.

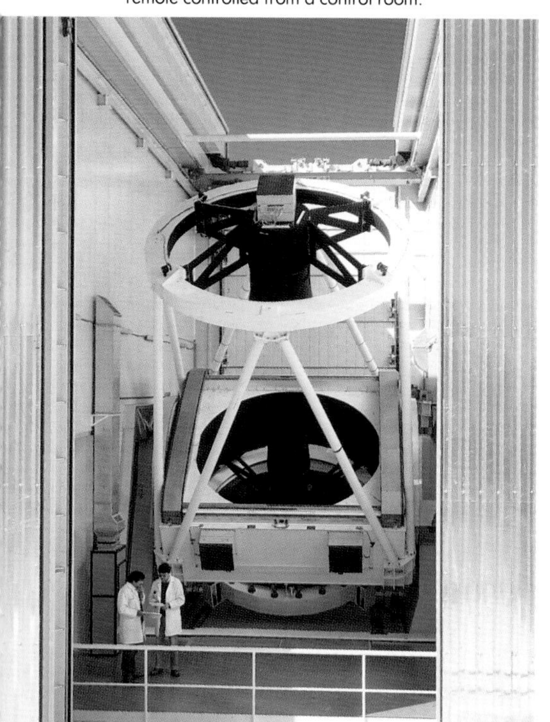

trum finding sources has been difficult, more than 60,000 X-ray sources in the universe have been discovered with the successful X-ray satellite ROSAT: stars, remainders of supernovas, galaxies, and a number of unknown objects, which are continuing to be examined now (see photo on p. 29). ROSAT was launched on June 1, 1990, from Cape Canaveral, and up to now it has been the largest German research-satellite. The project management was in the hands of the Max Planck Institute for extraterrestrial physics in Garching near Munich, Germany.

Optical Astronomy

Since astrophysicists strive to obtain the maximum amount of information about objects in outer space, the "classical" field of optical astronomy has been and still is interesting. Everywhere in the world, large observatories have been built and are being built, especially where the climatic conditions are favorable (e.g., mountain observatories). The technical progress has set new standards in regards to photo resolution and measurement precision. The 3.5-meter New Technology Telescope (NTT) of the European Southern Observatory (ESO) is equipped with an "active optics" system. That means the huge main mirror keeps its ideal form in each position with the help of computers so that small corrections can be made continuously. Another improvement is the "adaptive optics." While passing through Earth's atmosphere, a light beam of a star is slightly bent in each telescope, because of atmospheric disturbance, wind movement, and other similar factors. Therefore, the light wave reaches the telescope at an angle. The disturbed picture can be rectified in the computer through photos that are taken quickly in consecutive time periods.

These are procedures which further the construction of larger 8–10-meter telescopes. Indeed, one already thinks about the electronic linking together of a number of telescopes in order to achieve, with this mathematical total, apertures of up to 100 m (330 ft.).

Radio Astronomy

This field gets its most important data from the interstellar area. But radio astronomical observations reach from the bodies of the solar system to the galaxies. Many findings have been made since the discovery of the cosmic radio-radiation in 1932 by Karl G. Jansky and the building of large

The 100-meter radio telescope near Effelsberg in Eifel, Germany. It is presently the largest fully steerable radio telescope.

radio observatories. The netting of several instruments and the bunching toward a certain goal in the universe enhanced the productivity power astonishingly. Another improvement will come about when radio telescopes feed their data into the radio-telescope net on elliptic rotation orbits.

By the way, it is possible for the amateur astronomer to receive radio signals from outer space with simple antennas. Information about this is in *Astronomy: A Handbook*.

Infrared Astronomy

Infrared radiation is also called heat radiation. In its light, you can see into cosmic landscapes, which are invisible in the optical field. The expanded gas and dust clouds in our Milky Way system ("interstellar clouds") are one example. These clouds are several million suns in mass and some hundred light years in expansion. The dust consists of ice and silicate- and metal-compounds. The temperatures are between 10 Kelvin (−263°C/−441.4°F) and 100 Kelvin (−173°C/−279.4°F). Here, stars and planet systems are born!

Measurements depend on how well the measurement tool can suppress its own background radiation. To do so, the instrument must be cooled considerably (e.g., with liquid helium) so that its own heat does not cover up the searched-for signal.

In 1995, the European Infrared Space Observatory (ISO) opened. More research will be done from an airplane with a 2.5-meter telescope. The airplane will fly above the densest area of the Earth's atmosphere. With special telescopes from Earth, the infrared spectrum, which is directly adjacent to the visible spectrum of light, can be observed. In the graph on this page, you can see some spectral gaps in which the infrared radiation reaches all the way into the surface of the Earth.

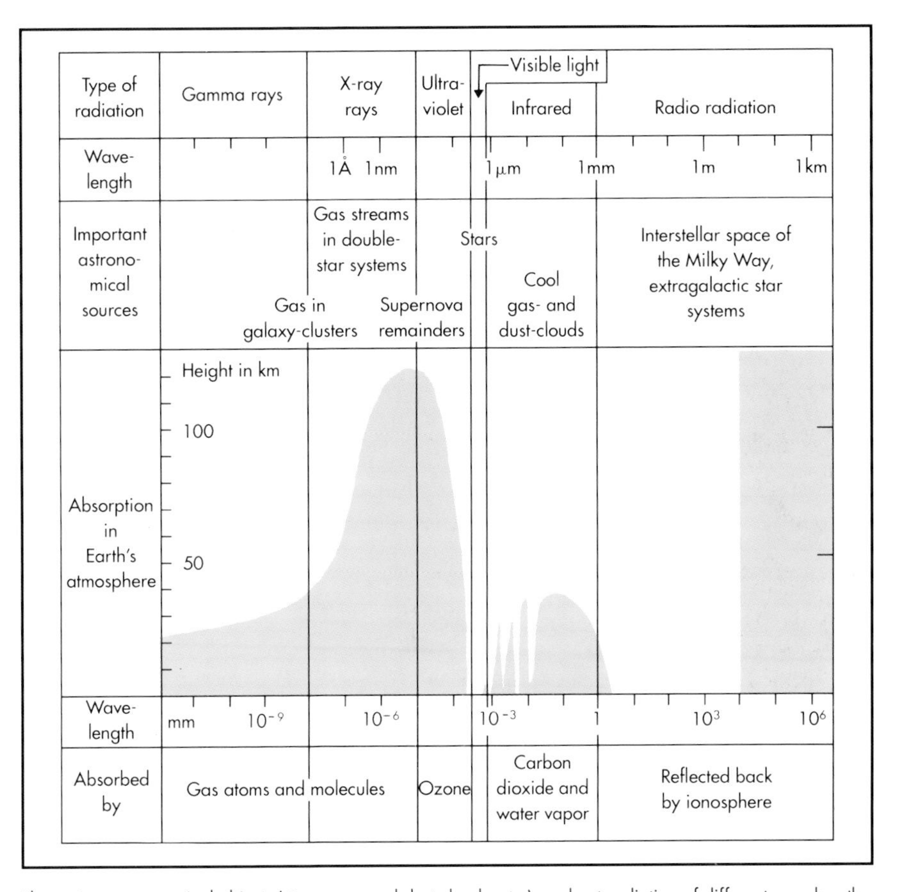

The various astronomical objects (stars, gas- and dust-clouds, etc.) send out radiation of different wavelengths. This radiation only partially reaches the Earth's surface (especially visible light and radio waves, which are between 1 cm and 20 m in wavelength) and it can be observed there with optical and radio astronomy instruments. The other parts—e.g., the gamma rays and X-rays—can only be measured when the instruments for measuring them are placed in outer space by rockets and space probes. The curve indicates in which height of the atmosphere the radiation decreases to half of its starting value.

The World of Fixed Stars

Science has learned a lot about fixed stars from observing the sun. After all, the sun is the closest fixed star, and its atmosphere can be examined rather precisely (see p. 89). The other stars are so far away that they can be pictured only as dots and only the light that they emit can be analyzed. Because they seem to be standing stationary in the sky, they are called fixed stars. In reality, they all move in space. English astronomer Edmond Halley was the first one to discover in 1718 the individual movement of stars. The first information about the distance of the stars was available in the middle of the last century. The first parallaxes were directly measured, for example, in 1838 the parallax of 61 Cygni by Friedrich Bessel.

The next fixed star is in the southern sky. It is Alpha Centauri A (see p. 78) with a parallax of 0.751″—less than one arc-second (″). From that, a distance of 4.3 light-years is calculated. A light-year is the distance that light travels in 1 year. In just one second, it travels 300,000 km (186,000 miles)! In a year, that is 9,500 billion km (5,880 billion miles).

The knowledge of the distance of the stars is an important prerequisite to learning more about the stars' magnitudes, stars' sizes, and the distribution of the stars in outer space. The number of stars is huge—into the billions. By observing even with a small telescope, you can see that the band of the Milky Way is covered with stars.

Nevertheless, in the last 100 years, astronomers have succeeded in learning a lot about the stars with the help of the spectrometer, photometer, and photography. There are stars with diameters several hundred-fold that of our sun, stars with mass more than a hundred times the sun's, but also stars with only 1/100 of the sun's mass. And with the help of newly developed measurement and observation methods, special stars have been discovered, such as stars with changes in magnitude or with special characteristics in the radiation they emit.

Formation and End of Fixed Stars

The different characteristic physical features of stars—e.g., illuminating power and surface temperature that are observed—are indications of the stars' age. Hot blue stars with high illuminating power are younger stars. Young stars often appear in the neighborhood of clouds in the interstellar matter—the birthplaces of the stars. Here, we can also observe the coming into being of a star. Measurements with radio- and infrared telescopes make that possible. That way, astronomers today have very concrete ideas about a star's formation: Through the effect of gravity, a cloud of interstellar matter condenses. In order to keep up this process continuously until a star is formed, interstellar matter of 1000–10,000 sun-mass must be available. This is enough mass for the formation of a star group, as we observe it in clusters. Within such clouds of matter, a disk is formed with a condensed center that is assisted by rotation. In rough contours, this is already the new star. While the condensed center develops into a star, planets develop out of the disk. Seen that way, it suggests that there are many stars within planetary systems in outer space. Radio astronomical observations have shown that areas of star formations are embedded in giant molecular clouds. These molecular clouds are the largest objects, the richest in mass, in our Milky Way system. They obviously play an important role in the star formation. Since interstellar matter is cold, the star formation takes place at the beginning in a cold state, which is difficult to register through observation. Nevertheless, astrophysical research concentrates on this, because that way a key can be found for understanding the coming into being of the universe.

Large to extremely large energy-turnover characterizes practically all life stages of a star. The age of the star is also closely connected to the energy production in the inside of the star. During the aging process, the decrease in the energy supply releases new forces momentarily.

M16, an open cluster surrounded by emission nebulae, is between the constellations Scutum and Sagittarius (see p. 64). It is already recognizable in the binocular. The stars of the young cluster can be seen on top and top left of the larger picture. The strong reciprocal action with the interstellar hydrogen produces the shiny nebulae (center of the picture). The smaller color photo on the left shows a greatly enlarged finger-shaped gas column in M16. At the upper end, this gas column is exposed to strong UV-radiation of hot stars, which stimulates it to shine. In this column of dense hydrogen gas, stars are created.

Crab Nebula (M1) in the constellation Taurus. Up until 1000 years ago, a star-shaped object was assumed to be at this spot. In 1054, Chinese and Japanese astronomers observed the appearance of a supernova. High-energy radiation (synchrotron radiation), a white fog in the picture, starts from this "remnant star" (pulsar).

This is called a supernova. For astrophysicists, it is the end of a star that has used up its supply of nuclear fuel. The fusion fire in the inside of the star does not withstand the gravitation anymore, which presses from the outside onto its center. The gravitation crushes the star shell in fractions of seconds so that the star explodes. Essential parts of the star matter thereby come back into the interstellar matter and fill these up with heavy elements. An unimaginably densely packed package of matter less than 100 km (62 miles) in diameter remains. Electrically negative electrons push themselves into the positively charged atomic nuclei. The matter of the "star corpse" now consists only of electrically neutral nucleus building-blocks called neutrons. On Earth, one cubic-centimeter neutron star would weight 100 million tons! A dwarf star has come about with an enormous attractive power and furious rotation as well as with temperatures of about 1 million degrees at the surface and 100 million degrees in the inside!

Neutron stars appear for the observer as dot-shaped radiation sources in the radiowave- and X-ray spectrum, which send out rhythmic impulses (pulsars). The Supernova SN 1987 A is considered an astronomical "event of the century." It was discovered in the early morning of February 24, 1987. A star, rich in mass, had exploded in the Large Magellanic Cloud (see p. 75). Its end, accompanied by a huge increase in brightness, was visible with the naked eye in the southern hemisphere of the Earth. Never before, since the invention of the telescope, had the appearance

Segments of the Large Magellanic Cloud. The picture on the left was taken on February 23, 1987, shortly before the appearance of the Supernova 1987 A. The picture on the right was taken on February 27, 1987, 4 days after the appearance of the supernova, which is clearly visible to the right of the center of the picture.

of a supernova taken place so close to the Earth: about 170,000 light-years away.

If the aged star does not reach stable balance as a neutron star, its matter is compressed to a few kilometers and the gravity field reaches such a force that the object can no longer give off any type of radiation to the outside world. It becomes, in the end, a black hole. Science does not yet fully understand this condition in detail.

From the end of a supernova eruption into a neutron star can only happen to stars that are rich in mass. Although stars lower in mass also reach an extraordinary state of matter, it is not quite so spectacular. They become small, overly dense stars called white dwarfs. This star is formed in the inside of a star in the red giant stage and only then becomes observable when the outer layers of the red giant have dispersed. One can observe this stage in planetary nebulae (see p. 149). A star is called a red giant when, as a result of the aging process, the temperature at the center increases so much that its external layers expand and the star increases in diameter (up to 50 times the original diameter). A strong increase in its illuminating power is connected to this. The density of a white dwarf is, with 1 ton per cubic centimeter, clearly below that of the neutron star. But compared with about 1 g per cubic-centimeter on Earth, it is still very considerable.

Illuminating power, temperature, and diameter are important characteristics of a star which can be observed. The Hertzsprung-Russell Diagram, named after Dane Ejnar Hertzsprung and American astronomer Henry N. Russell, is a well-known and simple diagram of the development stages of a star.

Planets and Moons Shape Themselves

The coming into existence of our solar system with the planets and moons is closely connected with what was said on page 134 about the coming into existence of the stars. The coming into existence of planets in the "protoplanetary disk," which forms around the young star—in our case, the sun—can be described as follows:

Many, many tiny dust particles collide and remain together. Small, loose conglomerates grow to become larger and heavier ones. Gradually they become matter lumps, which sink from gravitation to the center. The growth of these firm bodies continues, perhaps through collision or catching of other bodies that come into the gravitational field of the larger bodies. The prototypes of future planets and moons are formed. The larger protoplanets absorb quite a lot of gas from the disk to develop into huge gaseous planets, like Jupiter. Infrared and radio observations confirm that the temperatures in the disks around young stars are very low. They are 0° (32°F)–100° (212°F). But another astonishing observation is in contrast with this: Meteorites, which have not been changed through any kind of chemical or physical process since their formation 4.5 billion years ago, were found to contain melted silicate-balls called chondrules about a millimeter (0.04 inches) in size. But this material melts only at temperatures over 1500°C (2732°F). Chondrules constitute over 80% of a meteorite.

Tests in laboratories provided first hints at very short-lived processes that result from the melting. One is crystal structures from the chondrules. On Earth, short, highly energetic processes occur during a thunderstorm, such as electrostatic discharges in the form of lightning. There are theoretical tests which are trying to prove that such discharges also took place in protoplanetary disks and which happened again and again. According to these considerations, they are lightnings with diameters of some kilometers (miles) and lengths of 10,000 km (62,000 miles) and more. But the mysteries connected to the creation of the solar system are by no means solved. But the further the astrophysicists progress with their insights about the coming into existence of stars, the more clarity we will gain about the prehistory of the Earth.

More information is given in the following paragraphs about some special types of stars, whose conditions can be observed by the astronomer with his or her telescope.

The research of double stars and variable stars gained a new importance under the impression of modern astronomy of all wavelengths. One example is X-ray sources in double stars, which could be depicted only with the instruments of X-ray astron-

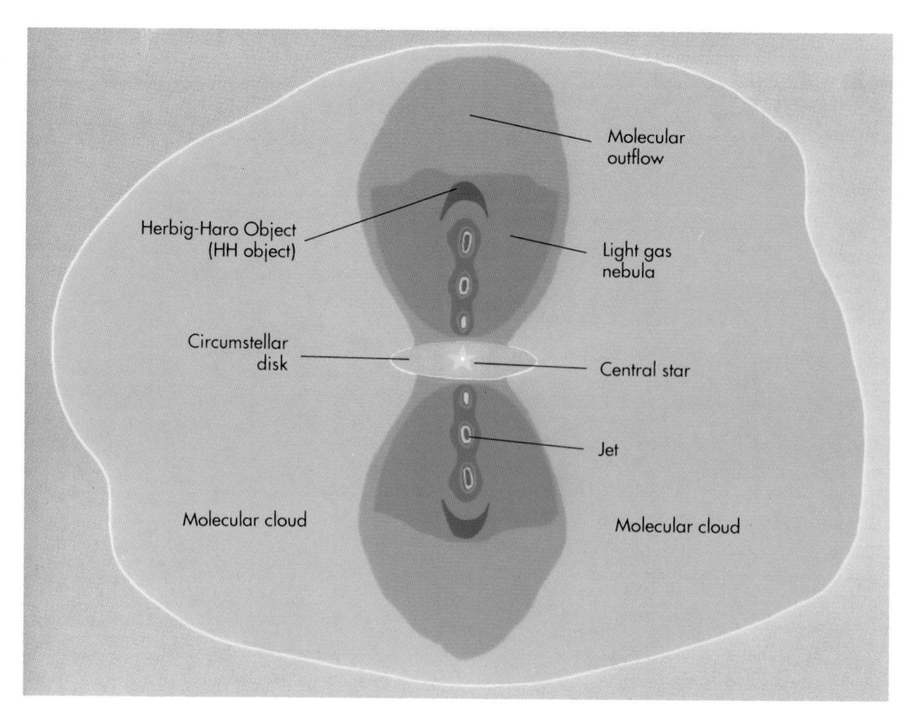

A schematic depiction of the typical appearances in the environment of a star coming into existence (according to J. Staude).

Right: A Herbig-Haro object, a small emission nebula with condensation appearances, which show streams of matter during the star formation.

Far right: Protoplanetary disk around the star Beta Pictoris. The outer layer, made of gas streaming in, which exists at the circumstellar disk, is now missing. Recognizable gaps point to the dynamic processes during a planet's formation.

omy. In the field of variable stars, since the introduction of radio astronomy and observation possibilities in outer space, magnitude changes have to be conceived more broadly. They are time-related radiation changes in the entire electromagnetic spectrum. The amateur astronomer can only use conventional visual and photographic methods, although there are technical advances now in electronic aids, such as a charge-coupled device (CCD). It is a light-sensitive electronic detector that has an almost professional quality (see p. 159).

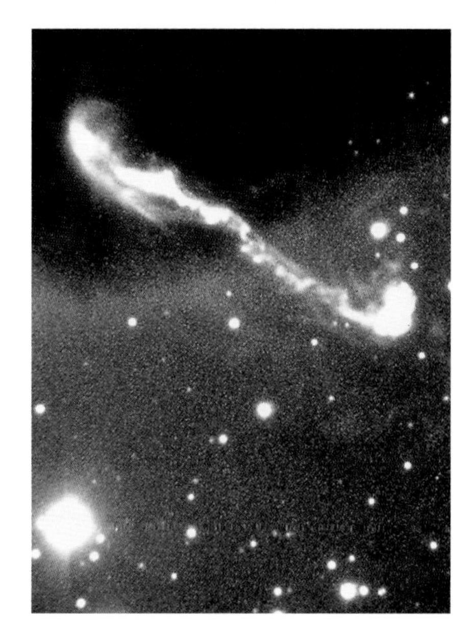

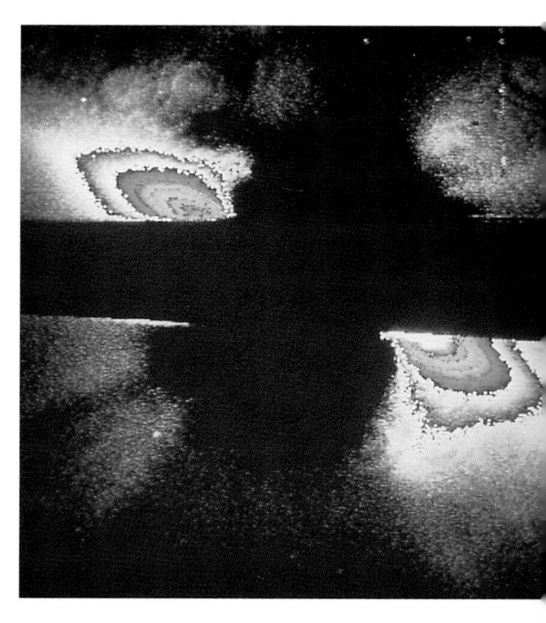

Double or Multiple Stars

As you scrutinize a sky area with binoculars or a small astronomical telescope and come upon two stars of the same or different magnitude standing strikingly close together, you should see spots. With the naked eye, you can also find stars that are relatively close. A nice example is the star Alcor next to the star Mizar in the constellation Ursa Major (see p. 45). Astronomers measure the apparent distance of the individual stars to each other with a protractor. The general rule is that we deal with an optical double star when the angle is not larger than:
20 arc-seconds (″)
for stars with an apparent magnitude of 4,
10″ for stars with 6^m,
5″ for stars with 9^m,
3″ for stars with 11^m,
1″ for stars that are less bright.

Optical double star? Two stars that stand close to each other can fall into one of two categories. The first one includes stars that stand optically close to each other but, in reality, have nothing to do with each other. They are called "optical double stars" and are of no further interest. The other group, on the other hand, consists of physical pairs of stars—under certain circumstances, even more—that orbit around each other. In other words, they describe their orbit according to the laws of gravity.

Then there are double stars which stand so close to each other that no telescope can resolve them. A spectrometer can prove that they are double stars. These stars are, therefore, called "spectroscopic binaries." Finally, you can also trace a double star by changes in brightness. When the two stars cover each other, it causes changes in brightness. Such is the case in so-called eclipsing binaries, whose most famous example is the star Algol in the constellation Perseus (see p. 37). Here the change in brightness can be effortlessly recognized.

Double stars are not rare. About one third of all stars are double or multiple. The rotation (period) lasts from a few hours to many centuries. The orbit of double stars with short periods is, in general, circular. Long-periodic double stars, on the other hand, often have eccentric orbits. Double stars are very important for science. When the movement of a star pair is not too slow, its orbit can be determined. In addition, if the distance can be calculated, then the mass can be determined. A star mass only reveals itself when two or more stars describe their orbit according to the laws of gravitation. Therefore, the observation of double stars is incredibly important for astrophysics.

Double star observation can also be spectacular when watching pairs that have a so-called "dark companion." That is, for example, the case with Barnard's star, which is the second closest to our solar system, or with 61 Cygni in the constellation Cygnus. Due to its relative closeness, for the first time in the history of astronomy, a trigonometrical parallax can be precisely measured at this fixed star. In addition to that, the star is an optical double star: The brighter partner has an apparent magnitude of 5.4, the weaker partner 6.2. The angle of 28 arc-seconds makes the observation with a small telescope possible (RA21^{h}07^m, decl. +38° 45′). Astronomer K. A. Strand was able to prove that another invisible celestial body exists, which circles around a star of the double star. Professor Albrecht Unsöld writes in his 1977 book *The New Cosmos*, "Our second-closest neighbor, Barnard's star, with the spectral-type M5V and a sun-mass of approximately 0.15, has a companion of only 0.0015 sun-masses or approximately 1.6 Jupiter-mass. We observe here a second planetary system at 1.84 pc distance." 1.84 pc means 1.84 parsec—the unit for the distance of fixed stars and star system. One pc corresponds to 3.26 light-years. Therefore, it takes the light of Barnard's star 6 years to travel to us. In cosmic dimensions, this is an astonishingly short time-period. So one can indeed call this fixed star our neighbor.

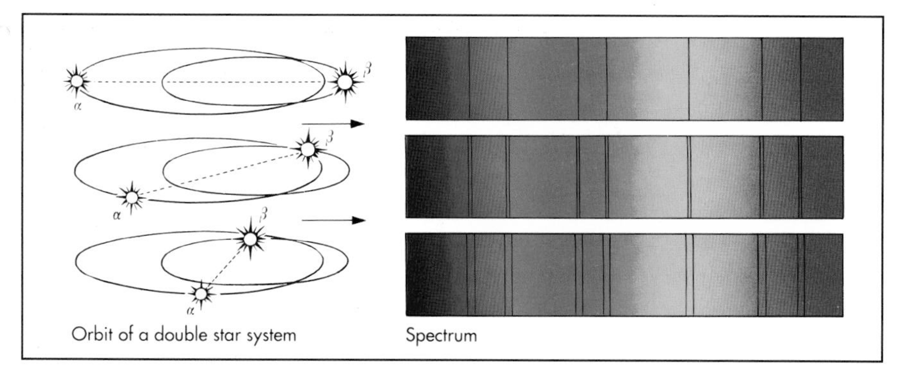

Astronomers have an interesting and reliable method for separating double stars that are standing very close together: spectral analysis.
Top: Alpha and Beta stars form a double star system. They move vertical to the line of sight. The spectrum does not show a shift yet.
Center: Alpha star moves in direction to the Earth (see arrow). In the spectrum, the lines show a shift towards violet. The so-called double-effect makes it noticeable: When a light source approaches Earth, Earth extends towards it and receives per second more vibrations. But more vibrations mean blue light, therefore, the spectral lines are shifted to blue and/or violet. At the same time, Beta star moves away (after all, the two stars move around each other). That means the red shifting of lines. Overall, that means a doubling of the lines in the united spectrum of both stars.
Bottom: The doubling seems to be the most pronounced when the directions of the movements of both stars are lying exactly in the line of view of the observer.

A second planetary system? The double star observation opens up new perspectives, because the expectation that there are a larger number of double stars with planet-like companions has been met.

Among the double stars are many multiple systems. Five percent of all double stars are multiple systems. The stars can be distinguished by their brightness, color, and size, which we can determine by visual observation. If their difference in apparent magnitude is large, the separation of the pair becomes difficult due to the outshining of the brighter star. Every telescope with a certain objective diameter has a certain separation capability for double stars. For that, a formula exists: Separable double star distance in arc-seconds (″) =

$$= \frac{11.7''}{\text{aperture in cm}}$$

According to that, a telescope with a 100-mm aperture still separates double stars with an angle altitude of 1.17″, provided that both stars (components) are equally bright. A one-magnitude difference in apparent brightness makes the separation more difficult. Observing double stars is a good way of testing the optical qualities of a telescope. But you must also take into account the cleanliness of air and the experience of the observer. The position of the companion star in relationship to the main star is for one thing characterized by the distance in arc-seconds, for another thing through the altitude, which is counted in degrees from north via east, south, west from 0 to 360. Lists of double stars that are suitable for binocular and small astrotelescopes are frequently published in the astronomy magazines. *Handbook for Star Friends*, which gives precise observation instructions including micrometer-measurements, also contains a compilation of 313 double and multiple systems and components up to the magnitude of 8.5 and distance between 1″ and 30″.

Albireo, the Beta star in the constellation Cygnus, is a very impressive example of the different colors of the individual stars of a double star. The lighter star is yellow, the weaker one is blue. Through binoculars with 15-fold enlargement, it is a beautiful sight! To find this star, see the map on page 48.

The fixed star Alpha, which is closest to our solar system at only 4 light-years distance, in the constellation Centaurus, is also a double star with components of 0.3^m and 1.7^m. It can be resolved with a telescope of 50 mm aperture (see p. 78).

Example of the resolution of a double star through a telescope with increasing enlargement (from left to right). That is the ideal condition! In reality, dust and turbulence in Earth's atmosphere disturb the observation and impair the resolution. Therefore, as a general rule, a clearly longishly depicted double star is considered resolved.

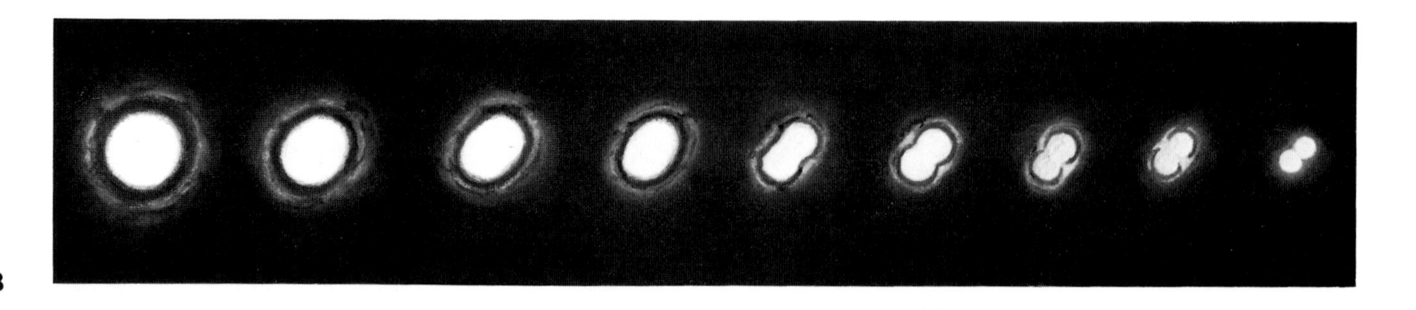

Simple Magnitude Estimation

Observing the light change of variable stars is possible by every astronomy buff with simple instruments, such as binoculars. Their relative magnitude is estimated by simply comparing the variable star to that of surrounding stars. For many variable stars there are maps with the environment of a respective number of comparable stars; for example, the map for the variable star Mira in the constellation Cetus. The numbers next to the stars mean their apparent brightness in magnitude. The field can be easily seen in binoculars.

The estimation goes as follows: The magnitude-difference between variable stars and comparable stars is indicated in 4 steps:

Step 0 = no recognizable magnitude difference;
Step 1 = a star is a bit brighter;
Step 2 = a star is clearly brighter;
Step 3 = a star is strikingly brighter.

The result of the estimations is recorded:

amV means that the variable star V is by m steps weaker than the comparable star a;
Vma means that the variable star V is by m steps brighter than the comparable star a.

An experienced observer achieves with this simple classification an accuracy up to 0.1^m, provided that the magnitude difference between the two stars, the step width, is not chosen too large. The choice of step 1 gives the most accurate values. With step 3, the dispersion grows. Appropriately, classification should be done with the help of comparing several stars.

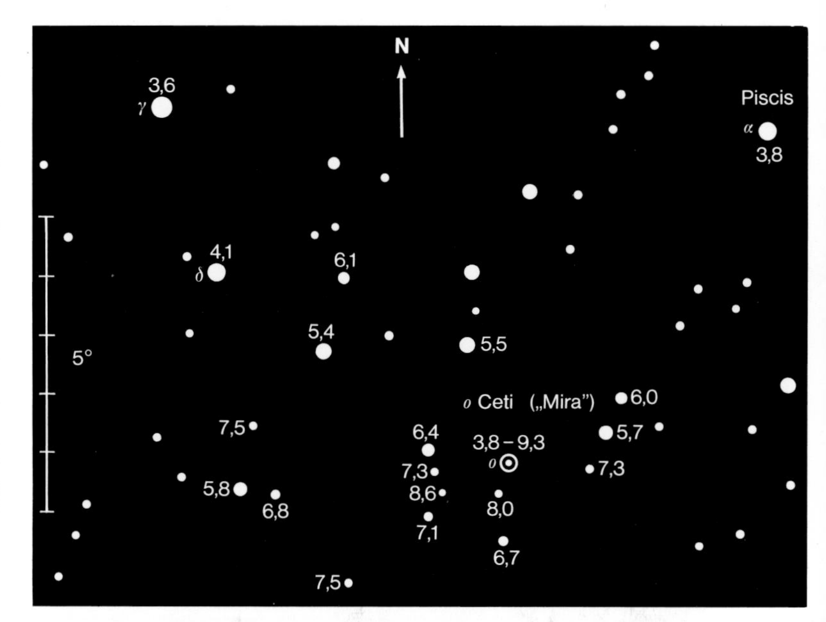

Variable Stars

Again and again one reads that this or that star has the magnitude of, for instance, 4. Magnitude has nothing to do with the diameter or circumference of a star. It refers to its brightness, that is, usually the so-called apparent brightness, which is labeled by an abbreviated, superscript m. Brightness impressions are perceived differently, for instance, by the human eye as compared to the photographic layers of films and film plates. Therefore, you should also distinguish the brightness impressions according to the visual and photographic apparent magnitude (see p. 28–29).

The classification into apparent magnitudes is based on establishing degrees of brightness. According to that, a star of the first magnitude is one hundred times brighter than a star of magnitude 6. In ancient astronomy, stars that were visible with the naked eye were classified as magnitude 6.

Orion Nebula (M42) on November 2, 1988, photographed by an amateur astronomer with a Schmidt-Cassegrain Telescope C 11. Exposed 50 minutes on Fujichrome 400. The well-known four-fold star "the Trapezoid" (see arrow) is in the Orion Nebula.

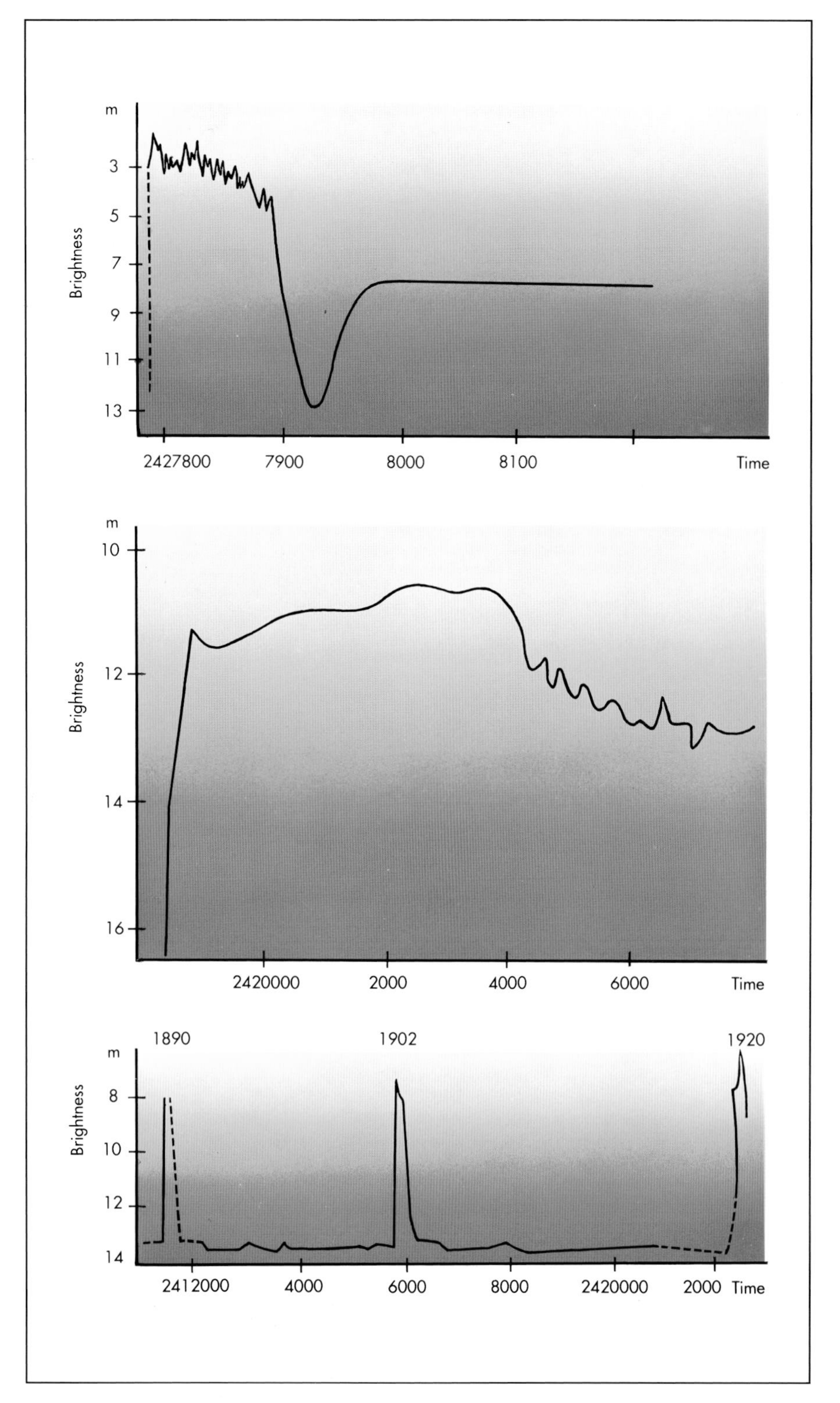

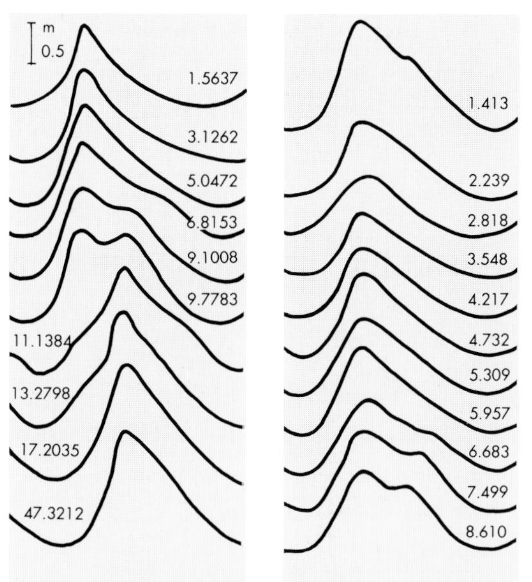

It is common to illustrate light curves of variable stars graphically. Above is the light curve form of Delta Cephei stars and the length of period that belongs to it (number indications in days). From: *Variable Stars* by Cuno Hoffmeister. Springer-Verlag, 1985.

Novae, old stars that suddenly increase in brightness, are fascinating objects in outer space. Often within 24 hours, their brightness can increase up to 10 magnitudes! A number of these interesting stars have been discovered by amateur astronomers, for instance the Nova Delphini by George Alcock of Peterborough, England, in 1967 (he used Tordalk 11 x 80 binoculars). The colored illustrations show three light-curves of novae: Nova DQ Herculis 1934 (top), Nova RT Serpentis 1909 (center), Nova T Pyxidis (bottom). After the brightness eruption (steep ascent of the curves), the descent of the brightness follows, quite differently, according to time and intensity. From: *Variable Stars* by Cuno Hoffmeister, Springer-Verlag, 1985.

The system of the continuous counting of the days is common, especially for determining the light-curve and period: The Julian date, which was introduced in 1582 by Joseph Scaliger, is presented on the time-line of such diagrams. The counting begins on January 1, 4713 B.C., noon-GMT. The number of days which have gone by since that day determine the Julian date:
for January 1, 1996 245 0083;
for January 1, 1997 245 0449;
for January 1, 1998 245 0814.

NGC1365, one of the largest known spiral galaxies in the southern sky. Because of its form, astronomers speak of a spiral beam. It is 50 million light-years away.

Yet, the relationship of the brightness-gradation from one magnitude to the following was astonishingly even, as modern control-measurements confirm. There is a mathematical equation for this: Two stars differ by $2.512 = \sqrt[5]{100}$ when they show the brightness difference of one magnitude.

The classification according to apparent magnitude takes place from Earth, depending on the appearance of the stars. But it does not yet tell us anything about the actual radiation of the stars. Therefore, the term "absolute magnitude" was introduced: "We define the absolute magnitude of a star as the apparent magnitude which the star would have if it was placed at a distance of 10 parsec (32.6 light-years) away from the solar system." (Otto Struve, *Astronomy: Introduction into Its Fundamentals*) Therefore, the issue here is to eliminate the distance factor.

As for variable stars, they are stars that change their magnitude regularly (=periodically) or irregularly. There are two very different causes for that:

1. Inner processes trigger the variability. Astronomers speak then also of "true" variability.
2. The mutual eclipsing of two stars, which orbit around each other, leads to the variability. Those are the mentioned eclipsing binaries, which are actually double stars (see p. 137).

Let's stay with the true ones. Inner processes? All fixed stars have their own radiation like our sun, which is also a fixed star. They produce energy through atomic processes in the core and bring this energy to the surface through radiation. Thereby, many things can happen: The atomic process does not proceed correctly, or disturbances can take place during the transport of energy. That of course influences the intensity of the radiation and, thus, the brightness of the star. True variability has different physical causes:

1. Stars pulsate so that their gas shells expand and contract again. This process happens astonishingly precisely (periodically). There are long-period and short-period variable stars. The typical example of a long-period variable is the star Mira Ceti (see p. 53). Short-period variables are stars with a period that is shorter than 5 hours!
2. Stars pulsate, but the process is not directly observable, because other influences contribute and trigger an irregular variability. The magnitude changes do not occur in strict periods.
3. Stars explode because accumulated energy suddenly breaks through the outer layers of the star's atmosphere and triggers extremely sudden magnitude increases. These stars are called novae, yes, even supernovae, because their original magnitude was so much less (see also p. 135).

These are the most important phenotypes of true variable stars. But there are other stars, for example, on whose surface a strong spot-formation causes magnitude-changes. Or stars on which magnitude-changes cause at the same time remarkable intensity-changes in the spectrum of light. Astronomers know by now several tens of thousands of variable stars and have classified them carefully according to characteristic features. Strictly speaking, probably all stars are somehow and sometime at one point variable. But, we do not notice that,

The two adjacent open clusters h and x Persei are still young and therefore consist of many bright stars.

even though the most modern measurement instruments reach only at 0.002^{m} the limit of accuracy.

The observation proof for the variable brightness of a star is determined through the ascertainment of a light curve. That is a graphical depiction of the apparent magnitude of the star in the course of a certain time period. The data for this light curve can be obtained with the help of visual estimates of the star's magnitude, the measuring of the blackness of star-dots on photos, or with light-electrical measurements. All magnitude estimates and measurements belong to the branch of science known as photometry. Important determination magnitudes of each light curve of variable stars are:

1. The time period between two consecutive largest and smallest magnitudes, which are called maximum and/or minimum.
2. The field of magnitude deviations between maximum and minimum (also called amplitude).
3. Shape of the curve—e.g., more symmetrical, more non-symmetrical.
4. Repetition of the curve shape and the amplitude following several maxima and minima.

There are a number of possibilities, and the pictures on page 140 show only a couple of typical examples. Any astronomer can make simple magnitude-estimates and light curves, which are based on those for brighter variable stars, by simply using binoculars. The estimates are done according to the reliable method of using adjacent stars whose apparent magnitude is known and constant (see example on p. 139). These estimates are, with some practice, quite accurate (approximately in the limit of 0.1^{m}) and constant enough to supply data for scientific evaluation. In a number of countries there are observation groups that execute these estimates systematically and pass on the results to internationally recognized collecting points. Seriously interested observers will find in the previously mentioned *Handbook for Star Friends* all information about the observation and evaluation. There are also described in it more demanding technical methods, e.g., magnitude measurements with the help of a photometer. The measurement accuracy can be essentially increased with electronic aids.

Knowledge of the physical causes for brightness deviations is very important in order to deepen knowledge about the construction of the stars. The determination of the mean apparent magnitude of certain variable stars results in distance-determinations of the star systems. One special example is the star Delta in the constellation Cepheus, which is known as the yardstick of the universe (see p. 51). Its discovery led to the first measurement of the distances to galaxies and, therefore, to our subsequent understanding of the scale of the universe.

Open Clusters

As you begin to observe the starry sky, you will soon realize that there are sections that are richer and sections lower in stars. An accumulation of stars occurs most distinctly in the Milky Way area (see p. 152). But there are also other groupings, which attract the observer, such as star clusters in the Milky Way system. There are two kinds of clusters that differ fundamentally by their size and construction: open clusters, or galactic clusters, and the globular clusters.

In this section we will be discussing open clusters, which include such famous objects as the Seven Sisters (The Pleiades; see p. 35 for details). The Pleiades were the first open cluster that drew the attention of humans. Formerly, though, these clusters were regarded from different viewpoints than they are now by modern astrophysicists. Huberta von Bronsart gives in her *Little Life Description of the Star Constellations*: "Some cultures calculated time according to Pleiades' year (just like moon year and sun year). The Incas in Peru began the year at the end of May with the early rise of the Pleiades as did the island inhabitants of the Pacific Ocean, for whom the year began in early summer. In Australia, the increase in warmth was attributed to the Pleiades. In Togo, West Africa, are festivals. These two festivals in June and November each last 3 weeks. Their beginnings are determined by the early rise of the Pleiades and/or their highest position in the sky at midnight. In some regions of the Arctic, the first month of the year, which occurs in June, is called "Pleiades rise." In tropical areas, the Pleiades are considered to be announcers of the rainy season, as they were, in early times, considered time markers for agriculture and navigation . . ."

As the sociologist analyzes the characteristics of human societies and interprets people's behavior, the astronomer analyzes the society of stars in order to better understand their origin, distribution, and movement in outer space. By no means do the stars move randomly through the cosmos.

Somewhere and somehow, each star belongs to a star system. Thus, the stars of an open, or galactic, cluster are not an accidental optical constellation in the sky. They belong together and show that to the observer in different ways.

The name "open cluster" already draws our attention to a characteristic feature: through binoculars or telescope the stars appear as a loose accumulation; sometimes it is so loose that it is difficult to recognize the cluster at first sight. Also, the arrangement in the sky is not necessarily circular, and there may not be an increased concentration of stars at the center of the cluster, which is the characteristic of globular clusters.

All open clusters are members of our Milky Way system. They are concentrated on the galactic plane. Therefore, the observer comes across these objects mainly in the Milky Way. Most of the open clusters have less than 100 stars, which does not preclude that also open clusters with 500 and 1000 stars have been discovered. The expansion in outer space is at a few parsec up to about 50 parsec. The open clusters are between 100 and 10,000 parsec away.

The closer an open cluster is, the more its stars are spread apart. The allocation to the same family then can only be determined on the basis of the speed and direction. The speeds of the stars that do not belong to star clusters, the so-called star field, have a broad range. The stars of an open cluster have approximately the same speed relating to space and the same direction. Star clusters, whose stars have no or little concentration but show the same movement, are called "moving clusters."

It was already mentioned that the Pleiades have been, since early times, a familiar sight in the sky for many civilizations.

The Pleiades, also called the Seven Sisters, is the most magnificent among the galactic clusters of the northern sky (see also p. 55).

The Messier Catalog

Simon Marius discovered the Andromeda Galaxy in 1612. Six years later, Johann Cysat discovered the Orion Nebula. The first famous discoverer of cosmic nebulae and star-clusters was Charles Messier (1730–1817). He was an astronomer working at the Marine Observatory in Paris to observe comets. He was one of the most successful comet discoverers of his time. He also put together a list of nebulae in order to avoid anyone confusing them with comets. Gradually a list of 110 foggy objects in the star sky was formed, of which Messier himself found 103. The list contains all kinds of possible galactic and extragalactic objects:

Open clusters (O)
Nebulae (G)
Planetary nebulae (P)
Globular clusters (K)
Galaxies (E).

In front of the catalog number, there stands an M for Messier.

The object M40 is a double star that was erroneously listed as a nebula. The objects M47, M48, and M91 were originally missing but, with the help of the observation books by Messier, they were filled in later on. M101 and M102 were originally listed as the same object. M104 to M109 were not listed in Messier's catalog, which was published in 1784, but handwritten remarks in his manuscript point to the fact that the objects were known to him.

Messier observed mostly with refractors of about 90-mm aperature and 1100-mm focal distance. An instrument of this size is available today to any interested amateur astronomer. The star maps on pages 34–87 contain the Messier objects.

NGC means "New General Catalog of Nebulae and Clusters," which is a catalog of over 6000 objects by J.L.E. Dreyer (1852–1926). A supplement to the NGC is called "Index Catalogue" and objects from here are abbreviated IC.

M	NGC	Coordinates		Type	Brightness	Remarks	Page
1	1952	5^h34^m	+22.0°	G	8^m	Crab Nebula	56
2	7089	21 34	– 0.9	K	6		68
3	5272	13 42	+28.4	K	6		62
4	6121	16 24	–26.5	K	6		64
5	5904	15 19	+ 2.1	K	6		62
6	6405	17 40	–32.2	O	5		82
7	6475	17 54	–34.8	O	4		82
8	6523	18 01	–24.4	G	6	Lagoon Nebula	64
9	6333	17 20	–18.5	K	7		64
10	6254	16 58	– 4.1	K	7		64
11	6705	18 51	– 6.2	O	6		66
12	6218	16 48	– 2.0	K	7		64
13	6205	16 43	+36.5	K	6	Globular cluster in Hercules	46
14	6402	17 38	– 3.2	K	8		64
15	7078	21 30	+12.1	K	6		68
16	6611	18 20	–13.8	O	6		64
17	6618	18 21	–16.2	G	7	Omega Nebula	64
18	6613	18 20	–17.2	O	8		64
19	6273	17 03	–26.3	K	7		64
20	6514	18 02	–23.0	G	9	Trifid Nebula	64
21	6531	18 05	–22.5	O	7		64
22	6656	18 37	–23.8	K	6		64
23	6494	17 57	–19.0	O	7		64
24	6603	18 20	–18.4	O	5		64
25	IC 4725	18 32	–19.3	O	7		64
26	6694	18 46	– 9.4	O	9		66
27	6853	20 01	+22.7	P	8	Dumbbell Nebula	66
28	6626	18 25	–24.9	K	7		64
29	6913	20 24	+38.6	O	7		48
30	7099	21 42	–23.2	K	8		68
31	224	0 43	+41.3	E	4	Andromeda Nebula	34
32	221	0 43	+40.9	E	9	Companion of the Andromeda Nebula	34
33	598	1 34	30.6	E	6	Triangulum Nebula	52
34	1039	2 42	+42.8	O	6		34
35	2168	6 09	+24.3	O	5		56
36	1960	5 36	+34.1	O	6		38
37	2099	5 52	+32.5	O	6		38
38	1912	5 28	+35.8	O	7		38
39	7092	21 33	+48.4	O	5		50
40	—	12 35	+58.2			Double Star	—
41	2287	6 47	–20.8	O	5		56
42	1976	5 36	– 5.4	G	3	Orion Nebula	56
43	1982	5 36	– 5.3	G	9		56
44	2632	8 40	+19.8	O	4	Praesepe	58
45	—	3 47	+24.2	O	2	The Pleiades	54
46	2437	7 42	–14.8	O	6		56
47	2422	7 37	–14.6	O	5		—
48	2548	8 14	– 5.8	O	6		—
49	4472	12 30	+ 8.0	E	9		60
50	2323	7 03	– 8.4	O	6		56
51	5194	13 30	+47.2	E	8		44
52	7654	23 25	+61.6	O	7		50
53	5024	13 13	+18.1	K	8		60
54	6715	18 55	–30.5	K	7		66
55	6809	19 40	–31.0	K	8		66
56	6779	19 18	+30.2	K	8		48

M	NGC	Coordinates		Type	Brightness	Remarks	Page
57	6720	18ʰ54ᵐ	+33.1°	P	9ᵐ	Ring Nebula in Lyra	48
58	4579	12 38	+11.8	E	8		60
59	4621	12 42	+11.6	E	9		60
60	4649	12 44	+11.5	E	9		60
61	4303	12 22	+ 4.5	E	10		60
62	6266	17 02	−30.2	K	9		64
63	5055	13 16	+42.0	E	10		42
64	4826	12 57	+21.6	E	7		60
65	3623	11 19	+13.1	E	10		60
66	3627	11 21	+13.0	E	9		60
67	2682	8 51	+11.8	O	6		58
68	4590	12 40	−26.8	K	9		60
69	6637	18 31	−32.4	K	9		82
70	6681	18 44	−32.3	K	10		82
71	6838	19 54	+18.7	K	9		66
72	6981	20 54	−12.5	K	10		66
73	6994	20 59	−12.6	O		Consists of only four stars	—
74	628	1 37	+15.8	E	10		52
75	6864	20 06	−22.0	K	8		66
76	650	1 42	+51.6	P	12		34
77	1068	2 43	+ 0.0	E	9		52
78	2068	5 47	+ 0.1	G	8		—
79	1904	5 24	−24.6	K	8		56
80	6093	16 18	−23.0	K	8		64
81	3031	9 56	+69.1	E	8		40
82	3034	9 56	+69.7	E	9		40
83	5236	13 37	−29.8	E	10		62
84	4374	12 26	+12.9	E	9		60
85	4382	12 26	+18.2	E	9		60
86	4406	12 27	+12.9	E	10		60
87	4486	12 31	+12.4	E	9		60
88	4501	12 32	+14.4	E	10		60
89	4552	12 36	+12.5	E	10		60
90	4569	12 37	+13.1	E	10		60
91	4567	12 37	+11.2	E	10		—
92	6341	17 18	+43.2	K	6		46
93	2447	7 45	−23.9	O	8		56
94	4736	12 51	+41.1	E	8		42
95	3351	10 44	+11.7	E	10		60
96	3368	10 47	+11.8	E	9		60
97	3587	11 15	+55.0	P	12	Owl Nebula	42
98	4192	12 14	+14.9	E	11		60
99	4254	12 19	+14.4	E	10		60
100	4321	12 23	+15.8	E	11		60
101	5457	14 03	+54.4	E	10		44
102	5866	15 06	+65.8	E	11		44
103	581	1 33	+60.8	O	7		34
104	4594	12 40	−11.7	E	9	Sombrero Galaxy	60
105	3379	10 48	+12.6	E	10		60
106	4258	12 20	+47.3	E	9		42
107	6171	16 33	−13.1	K	9		64
108	3556	11 12	+65.4	E	11		42
109	3992	11 58	+53.4	E	11		42

The Seven Sisters are also the best-known open cluster for scientific research. A spectroscopic procedure helped determine that their distance is approximately 120 parsec. Several hundred stars belong to the Pleiades cluster. The weakest stars have an apparent brightness of 17^m. The diffuse reflection nebula, which reflects the light of the brighter Pleiades stars, is very interesting. The existence of such a nebula suggests that maybe it is here, among remainders of dense matter, that stars are formed. The systematic research of open clusters contributed much new knowledge about the construction of our Milky Way and about the existence of the interstellar matter. In order to determine the distance of galactic clusters, it is possible to compare their apparent and absolute brightnesses (see p. 141). One can also establish a relationship between the number of stars in a cluster and their density. With increasing distance, this relationship deviates more and more from the norm, supporting the assumption of light absorption through interstellar matter.

Open clusters are well suited to show the stars in different development conditions. When the stars are embedded in nebulae (for example, the Pleiades), they are an accumulation of relatively young stars. The Pleiades, the most well-known open cluster, is between a few million and 5 billion years old. Stars come into existence from a concentration of interstellar gas, which occurs especially frequently in the galactic plane. Thereby, in addition to high density of the matter and strong magnetic fields, sufficient mass and only weak inner movement of the matter are the necessary preconditions for the process of star formations (see p. 134).

In comparison to the age of the Milky Way, which is indicated to be 10 billion years old, most open clusters are young star systems. At least a large part of the galactic clusters represent the first stage of the stars after the star-formation process. The life span of open clusters is limited—most likely one billion years. The individual clustered star escapes little by little. The stars, which belong together, are subject

to mutual attraction, which can become so effective that two stars throw each other out of the orbit. Thereby, an energy exchange takes place, which gives the necessary flight speed to one of the stars.

The constellation Perseus is one of the most beautiful double clusters (see pages 36–37 for more details).

Globular Clusters

We see the brightest globular clusters in the southern sky in the constellations Centaurus and Tucana (see maps on p. 79 and 71). With the naked eye, the globular clusters Omega Centauri and 47 Tucanae can be perceived as "little fuzzy stars." The brightest globular cluster of the northern sky is the object M13 (see p. 47) in the constellation Hercules. Through binoculars, the resolution into individual stars can be made out along the edge. But what attracts attention immediately is how pronounced is the symmetry of the star accumulation is and how much denser the cluster appears toward the center. Even a larger telescope cannot resolve the individual stars in the center. Huge telescopes in American observatories could not do it as well. The photo on this page gives us an idea of the enormous density of stars, which must be prevalent in a globular cluster.

Seen that way, globular clusters already have some simple external differences from open clusters. There are between 100,000 and 10,000,000 stars in a globular cluster. The linear diameters are in general between 10 and 20 parsec. About 150 globular clusters are listed in detail. A similar number are assumed to be hidden from the telescopes of astronomers behind clouds of interstellar matter.

In contrast to galactic clusters, globular clusters contain no reflective nebulae. S. v. Hoerner and K. Schaifers in *Meyer's Handbook about Outer Space* explain: "One may assume that all the gas, which had remained in them at the beginning during the star formation or which in the course

A color photo of the globular cluster M13 in the constellation Hercules. This degree of resolution into individual stars can only be done with very large telescopes. In general, it is only the edge of the stars that can be effortlessly seen with a small telescope.

of the progressed development of stars had been repelled again, was swept out of the cluster (through the gases of the Milky Way) during each passing swing of the cluster through the level of the Milky Way." In other words, a gathering of older generations of stars?

Globular clusters are, in contrast to open clusters, not restricted to the galactic plane. Rather, they exist in all galactic latitudes. They form a ring around our Milky Way system and thus represent something like the outposts of the Milky Way system. Globular clusters lead their own life within our galaxy. The ring of the globular cluster hardly takes part in the rotation of the galaxy. Of course the globular clusters in the Milky Way environment are exposed to the attraction of the nucleus of the Milky Way system. Therefore, they have a galactic rotation-time and they have already "dove through" the plane of the Milky Way more than sixty times since the existence of the Milky Way system.

We must base our considerations on the fact that at the very beginning, at the time of the birth of our Milky Way, a huge hydrogen cloud existed, which rotated around its center and in the course of time dissolved into smaller gas swirls, in which the stars then formed. Most of the gas swirls were near the ecliptic of our galaxy. Here, they united to the well-known lentil-shaped spiral system. But not all gas swirls ended up there. Although the force was not sufficient enough for the formation of an independent galaxy, the attraction of the nucleus of the Milky Way system caused the gas swirls, which were located at the edge, to begin galactic rotation together with the stars that were gradually formed. When immersing into the plane of the Milky Way, the free gas was "swept out," and the stars continued their galactic rotation as globular clusters. Thus, these clusters consist of the oldest stars of our galaxy, because hydrogen gas, the material for new star formations, does not exist anymore in globular clusters. Thus, globular clusters are also about 10 billion years old—the same age as our Milky Way system.

In contrast to open clusters, the accumulation of stars in globular clusters is stable; only sometimes do stars all the way on the outside at the edge get "lost."

By the way, astronomers have proved that globular clusters in other galaxies form a similar wreath around the system as in our Milky Way system. The famous galaxy in the constellation Andromeda, M31 (see p. 30), which is the galaxy closest to us, has about 200 globular star clusters.

M3, the most beautiful globular cluster in the northern sky, after M13, is not far away from the bright star Arcturus in the constellation Boötes (see p. 62). It has an apparent brightness of 6.38^{m}. With a 4-inch telescope, you can see individual stars at the cluster's edge.

Planetary Nebulae

Many astronomical objects are named purely by their appearance and then retain that name even after it turns out that their actual physical nature is not what the name implies. Thus, we continue to speak about the moon's oceans (lat. *Maria*, plural *Mariae*) and, too, the planetary nebulae, the latter which have not the least bit to do with either the large or small planets. The ring of light in the constellation Lyra (see p. 149) is a very beautiful example of a planetary nebula. Through a small telescope, you can see a small disk-shaped object that is comparable to the sight of a far-away planet, for instance, Uranus. Planetary nebulae owe their misleading name only to this seeming similarity in looks. Through a larger amateur telescope, from about 200-mm aperture on, the sight changes: The ring of light in the Lyra now really seems to be ring

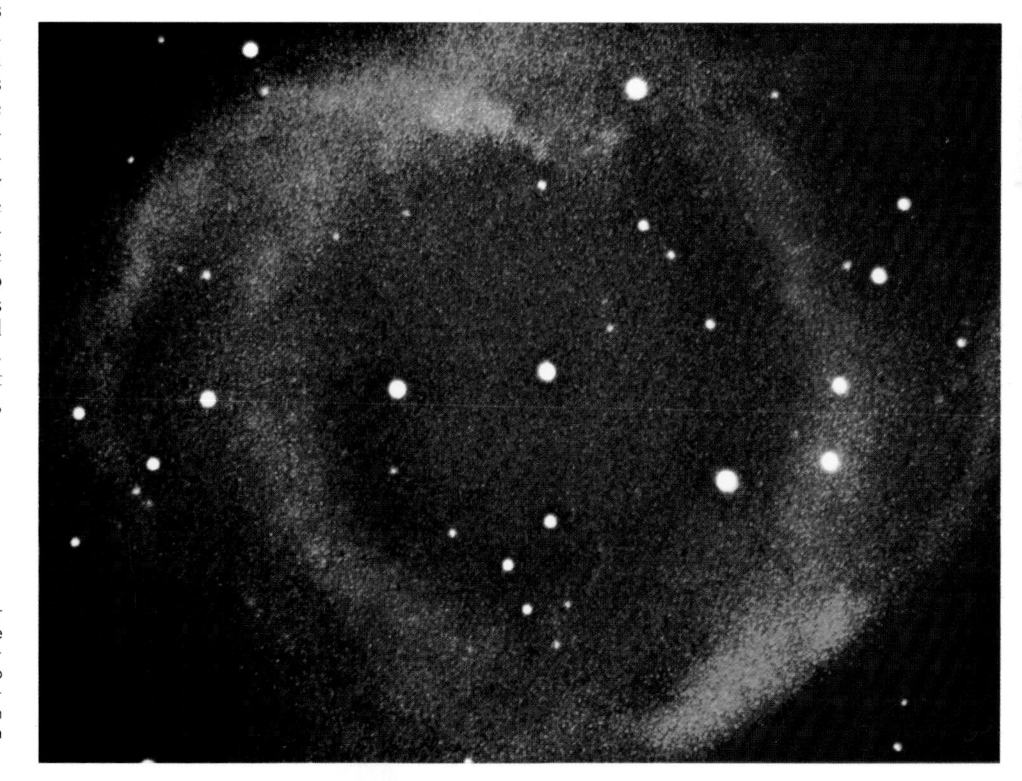

This cosmic formation, whose official catalog number is NGC 7293, is called Helix Nebula. It belongs to the most interesting planetary nebulae, which have nothing to do with planets. Rather, they are stimulated to shine by a central star. The nebula consists of hydrogen, helium, oxygen, nitrogen, and neon gas, which the central star has thrust out. Is this proof for the birth or death of a star?

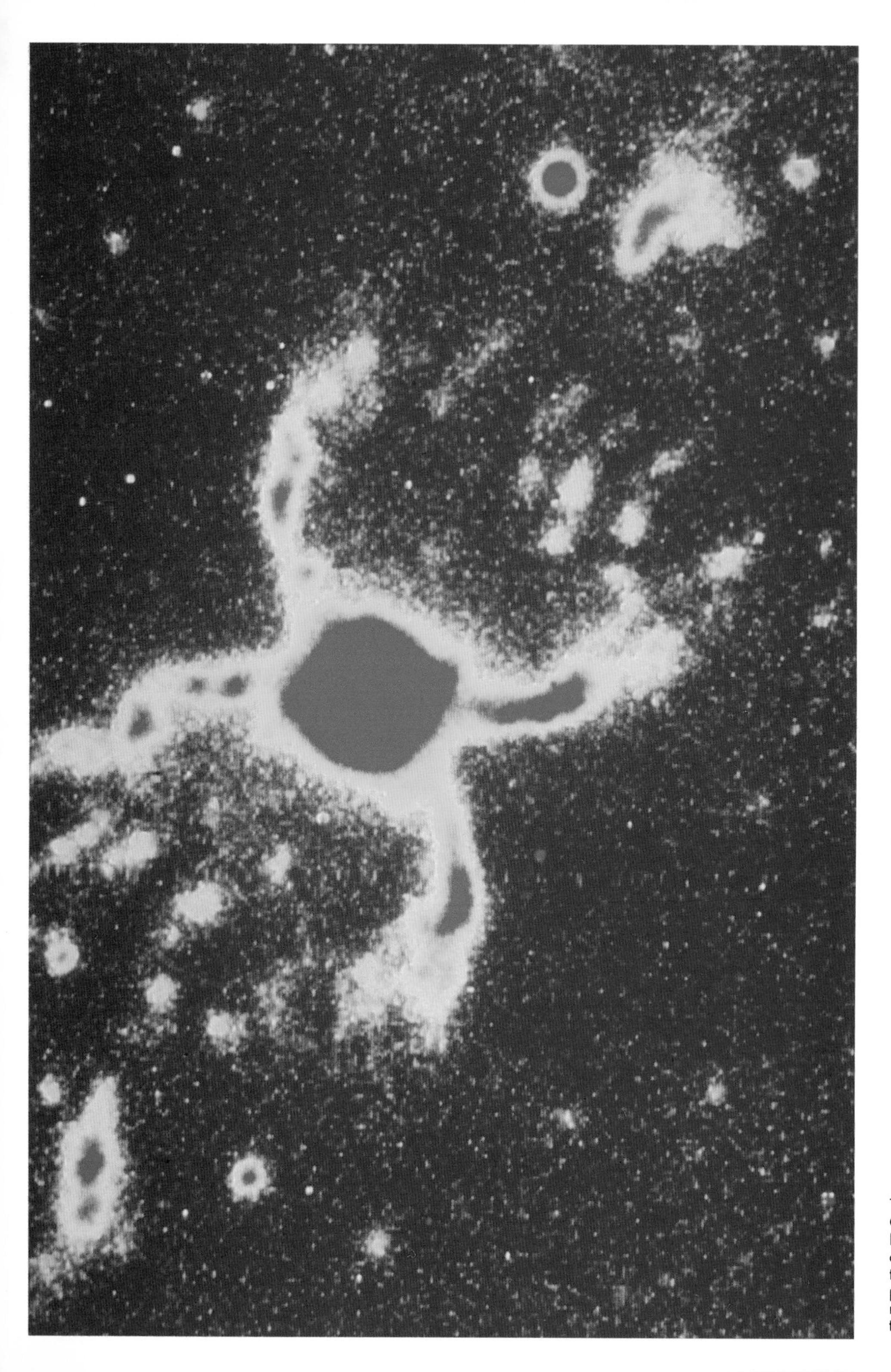

shaped. The center of the ring appears dark. In the center, a so-called central star has the apparent magnitude of 15 and shows itself only in essentially larger telescopes. In any case, with closer inspection, there is nothing left anymore of the "planet."

Planetary nebulae make their appearance in many different forms. In our Milky Way system, about 1500 such nebulae are known. Even when the forms vary, all planetary nebulae recorded until now are ring- or hourglass-shaped. Frequently, a form of ring occurs in whose center is a star. The planetary nebulae reflect light from that central star, which is the cause of the entire phenomenon.

The central stars in the planetary nebulae are not very large—smaller than our sun. But they are extremely hot; their surface temperatures are between 50,000 and 100,000°. Nevertheless, the central star is less conspicuous to the observer, because the ring of light is much brighter. Yet, all light of the nebula is only light thrown off the surface of the star in the center. The nebula represents a very thin gas shell, which is stimulated to shine by the gently erupting central star.

Planetary nebulae demonstrate to us a gigantic cosmic event. The gas shell expands with speeds of 10–50 km/s (6.2–31 miles/sec). Evidence indicates that these gas masses once belonged to the central star. An explosion, or something similar, must have taken place. What kind of stars are these, that such a thing can happen? Does the planetary nebula present to the observer a stage of birth or death of a star?

This strange cosmic object, known as the "Southern Crab Nebula" (He 2-104), deserves the name Crab Nebula more than the famous Crab Nebula in the constellation Taurus (see page 135). The object is in the southern sky. This photo depicts the birth of a planetary nebula. It was taken with a 2.2-meter mirror telescope from the European Southern Observatory (ESO).

A splendid example of a planetary nebula is the ring of light in the constellation Lyra (M57). This color photo shows the central star as well as the gas shell. With the new astronomical aids for the amateur astronomer, it is by all means worthwhile to observe this interesting object.

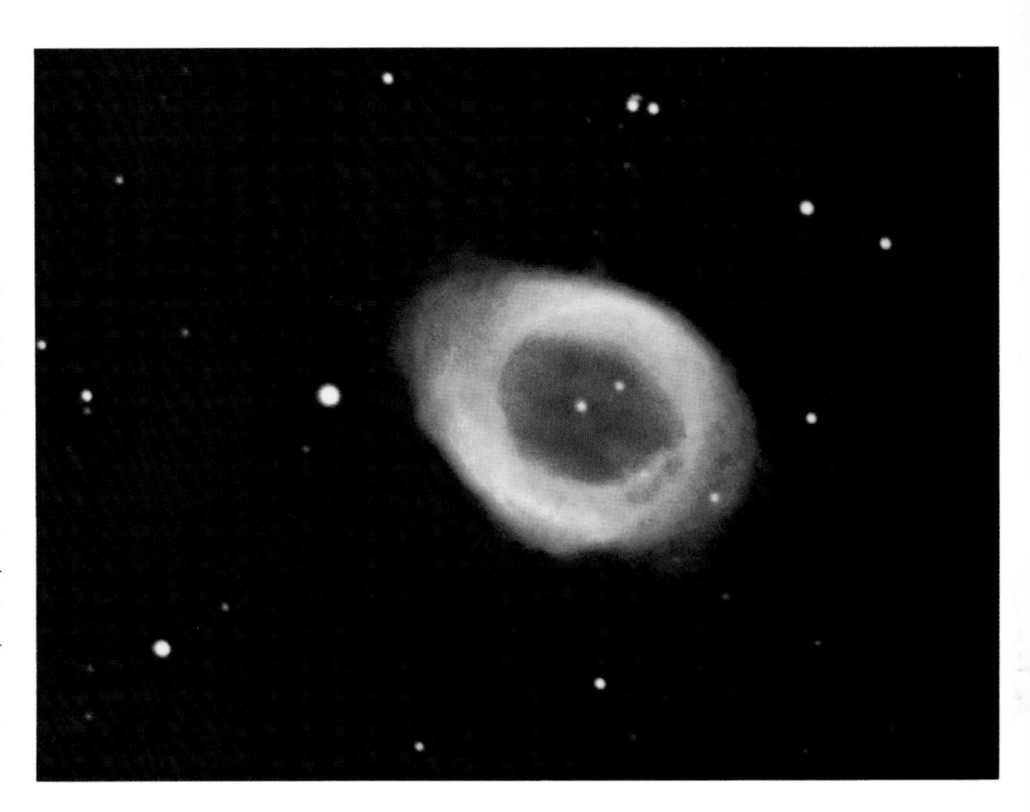

The central star blows up and discards a thin layer of its atmosphere. It is a condition like that of the variable stars. It is often said that there is a relationship between the planetary nebulae and novae, or supernovae. Strong magnetic fields dominate the planetary nebulae, and the life span of such nebulae is one hundred thousand times shorter than that of a common fixed star. Therefore, it is also believed that planetary nebulae were remainders of star-formation processes. The general consensus is that they are of short duration and are formed by dying stars. Many central stars are so-called "white dwarfs," which are a type of star of very small mass nearing the end of its life. Their energy resources are coming to an end. In any case, it seems to be certain that planetary nebulae are relatively "short-lived" objects. The expanding gas hits the interstellar matter that is being swirled up. But that slows down the process of the expansion, and the outer parts of the nebula condense. The observation of relatively sharply marked-off, bright outside edges of the planetary nebulae reveals this. The gas shell does not yet belong to the interstellar matter. But for how long? The dispersion of the gas leads, when the energy supply of the central star ends, to the disappearance of the planetary nebula, that is, to the breaking up of the nebula matter into interstellar matter . . . Opinions are not unanimous, so planetary nebulae are still objects of research. Their high expansion speed can be measured with a spectrograph. The study of the light spectrum omitted by planetary nebulae also led to the discovery of so-called "emission" lines of the component gases: An example is the characteristic green of the inner region due to doubly ionized oxygen, which can only occur in strongly stimulated and very thin gases. Planetary nebulae do not have a continuum because they have emission lines.

Reflection Nebulae

The most frequent nebulae are the diffuse clouds of dust and gas that are in interstellar matter. To the south of the three belt stars of the constellation Orion is a nebula that can immediately be seen without any visual aids. With the help of binoculars, an interesting cosmic landscape unveils on a dark, moonless night: We observe the famous Orion Nebula, M42, with its neighbor stars (see p. 139). In the constellation Sagittarius is another opportunity for a similar exciting sight: The nebulae M8 and M20 are in this part of the Milky Way, which is interspersed with numerous dark clouds. M8, also known as the Lagoon Nebula, is an impressive object through binoculars.

Nebula. What is that? At the edge, where globular clusters as well as galaxies can be resolved into individual stars, the nebula keeps its milky look, although it is not a circular or otherwise regularly-shaped form. The edges are fuzzy, dark inclusions (they look like dark clouds) that obscure the bright shine of the background stars. These diffuse nebulae are nothing else but finely distributed interstellar gas and dust which shine because of the ultraviolet radiation of stars standing close by. If the gas and dust absorb light, it is called a dark nebula; if it reflects light, it is called

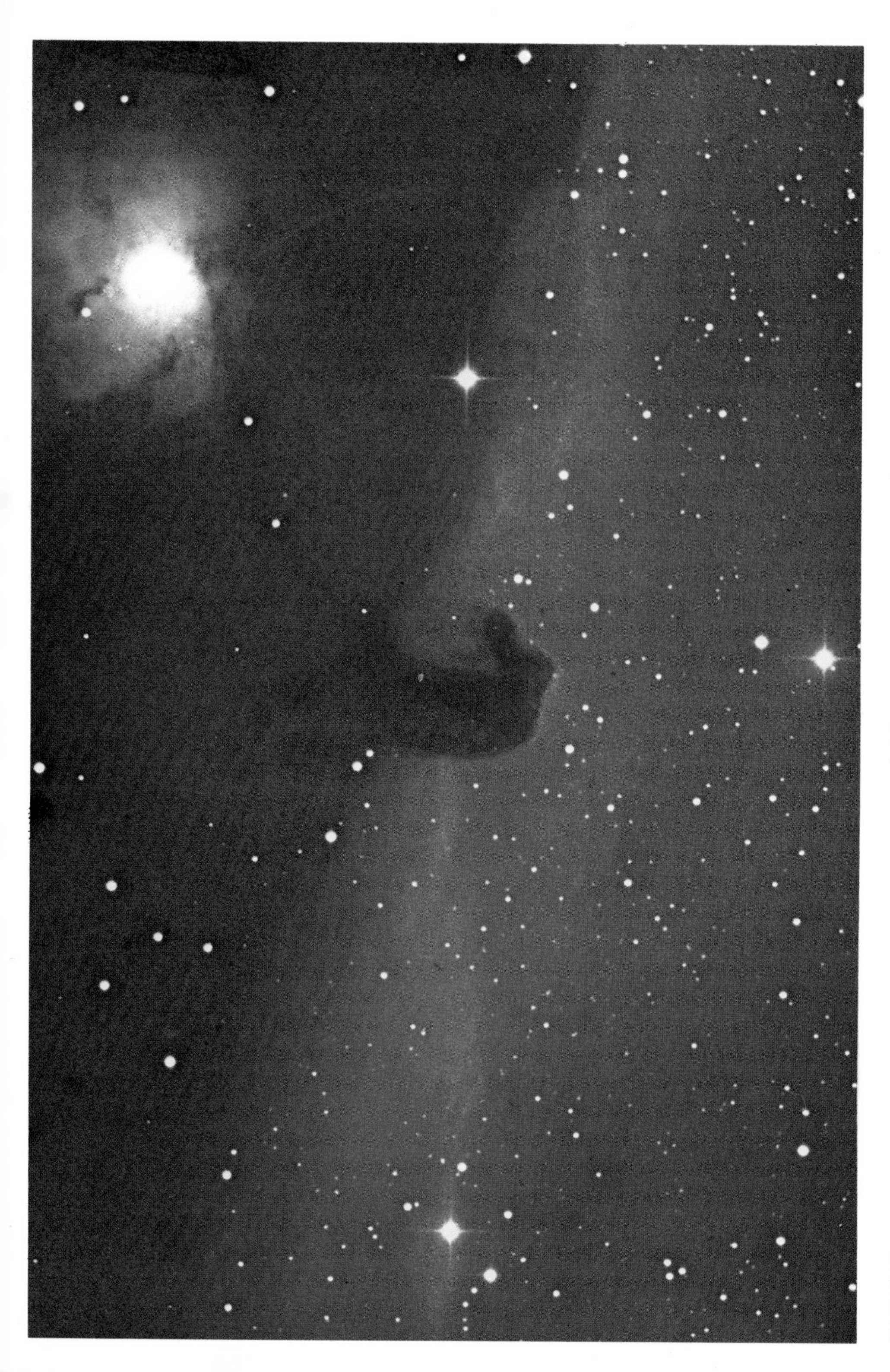

a reflection nebula. Reflection nebulae and dark nebulae often stand close to each other, for example in the environment of the eastern belt star of Orion (ζ Orionis, see p. 56). The most striking example of a dark nebula is in the constellation Southern Cross (see p. 152): Coalsack Nebula is an expanded dark cloud in interstellar matter that covers up the light of many Milky Way stars.

There is an enormous amount of gas and dust in space. From this substance, when certain physical conditions are fulfilled, new stars are still being formed today—10 billion years after the formation of our Milky Way system. All stars and their planets are products of a process in which interstellar gas and dust condense from the gravitational attraction of individual particles. The globe shape comes about because the gravitational force is aligned towards the center point of the attracting body. When the body rotates at the same time, the globe flattens, more or less. This is noticeable by looking for some time. Jupiter's rotation causes it to bulge (to be oblate) at the equator. Sir Isaac Newton (1643–1727) discovered this theory of gravitational attraction, which two masses exert onto each other. In his most important publication, *Philosophiae naturalis principia mathematica* (1687), he demonstrates that the curved orbit of a planet around the sun consists of two elements:

1. of the planet's own movement. In other words: A planet continues in a state of uniform motion in a straight line until that state is changed by the action of a force on the body.

In the area around the left belt star ζ Orionis, in the constellation Orion, is a large mixture of interstellar matter. Bright and hot stars stimulate gas to shine. The Horsehead Nebula is a dark nebula, which is surrounded by hot gas.

A typical emission nebula is the Lagoon Nebula (M8) in the constellation Sagittarius. In contrast to the reflection nebula, here is sufficient ultraviolet in the radiation of the stimulating star to cause the gas to shine on its own.

2. from a force that is directed towards the center movement, that is, having the inverse proportional effect to the square of the distance, equal to the force that gives the bodies their weight on the Earth.

This is how Newton first mathematically formulated laws of gravitation. They state that the sun attracts each planet and each planet its moons with a force that is proportional to the mass of the attracting body and inversely proportional to the square of the distance. The laws of gravitation, named after Newton, solve all important problems of space science, not only within our solar system, but also outside in outer space, in the Milky Way system as well as in other galaxies. Yet, the physical nature of the gravitational force is still today an unsolved puzzle relating to natural science.

What makes up the interstellar matter? As always, it is mostly made up of hydrogen, which filled the largest part of the outer space when our Milky Way system came into existence. The stars, which formed at that time, gave off and are still constantly giving off energy into space. Nuclear processes from the inside of the star provide the supply. There, a constant transformation of hydrogen into helium takes place. But inevitably, the hydrogen supply of stars will come to an end at some point. Something happens in the stars that is like aging. The helium is transformed again, in most probability, into heavy elements. Therefore, it is safe to say that the chemical makeup in outer space is constantly changing.

These wisps of clouds belong to the Veil Nebula in the constellation Cygnus. It is believed to be the remnants of a supernova. With the high beginning-speed of 100 km/sec (62 miles/sec), the gas was thrust from a center into outer space and was slowed down by the interstellar matter. The explosion would have to have taken place 30,000 years ago.

At the same time, a succession of star births and deaths and new star births take place. Thereby, interstellar matter plays an important role. With the energy that our sun and all other stars emit day after day into outer space, which is often even enhanced by explosion-like processes in the outer layers of hot gas from the globe, material that is already chemically transformed is released from the stars into the gas-formed interstellar medium. In some ways, we can think of it as space pollution. The interstellar matter is constantly enriched with helium and heavy elements. Although the stars formed billions of years ago out of almost pure hydrogen, all new condensations of interstellar matter contain a chemical composition: 60% hydrogen, 38% helium, 2% heavy elements. The physical condition of the interstellar matter hints at a portion of 99% gas and 1% dust, whereby the interstellar dust is identical with the heavy elements that are made up of mini-grains of a ten-thousandth millimeter or even smaller. The heavy elements in the cosmos weigh more than helium. They are made up of lithium, berylium, boron, carbon, nitrogen, oxygen, neon, aluminum, sulphur, and iron.

When looking at the Orion Nebula or the Lagoon Nebula through binoculars or a telescope, keep in mind that the components of the interstellar matter trigger reflection of light. Thus, it creates the impression of a reflection nebula. The typical examples for that are the nebulae that surround the Pleiades stars (see p. 143). Emission Nebulae consist of interstellar gas, which is stimulated by the radiation of adjacent stars to shine and radiate a line spectrum. In this case, the stars contain enough energy to ionize the nebula.

The interstellar matter also presents the observing astronomer with many technical problems. The light absorption, which is connected with it, makes it necessary to correct the distance determinations in outer space, which are based on the apparent brightness of stars. The bright and dark gas and dust clouds are also strong transmitters of radio waves. They are the objects thus of research for a young branch of astronomical science called radio astronomy.

Milky Way and Other Galaxies

To observe the Milky Way far away from all disturbing sources of light—in the country or high in the mountains—is one of the most impressive experiences in astronomy. The sight of the overcrowded regions of stars is simply stunning! We find such overcrowded regions of stars, for example, in the constellations Sagittarius and Cygnus. With the help of binoculars or a small astronomical telescope, the blurry, shiny band of the Milky Way, which expands alongside a large circle all over the celestial sphere, will resolve into thousands of individual stars in which, in between, are dark and reflection nebulae of bizarre forms.

What actually is the Milky Way? It is the perfect example of a star system—i.e., our star system to which the sun and planets belong and around which, in outer space, move the stars. Space is also filled large star systems, which astronomers refer to as galaxies. Our Milky Way is a huge star system, with no less that 200 billion objects.

The Milky Way has stirred the imagination of people of all civilizations. In many cultures, there is the notion of the Milky Way as a road. At one time, it was believed to be where the souls of the dead and unborn appeared. It was also once regarded as a fodder path in which the stars were corn for the chickens. The ancient Greeks wavered between belief and physics. Aristotle spoke of an accumulation of flammable vapors of Earthly and cosmic origin, while the majority of his contemporaries believed in a gathering of heroes. The Arabs finally came up with the idea of an accumulation of stars.

In the meantime, telescopes, astrophotographs, and radio telescopes have resolved this matter and confirmed the Milky Way as a star system. By the way, a star system with the same spiral character as other galaxies impressively show up in photos. We also would be able to see this immedi-

The Milky Way together with the constellation Crux (Southern Cross). In the center of the picture is the Coalsack Nebula, an expanded dark nebula in outer space.

A view into the Milk Way with its vast quantity of individual stars, including our sun, in between dark clouds of interstellar matter. The trace of the balloon Satellite *Echo I*, which flew during the exposure of the photo through the visual field, shows up in this photo as a line. With the increasing number of artificial Earth satellites, such "embellishments" of astrophotos are becoming more and more frequent (see also p. 128).

ately if it was possible for us to observe the Milky Way from outer space. But as it is, we are observing from the level of a flattened, lentil-shaped system. That causes the concentration of the stars onto that shiny band, which moves around the sky. Besides stars and nebulae, there are other interesting galactic objects in the Milky Way: planetary nebulae, open clusters, globular clusters (see pages 142 and 146–147).

A star system has all possible celestial objects, which raises the question: Why not planets with living creatures?

We can observe the Milky Way and its cosmic landscapes with binoculars. It becomes more difficult when we search this way for adjacent galaxies. The best known one is in the constellation Andromeda. Known as M31 in the Messier Catalog, it is better called the Andromeda Galaxy. This spiral galaxy is a must-see object—provided that geographic latitude allows its observation. The Andromeda Nebula is 1,500,000 light-years away from Earth. Its center is dimly visible to the naked eye but it is better seen with a small telescope. Because of spectacular photos taken, many observers may be disappointed when viewing the Andromeda Galaxy, or other spiral galaxies, themselves. The observation of such objects depends a lot on the environmental conditions at the location of observation and on the adaptation of the eyes of the observer to darkness and the telescope. It is very impressive to realize how far the Andromeda Galaxy expands in 14 x 100 binoculars under favorable conditions. The binocular observation with a modern refractor 1:8 to 1:5 or a mirror telescope 1:6 to 1:4 is also recommended. You can use wide-angle lenses to bring enlargements between 30- and 60-fold.

Outside the Milky Way, star systems (called galaxies) exist in many forms. There are elliptic and lentil-shaped galaxies. Each of these galactic systems consists of many billion individual systems. The galaxies form double and multiple systems in outer space. Our Milky Way system forms with other galaxies (the Large and the Small Magellanic Clouds, the Andromeda Galaxy and others) the Local Group.

The galaxies drift apart in outer space. These galaxies are billions of light-years away. Astronomers detect flight speeds that come close to light-speed. A center point in outer space cannot be detected. New research results of radio-, X-ray-, and infrared astronomy suggest explosion-like occurrences in a part of the galaxies from above-average high-energy radiation. Astrophysicists refer to them as "active galaxies," that include quasars, which are radio sources that look like a fixed star through a large telescope. In reality, these star-like objects are energetic cores of young galaxies, which are at the beginning of their development.

The Big Bang

When did the large star systems, the galaxies, come into existence? The larger the telescopes become, the more they discover distant Milky Way systems. We already know more than 100 million objects in the sky. And in each of these galaxies there exists billions of individual stars—truly gigantic. It is the desire of astrophysicists to observe more and more distant objects and, thus, look back deeper and deeper into the past of the outer space. They hope to prove events that, as a result of the uncoupling of natural forces, have remained as visible signs.

Observations hint at the drifting apart of galaxies. Formerly, they were closer together. But astrophysicists see in so-called background radiation the best evidence for the beginning of an expanding outer space in the stage of high matter-density and high temperatures. The expansion of the observable universe began between 10 billion and 20 billion years ago. The radiation that is richest in energy is gamma ray radiation. Here, the energy was stored, which set the expansion into motion ("the big bang"). The "time witness" of this inferno is the so-called cosmic background radiation. It was discovered in 1965 but was suspected of existing since the 1920s. It stems from a time when outer space was only some 100,000 years old. At that time, galaxies did not yet exist. The strength of the background radiation is with high accuracy completely independent from the direction it comes from. None of the interstellar objects that we know can be traced as source for the background radiation. It separated itself some 100,000 years after the big bang from the matter and energy of the universe. Since then, the universe started to expand in all directions. The energy distribution measured corresponds astonishingly well to Planck's radiation-law which describes how objects in space emit radiation at different temperatures.

Processes that occurred during the formation of the stars also accompanied the formation of the galaxies. As the expanded matter accumulated, bodies of large masses formed. They condensed and combined through their own gravitation. Fully developed galaxies extend back about 2 billion years—to the point in time of the original big bang. But those are isolated cases. Overall, any observational evidence for the first billions of years after the original bang are very sparse. Thus, the formation of galaxies—from the birth of a simple organized outer space to the diversity of galaxies with large formations of clusters—will occupy science for many years to come.

Galaxy M33 shows its spiral form. It both looks like and is a rotating cosmic system.

A Milky Way system at the edge of the universe. Photo taken with the 3.5-meter telescope (NTT) at the European Southern Observatory (ESO).

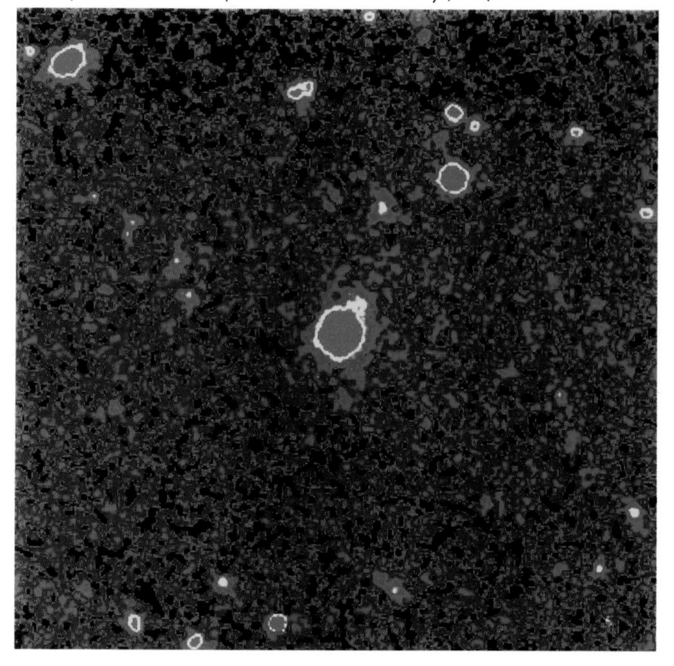

Telescopes for Astronomers

Some people may think that you need large and expensive telescopes for astronomical observations. That is not the case. Binoculars and small astronomical telescopes up to 100-mm objective aperture can achieve a lot. By nature, the human eye is an observation instrument. Light, whose quantity is decisive for stimulating the retina, enters through the pupil. From youth through middle age, the pupil's diameter is about 6.3 mm when the eye is fully adapted to darkness. On clear, moonless nights the naked eye can perceive stars of the 1st magnitude. A number of companies offer small astronomical telescopes with apertures between 50 and 100 mm (including mounting and eyepiece).

Paul Ahnert writes in the magazine *The Stars* about the efficiency of a telescope with a 63-mm opening; "It may be a small telescope, but its aperture has one hundredfold the surface of an eye's pupil. That means it collects a hundred times as much light and, thus, shows objects that are a hundred times weaker in light than those that the naked eye can see. Since the apparent magnitude of a star decreases with the square of the distance, we can still recognize a star like the sun from a 10-fold distance (over a distance of 550 light-years). But an extension of the reachable distance to ten times the amount means an expansion of the surveillable space to the 10^3 (1000-fold). Of course, one sees absolutely bright objects also from a much greater distance. Many of the galaxies in the constellations Coma Berenices and Virgo, which are completely invisible for the naked eye, can be well seen from a distance of about 30 million light-years. That's what the small telescope can achieve."

Small telescopes allow you to view moon and planetary occurrences without problems. Such typical phenomena as the phases of Venus, the polar caps of Mars, the stripes of Jupiter, and the rings of Saturn can be well seen under the right observation conditions.

Of course there are larger telescopes for astronomers. They cost, in general, several thousands of dollars. But, besides the costs, telescopes with more than a 100-mm aperture are heavier and more unmanageable with the necessary mounting. They need a firm setting up on a balcony, in an attic, or, better, in a garden to be at their optimal use. But you must be sure that the foundation is stable (e.g., concrete floor) and that you have enough unobstructed view of the sky. Most of the time, you will need weather protection gear (e.g., protection hut, dome) for the firmly set-up telescope.

A transportable amateur refractor from Astro-Physics, with a 150-mm aperture, parallactic mounting, and tripod.

Modern telescopes are now made to be more mobile: short tube, low weight, and small volume for transport. Mountings with electronic motors make the tracking easier. There are instruments with 15–30-cm apertures that can still be considered easily transportable. For travel with larger instruments, suitably built trailers are available. Since fancier and larger telescopes can become quite expensive, we recommend in this book binoculars and small telescopes. Anyone can look for more optical aids. But in many cases the small instrument will fit its purpose, according to the motto "Each telescope has its sky."

Parts of the Telescope

Here is some basic information about astronomical telescopes:

1. An astronomical telescope consists of a light-gathering objective and light-gathering ocular, or eyepiece.
2. Objectives are made of lenses or mirrors, which are referred to as refractors and reflectors, respectively.
3. The eyepiece is a magnifying glass that increases the visual angle under which the picture, which is created by the objective, is being perceived by the eye.
4. The image in an astronomical telescope is upside down and mirror image.
5. The amount of a telescope's magnification depends on the size of the objective's aperture in relation to its distance from the eyepiece (called focal length).

Usually, astronomical telescopes are viewed with one eye. Therefore, binoculars are easier to view through. There are affordable binocular attachments for astronomical telescopes (e.g., from Pentax) with an aperture ratio of 1:10 to 1:5 to give sufficiently bright pictures.

Objectives

Here is some important information about objective lenses:

1. One objective lens for small openings and long focal lengths (1:20). Disturbing edges of color. Cheap optics for experimenting (self constructed).
2. Two objective lenses cemented (Achromat) for aperture up to 80 mmm (1:15). Secondary spectrum noticeable, but hardly disturbing. Affordable optic for amateur astronomers.
3. Two objective lenses with air distance between them (Fraunhofer USA). Standard objective lens for astronomical purposes (1:15).
4. Two objective lenses with air distance and reduced secondary spectrum (AS-Objective by Zeiss, HA-Objective by Lichtenknecker). Meets high demands in regards to correction (1:15 to 1:10).
5. Two objective lenses with air distance. Second lens made of fluorspar, or a special glass. Made by Meade, Pentax, Takahashi, Vixen. These objectives are apochromatic (1:9 to 1:8).

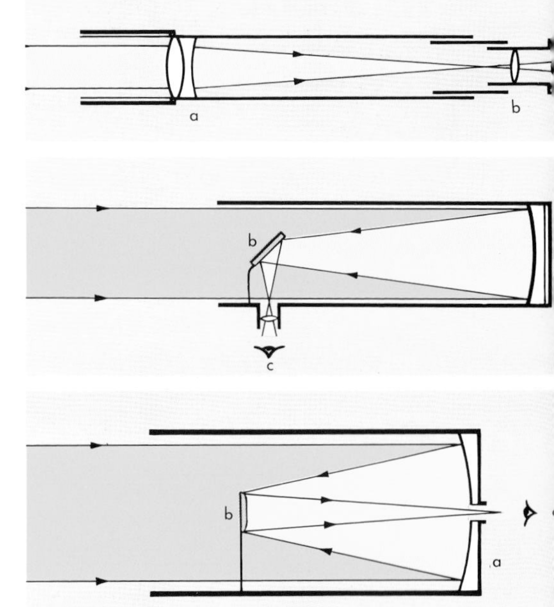

Top: Ray movement in a refractor (lens) telescope. Achromatic objective (a), eyepiece (b), observer (c). Center: Ray movement in a reflector telescope, according to Newton: Primary mirror (a), secondary mirror (b), view into the eyepiece for the observer (c). Bottom: Ray movement in a reflector telescope, according to Cassegrain. Primary mirror with a central hole (a), secondary, convex mirror (b), view into the eyepiece for the observer (c).

6. Three objective lenses with air distances between them. The lenses are made of special glasses (e.g., CA-Objective by Lichtenknecker) or with bound lenses made of special glasses (e.g., fluorspar). Manufacturers include Astro-Physics, Takahashi, Zeiss. All these objectives are apochromatic and deliver practically color-free pictures (1:9 to 1:5). Individual manufacturers use more than three lenses for their apochromats, e.g., Tele Vue. All objectives are equally suitable for visual observation and photography.
7. Mulitiple objective lenses for photographic usage that are similar to Tessar lenses (1:6 to 1:2.8).

The Schmidt-Cassegrain telescope is very popular because of its easy usability. Here is a Celstron 8 with 20-cm aperture and 2-m focal length. The same type is made by Meade.

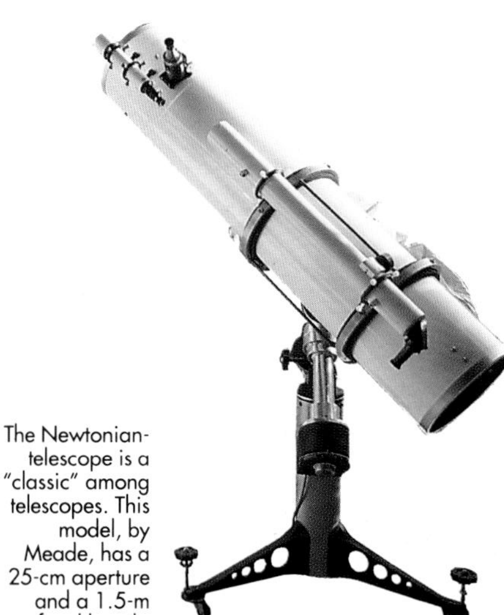

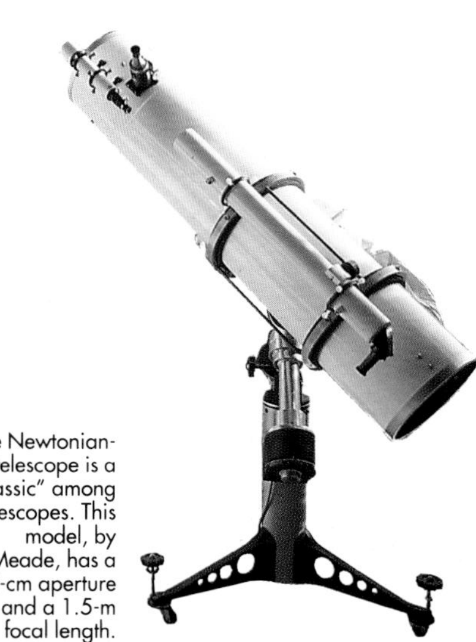

The Newtonian-telescope is a "classic" among telescopes. This model, by Meade, has a 25-cm aperture and a 1.5-m focal length.

When a mirror takes the place of the lens, then the expenditure of getting rid of color deviation (chromatic aberration) is avoided. Globe and parabolic reflectors deliver color-pure pictures and they do not know a difference between visual and photographic focal lengths. But the globe surface is not free of picture errors. The simple globe mirror can therefore only be used with an aperture ratio between 1:10 and 1:20. In the case of the more sophisticated parabolic reflector, one can choose between 1:5 and 1:10. The most common mirror/reflector systems are:

1. Newtonian telescope. The picture, produced by a concave reflector, is caught by a flat mirror and diverted to the eyepiece. Simple optics, easy construction, low cost. The picture contrast depends on the size of the collecting mirror (silhouetting). A collecting mirror that is no more than 15% of the primary mirror in diameter will bring a high picture-contrast. A tube, which is turnable, makes it easy to look into the eyepiece.
2. Cassegrain telescope. A compound reflecting telescope system. In front of the focal point of the pierced paraboloid reflector is a convex, small secondary mirror, which also corrects optically and stretches the focal length. Short construction time, very good definition in a small visual field for longer focal lengths. Disadvantages: The mirror is in the movement of rays; it demands optics of high quality, and experience is necessary for centering the optical system.

Another reflecting system used by amateur astronomers is the slanted mirror. The quality of the picture and picture contrast are excellent. Like all reflecting systems, it is free of color aberrations, but it is also prone to thermal influences of the environment (open tube). Never choose a primary mirror smaller than 100 mm (4 inches) for reflecting telescopes. In any case, the refractor is easier to use and optically more efficient with small telescopes (aperture below 100 mm). According to the newest developments, lenses and mirrors are combined with objectives (catadioptric systems). Three types are worth mentioning:

1. The Schmidt-Cassegrain system with correcting-plate and convex, or paraboloid reflector. Almost free of color-aberration. Very compact method of construction, closed optical system, therefore less prone to thermal influences of the environment. Disadvantages: The obstructed movement of rays and outer-axial picture errors. An affordable telescope. Well-known makes include Celstron and Meade.
2. Maksutov with meniscus lens and globe mirror. Almost free of color aberrations. Very compact method of construction, closed optical system, therefore less prone to thermal influences of the environment. Obstructed movement of rays, but better outer-axial picture correction in comparison to the Schmidt-Cassegrain. Manufacturers include, among others, Intes, Questar, and Zeiss.
3. Schmidt camera with glass correction-plate (flat with curved sections) and spherical mirror. A device only for astrophotography. Sharp pictures and very light-strong (1:2 to 1:4); in other words, short photograph exposures are required. Very sensitive to scattered light. The best modification is the Flat-field camera, which is at present only built by Lichtenknecker.

Also the Maksutov telescope with the meniscus lens is a compact telescope. In the picture, a model by Zeiss with 18-cm aperture and 1.8-m focal length.

A 3-inch refractor (Vixen SP 80 M) with the Baader 10-A prominence attachment, a special device that makes loops of gases visible.

Firmly placed telescope with a removable roof — the dream of many amateur astronomers.

Eyepieces

The quality of a picture depends to a great degree on the quality of the eyepiece. The most important types of eyepieces are:

1. Huygenian. Affordable two-lens design. Good for focal lengths of 16 mm and a visual field of ±30°.
2. Mittenzwey. Two-lens design with large visual field (about 50°). For medium focal lengths.
3. Kellner. Three-lens design with achromatic eye lens. Good correction for long and medium focal lengths. Suitable for a visual field up to ±40°.
4. Monocentric (achromatic). Three-lens cemented. For focal lengths of 10 mm and shorter. Practically free of reflections. Visual field ±15°.
5. Orthoscopic ocular (achromatic). Four lenses. For short and medium focal lengths. Visual field up to ±45°.

Besides these "classic" eyepieces, there are also wide-field eyepieces, which combine good optical drawing with a large visual field. They are good for almost all observations. These eyepieces have at least four lenses and are designed by Erfle and Plossl. Their construction and focal length are comfortable for the observer after some getting used to. Suppliers include, among others, Baader Planetarium, Celestron, Meade, and Tele Vue.

A set of three eyepieces (for focal lengths: 40 mm, 20 mm, 10 mm) are usually all that an amateur astronomer needs. Anyone who frequently observes double stars, the moon, and planets will occasionally need greater magnifications and can look for eyepieces with 5-mm and 7.5-mm focal length.

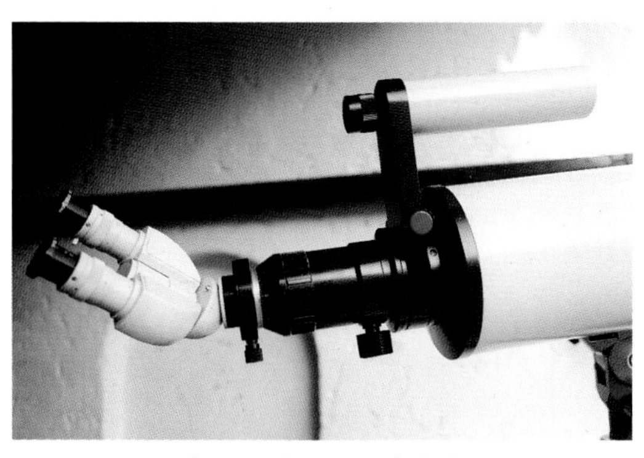

Binocular attachments and finder are important accessories for a telescope.

Accessories

Telescopes usually come with finders (standard 7 × 50). Finders are small, low power refractors with wide visual field. Using half of a binocular is also suitable as a finder.

Sun filters, either made of foil (special foil) or glass, are indispensable accessories for observing the sun. Foil or glass are encased and stuck on the front onto the telescope tube. You can make foil filters yourself.

A special sun filter is the H-Alpha filter. The DayStar H-Alpha filter can, at the same time, make the sun's surface and the prominences observable. The sun's chromosphere, which is a part of the sun's atmosphere, as well as the sun spots, solar flares, spicules, and granulations can be observed. Besides the atmospheric conditions, the visibility of the objects depends on the half-value width of the filter.

The nebula and comet filters are another type of filter that is simply stuck or screwed onto the eyepiece. The purpose of these filters is to enhance the contrast, especially in the visual and photographic observation of reflection nebulae and planetary nebulae.

Mountings

Telescope mounts come in two designs: alt-azimuth and equatorial. With the equatorial mounting, the tracking of the stars is simpler. Tracking is necessary for magnifications, since the objects disappear too fast from the field of view, due to the spin of the Earth. It is important that the mounting be strong and stable enough for the telescope and any movement by the wind or your hand. Many mountings today are homemade. But the equatorial mountings made by companies usually have fine movements, electronic tracking, and, most recently, computer-supported focusing.

The equatorial mounting is designed with a polar axis, which compensates for the Earth's rotation. That way, the setup is essentially made easier. If the mounting is equipped with a computer tracking, the setting of the stored celestial objects is done by pushing a button.

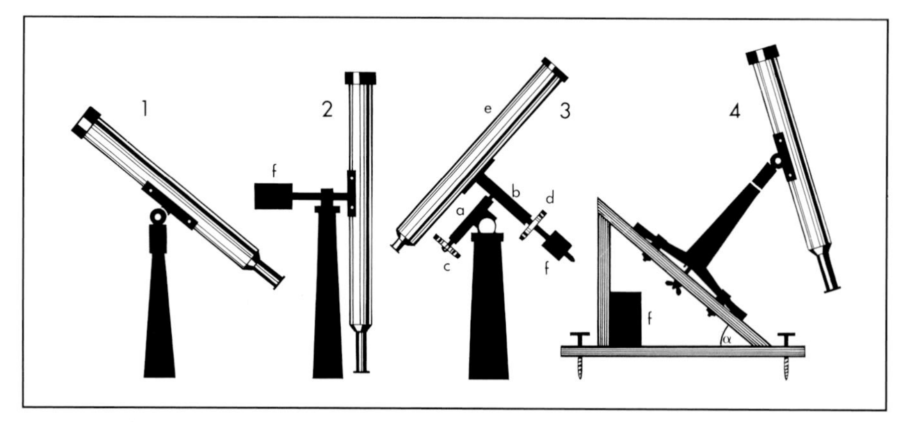

Pictures 1 and 2 show the mounted telescope, which can turn around two axes (one axis horizontal, one axis vertical). They are called alt-azimuth or horizontal mounting. Picture 3 depicts the equatorial setup of the telescope: a = polar axis, b = declination axis, c = polar circle, d = declination circle, e = tube, f = balancing or counterweight. In the picture, the telescope is aligned towards the celestial pole. If you tilt the vertical axis of an alt-azimuth-mounted telescope in such a way that it points toward the celestial pole, the setup shifts into an equatorial one. A small telescope with alt-azimuth mountings can in this way be set up equatorially (Picture 4). The angle, α, corresponds to 90° minus the geographic latitude. The counterweight, f, makes sure that the telescope does not tip over.

A very popular alt-azimuth mounting was developed by John Dobson. These instruments are easier to handle, light, and cheap. The optical system used is generally a Newtonian mirror. The Dobson mounting is a good example of a wood construction.

With the help of computer tracking, it is possible to track an alt-azimuth mountings "equatorially." This happens with two computer tracking engines at both axes. An additional tracking engine, a clock drive, can be attached to the polar axis to make the tracking almost automatic. This way, celestial objects can be photographed without any adjustment of the polar axis.

Look in astronomical magazines for the most up-to-date models for telescopes, mountings, and accessories. You can also get information from state observatories or astronomy stores.

It is always useful for any amateur astronomer to talk with an experienced astronomer, before buying a telescope as well as accessories, which can be, depending on size, quite expensive. Most of the time, affordable basic equipment is sufficient.

Newtonian model ICS with lattice tube and Dobson mounting; 23-cm aperture and 1.5-m focal length; light (in weight), capable of being disassembled and handy to use.

Photographing the Stars

It is easy to take good photos of the sun, moon, and other celestial bodies with small telescopes. Even though the refractor-objectives are usually corrected for visual observation (except for modern achromates, in which the visual and photographic levels are identical), they are suitable for astrophotographs, just as the mirror is. Thereby, the possibilities are

1. to photograph in the focal point of the telescope (without the eyepiece) or
2. to photograph the picture, which is projected by the eyepiece.

This is very easily done with the help of the casing of a single-lens reflex camera after removing the camera optics. You can experiment using the slower-speed film that is commonly sold. You can get snapshots of the moon from a standing telescope with exposure time between 1/5 and 1/25 seconds.

Cameras of all formats can also be used as mini-astrocameras for taking photos of the starfield. They are fastened onto the equatorial mounted telescope, parallel to the optical axis of the telescope and tracked after the stars according to the time of exposure. You can get pretty pictures using light-strong small-picture objectives (about 1:2) with highly sensitive film (ISO 400) exposed for a short amount of time (1–10 minutes). The short focal length of small-picture objectives (50–70 mm) is relatively insensitive to after-tracking mistakes, especially in the case of short exposure times. For after-tracking, the telescope serves as lead tube. Focus the eyepiece onto a bright star so that it is the center of the visual field and makes sure, either manually or with an electronic motor, that the star always remains in the center of the field during time of exposure.

Motor-operated tracking in modern mountings are better than manual ones. Exact corrections during long exposure times are done by pushing a button.

With the CCD-technique, the range of amateur telescopes has also improved enormously.

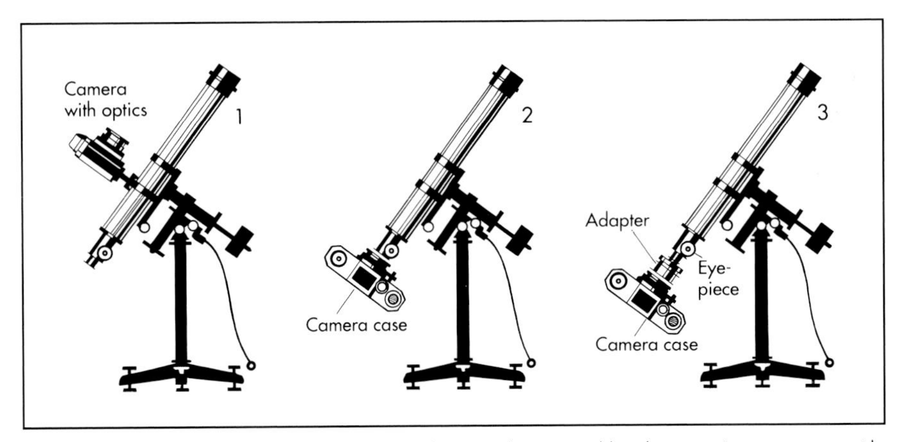

Possibilities for astrophotographs with an equatorial-mount telescope and hand camera (camera system with exchange optics): 1 = camera is screwed on parallel to the optical axis of the telescope. Telescope is the lead tube for long exposure times. 2 = Camera case (without optics) is attached to the eyepiece holder instead of the telescope (via adapter). This way, photos are possible through the telescopic focal point. 3 = Camera case is attached behind the eyepiece.

Binoculars

Binoculars, or field glasses, should not be used like a telescope for small details and large magnifications. They should be used for observing the fixed star world with its numerous star fields, star clusters, nebulae, and variable and double stars. Good field-glass optics can also show moon craters and sunspots, Jupiter's moons, and the phases of Venus.

Binoculars have four specific advantages:

1. Easy to handle, even the larger ones (in comparison to the telescope);
2. Considerable light gathering efficiency;
3. Both eyes for observation;
4. Upright pictures, usable also for Earth observations.

There are many kinds of binoculars available today. For astronomical use, you should pay attention to several things when purchasing a new one. Binoculars are made in different degrees of efficiency. Hunting, night, or marine binoculars not only hint at certain fields of use, but also their optical efficiency. The numbers 11 × 80 or 10 × 40 engraved on each binocular indicate the magnification (1st number) and the aperture diameter (2nd number). Magnification and aperture diameter are the criteria for the optical-efficiency capability of a binocular. The larger the aperture, the greater the amount of light entering the eye of the observer. In order for the observed object to appear as bright and clear as possible, you need as large an aperture as possible. For astronomical purposes, your aperture should be between 50 and 100 mm. Besides the aperture (opticians call it enter pupil), magnification plays a role. If the aperture remains the same and the magnification increases, the light strength of the binocular decreases. Even though the angle at which the observer sees the object becomes larger, the picture becomes paler and loses its brilliance. The exit pupil (the diameter of the cone of light that leaves the eyepiece

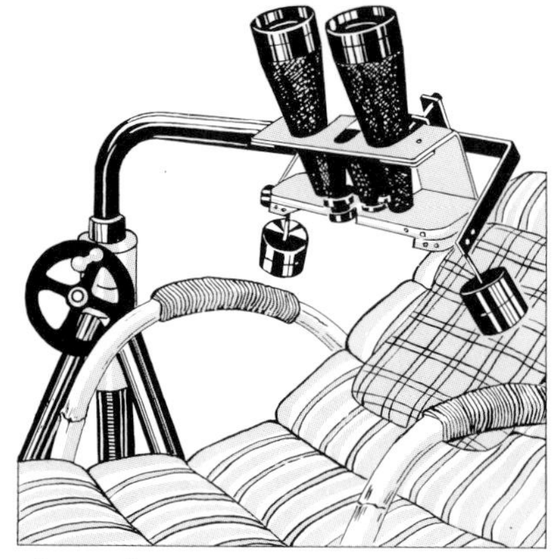

Alt-azimuth mounting for large binoculars to observe from a deck chair.

and enters the eye) becomes smaller. Only the aperture is duplicated by the eyepiece. When you hold the binoculars against a window and observe the eyepiece from 20–30 cm away, you will immediately recognize the exit pupil in the eyepiece as a light disk.

The following relationship is true:
Aperture: Magnification = exit pupil in mm.

In the table below are the exit pupils of four binoculars. The column "geometrical light strength" gives the squared number for each exit pupil, showing the difference in light strength between the four designs even more clearly. Even though the human eye can maximally reach a pupil diameter of 8 mm (⅓ in.), this value is rarely reached. With increasing age, the pupil diameter decreases. A more realistic value is, on average, around 5 mm (.2 in.). Since the pupil does not let any more light in than the diameter allows, binoculars with exit pupils of 4–6 mm for astronomical observations are optimal. But it would be wrong to judge binoculars exclusively according to light strength. A too-large light strength brings in too much of the sky's brightness, which can either blur or cover up fine light impressions. This can also happen with astronomical telescopes when a very small magnification is chosen. It can happen not only for observations at dawn or dusk, but also at night, since the sky is always brightened up by stars or city light. The magnification brings in the balance. The overall efficiency of binoculars, therefore, depends on a balanced relationship between light strength and magnification. For that, there is an equation (dawn/dusk number = DN):

$$\text{DN} = \sqrt{\text{aperture} \times \text{magnification}}$$

When purchasing binoculars, you will also have to consider how easily it can be handled, which, under certain conditions, can be more important than extreme optical values. Binoculars between 10 × 40 and 10 × 50 have easy focusing and good optical efficiency.

You should not underestimate simple binoculars with 8 × 30. They can be used for astronomical observations. In addition, they have the advantage in that you can observe with them, in general, freely. The much heavier binoculars with 11 × 80, 15 × 60, 14 × 100 need a tripod if you want to spend hours viewing the stars.

There are binoculars with picture stabilization, such as the Canon 12 × 36 IS. It has a 12-fold magnification at only 36-mm aperture, but it is impressive with its stable pictures. Details can also be seen much better, and the observer does not get tired so fast. Despite its 12-fold magnification, a tripod is not necessary.

There are also monoculars, or so-called spectives. Latter ones (e.g., the FV 63 spective by Minolta with 63-mm aperture and magnifications between 20- and 50-fold) are often equipped with a zoom lens, which makes transitionless magnifications possible. Spectives with these magnifications depend on tripods even more than binoculars.

Optical Data of Binoculars

Brand	Enlargement	Aperture opening	Exit pupil	Geom. light strength	Dusk/dawn number	Visual field in degrees
Leica 10 ×42 BA	10x	42 mm	4.2 mm	17	20.0	7.0°
Zeiss 15 × 60	15x	60 mm	4 mm	16	30.0	4.6°
Fujinon FMT-SX 16 × 70	16x	70 mm	4.4 mm	19	33.5	4.0°
Vixen Deluxe 14 × 80	14x	80 mm	5.7 mm	32	33.5	3.5°

The History of Astronomy

Antiquity and the Middle Ages

Astronomy is one of the oldest of the natural sciences. Ever since there have been people on Earth, the starry sky has been a focus of their thirst for knowledge. The systematic observation of the stars began at the moment when the stars were given cultural and religious significance and, at the same time, helped to determine time and place in people's daily lives. The daily measurements of time eventually led to the development of yearly calendars.

The oldest astronomical observations are handed down through writings and culture documentations of the old civilizations of the Near and Far East. There are Chinese documents about solar eclipses from the 3rd century B.C. Hindu and the Babylonian civilizations also have documents dating as far back in time. The Mayan civilization of Central America made sky observations regularly in the 4th century B.C. The interpretation of an old Mayan manuscript called the Dresdener Codex hints at the observation of a total lunar eclipse on February 15, 3379 B.C.

In antiquity, the stars played the roles of gods. Each celestial scientific observation was an aid for religion and myth. Thus began the basis for astrology, the belief in the influence of the stars on human destiny. Gradually, mankind began to study the physical reality of the cosmos. In addition to the religious ties of astronomical observations, the mathematical knowledge of the people of antiquity is stunning. Studying the orientation of the sun goes even back to the Stone Age—i.e., the ruins of Stonehenge in England and the great stones of Carnac in France. In the Alps are the Uhren Mountains, whose peaks, with such names as Neuner (niner), Zehner (tener), Elfer (elevener), Zwolfer (twelver), and Einer (oner), were once used as gigantic, natural time-markers.

For more than 5000 years, the events of outer space have been followed from Earth—only 390 years of those with the telescope. Astronomers of Antiquity and the Middle Ages were dependent on their eyes alone. Simple measurement and visual devices were their only aids. But the fundamentals of spherical astronomy were already recognized by the ancient Egyptians and Babylonians. The determination of the length of the year stood in the centerpiece of interest. The seeming movement of the sun through the constellations of the zodiac was the obvious natural basis for time measurements. In contrast to the orbits of the planets—Mercury, Venus, Mars, Jupiter, and Saturn were known in Antiquity—the orbit of the sun was not complicated. Those with knowledge about the sky called it ecliptic (the "line of the darkness") because there, striking phenomena like solar and lunar eclipses took place. They knew that solar eclipses only took place during a new moon, and lunar eclipses only occurred during a full moon. The periods of the eclipses, the so-called Saros cycle (18 years 11 days) were well known to Babylonians 1000 years B.C. With the help of this cycle, Thales von Milet predicted a solar eclipse, which promptly took place on May 22, 585 B.C.

All important points on the ecliptic were known in Antiquity: vernal and autumnal equinox, and summer and winter solstice. From repeated observations of the solstices came the definition of the tropical year (the name has nothing to do with the tropics but stems from the Greek word "tropos" = turning point). The repeated observations of a certain bright fixed star in the dawn served as basis for the sidereal year. The Babylonians missed the actual length of this year with 365.256 days by only 4½ minutes, and the duration of the day (i.e., a rotation of the fixed star sphere as a result of the Earth's rotation) by 3/100 more.

Ancient astronomers also occupied themselves with the shape of the Earth. Many ideas were discussed: the Earth as a flat disk or a cylindrical column in which one can swim on the water of the ocean, and so on. Plato (427–347 B.C.), a pupil of Socrates, expounded the global shape of the Earth, as did the scholars Pythagoras (580–500 B.C.) and Zenophenes (565–480) before him. For Aristotle (389–322 B.C.), the global shape of the Earth was a matter of fact. The fact that on the high sea the tip of the mast of the ship was first visible and that the height of the celestial pole rose towards north and fell towards south served as proof of the Earth's curvature. Aristotle also drew attention to the fact that during a solar eclipse, the edge of the Earth's shadow always had the form of the arc of a circle with the same curve. A globe is the only geometrical body whose shadow upon a projection surface is always circular.

The gnomon of the sun observatory in Machu Picchu (Peru), estimated 1500 years old, is an example of the observation of the shadow cast by the sun for determining the solstice.

The orbits of the planets were a problem for ancient astronomers. They did not fit into the dogma of the uniform circular movement of all stars. The dogma was tightly connected to religious ideas of this epoch, which believed in a harmony on Earth and in outer space. Ptolemy (A.D. 100–170), the famous astronomer of late antiquity, explained the nonuniformity of planet movement with a theory involving epicycle and eccentricity. An epicycle is a circle on which the planet moves at a constant speed. The central point of this circle wanders itself on another circle, in whose center, or eccentrically from that, is the Earth. Ptolemy developed his system, which only 1300 years later seemed to need improvement by the scholarly world.

The time between Ptolemy and Copernicus is often called the rest period, in which there was a decline in the advancement of astronomy. But, even though there were no new theories, numerous innovations and refinements of mathematical methods and instrumental tools were made. The Arabs continued to develop mathematics and observation techniques. Al Battani (died A.D. 929) formulated the cosine theorem of spherical trigonometry. His countryman, Abul Wefa (A.D. 940–998), formulated tables of the sine and tangent functions. It is also the Arabs who handed down ancient astronomical knowledge to the West. There, the occupation with astronomy became a part of every academic education. The acceptance of Arabic astronomy was possible because of the interest in science relating to nature. Famous theologians, such as Thierry von Chartres, Hugo von Sankt Viktor, Albertus Magnus, and Thomas of Aquinas, were opposed to the theology of the Middle Ages as being the big obstacle to the development of the natural sciences. They praised astronomy for its help in realistically perceiving god and the world.

Summary: Ancient astronomers endeavored to interpret the phenomena of the stars, which could be seen with the naked eye, and to bring them into a correct relationship with their view of life. The result was a geocentric view of the Earth.

Nicolaus Copernicus, astronomer, 1473–1543.

Modern Times

The most important event at the beginning of modern times was the transition from a geocentric to a heliocentric view of life. It began at the end of the Middle Ages. Ancient knowledge and spiritual belief were incorporated into Scholasticism for a critical spirit of modern scientific thinking.

The new view of life and the final explanation of the movements of the sun, moon, and planets are inseparably connected with the names Nicolaus Copernicus (1473–1543), Galileo Galilei (1564–1642), Johannes Kepler (1571–1630), and Isaac Newton (1642–1727).

Johannes Kepler, astronomer, 1571–1630.

Copernicus disagreed with the geocentric view of life through his own astronomical observations and by rethinking Ptolemy's theory. Astronomers Tycho Brahe (1546–1601) and Kepler, and university professor Galileo Galilei from Padua (Italy), created the empirical foundation for the new doctrine through new and precise observations. The theory of Ptolemy was for about one and a half centuries the scientific belief as well as the fundamental philosophy of life for European society during the Middle Ages. This social world was dominated by the *ordinato ad finem* (the thinking of order), where religion influenced politics, economics, and science. Into this seemingly so unshakable structure, Copernicus expounded his idea that the sun was the center of the world with the Earth as one of its satellites, in his book *De Revolutionibus Orbium Coelestium* (1543). His assertion was enormously revolutionary at the time. Copernicus' idea occurred when the Earth known to Europeans was becoming larger and the seaways to India and America were being discovered.

The dramatic moment for this new theory came only 100 years later with the discovery of the telescope. Thus, it became possible to examine the stars more closely and make astonishing discoveries. Galileo saw the rough surface of the moon in the telescope. In 1610, he discovered the four bright moons of Jupiter and, with the observation of the phases of the Venus, confirmed the heliocentric view of life. In 1609, Kepler published the first two of his planetary motion laws, which are based on observation of the planet Mars that Tycho Brahe had made without a telescope. In 1619, he published his third law (see page 107). Kepler had the personal advantage of publishing his work in Germany without punishment. The world had to accept this new astronomic concept. In 1616, *De Rev-*

Galileo Galilei, astronomer and physicist, 1564–1642.

olutionibus Orbium Coelestium was put in the index of the Roman Catholic Church (73 years after it was published), where it remained recorded until 1757. Galileo was put to trial and sentenced to house arrest for the rest of his life.

The telescope opened to astronomers the possibility of physical research of the individual celestial bodies. Before then, their work had been based upon theories.

Another important moment in explaining the movements of the stars was the formulation of the laws of gravity by Isaac Newton (see page 150). It gave the solution to the problem of the movement of a planet around the sun. More and more precise precalculations of the orbits of the celestial bodies of the solar system characterize astronomy at the turn of the 18th to 19th century. In addition to that, the investigation of outer space continued. The exploration of the world of the fixed stars advanced with the construction of highly efficient mirror telescopes in the second half of the 18th century. William Herschel (1738–1822) catalogued and measured many of the fixed stars. He was originally a trained military musician in England. Friedrich Wilhelm Bessel (1784–1846) and other astronomers of the early 19th century thought the sole purpose of astronomy was "finding rules for the movement of each star, from which its location for any time is concluded." But shifts to a new direction had already begun. In 1814, Joseph von Fraunhofer analyzed sunlight in a prism and found the spec-

Sir Isaac Newton, astronomer, physicist, and mathematician, 1642–1727.

trum of the sun. In 1859, Kirchoff and Bunsen explained the foundations of spectral analysis. From then on, the research of the physical property and composition of the stars was as important as the knowledge of their movement and arrangement in space. It was the beginning of astrophysics.

The 19th century also brought decisive progress in observation techniques. The quality of the astronomical telescopes and aids (spectroscopes, photometers) was constantly being improved. But the introduction of photography into astronomical observation was of revolutionary importance. The photographic plate made it possible to gather up light impressions and

Joseph von Fraunhofer, astronomer and physicist, 1787–1826.

thus, with the help of long exposure times, made celestial bodies, which were unseen by visual observation, visible. The photographic plate is also a relatively durable document that can be viewed again and again. Astrophotography especially furthered the exploration of the Milky Way (with its star clusters and nebulae) and spiral galaxies. The first decades of this century marked the arrival of mirror telescopes, especially in the United States. As a result, American astronomy gained international recognition.

Summary: The 16th and 17th centuries shifted the view of the Earth. The heliocentric view of life replaced the geocentric one. The next two centuries brought about mathematical refinements for the movement doctrine of the stars as well as new theoretical and technical foundations for astrophysical exploration of outer space.

The 20th Century

The events in this century follow in rapid succession. As in all sciences, the number of discoveries in astronomy grows rapidly. Besides optical astronomy (which controls, with the help of the human eye, photographic plate, light-electrical cells, visible spectral field, and the adjacent ultraviolet and infrared), radio astronomy has for some decades helped in many of these discoveries. Large mirrors and antenna systems with very sensitive receivers are some of the equipment used. With the landing of human beings on the moon for the first time (Americans Neil Armstrong and Edwin Aldrin of Apollo 11 on July 21, 1969), not only has an old dream of mankind been fulfilled, but also the opportunity has been created to study a celestial body up close and to bring samples of it back to Earth.

The Viking mission (with soft landing of Viking 1 and 2 on July 20, 1976 and September 3, 1976 on Mars) and the flights of the Voyager probes (to the planets Jupiter and Saturn in 1979, 1980, and 1981) as well as of the Voyager 2 probe (to the planets Uranus in 1986 and Neptune in 1989) supplied astrophysicists and geologists with further material for the better understanding of our solar system.

Two focal points characterize the astronomy of our century: First is the physics of the individual star from which we get information about the development of the star. Second is the exploration of the Milky Way and galaxies.

In the 19th century, voluminous catalogues were put together which contain the positions of the stars and their magnitude—e.g., the Bonner Durchmusterung (around 1860) and, for the southern sky, the Córdoba Durchmusterung (around 1930). In 1904, Karl Schwarzschild founded photographic photometry in Göttingen. Photocells and photomultipliers produce more and more precise magnitude measurements. The measurements are done in precisely chosen wavelength fields and the color indices for individual stars are derived. Parallel to that is progress made in the field of spectroscopy. An important diagram is the Hertzsprung-Russell diagram, which was created by two astronomers: Ejnar Hertzsprung (1873–1967) and Henry Norris Russell (1877–1957). It depicts the relationship between the absolute magnitude and the spectral type of stars (discovered in 1913). It provides an overview of the physical properties of the individual stars in outer space. It also shows a mass-illumination relationship: the higher the absolute magnitude of a star, the more mass it has. Spectral analysis confirmed the theory that hydrogen is the most important element of each fixed star. If the helium portion is known, the development of a star in connection with the energy it produces from its core can be studied from this ratio of components and a statement about its age can be made.

In our century, outer space, in its development relating to time, has become an overall area of astronomical research. In 1918, another step was taken with the development by Harlow Shapley of photometrical distance-measurement of variable stars, such as Delta Cephei (see also p. 51). In the 1920s and 1930s, we developed concrete concepts about the construction of our Milky Way system. In 1924, Edwin Hubble resolved the outer parts of the Andromeda nebula into individual stars with a 100-inch mirror telescope from the Mt. Wilson Observatory. The similarity between the Milky Way and Andromeda nebula was determined. In 1929, Hubble discovered in the spectrums of the galaxies a red shift that was proportional to their distance. This was interpreted as an expansion of outer space.

In 1931, K.G. Jansky discovered in the Milky Way a type of radio wave: the non-thermal component of radio-frequency radiation. Meanwhile, many radio sources in and out of the Milky Way were noted. Radioastronomy made new discoveries about the energy resource of outer space. It is here that radiation, rich in energy to build matter in the cosmos, can be attributed. Development in space travel made the transport of scientific measuring equipment outside the Earth's atmosphere possible. Thus, in 1946, the ultraviolet spectrum of the sun was photographed for the first time (V2-Rocket as carrier). Without distinct absorption, only visible light and radio waves between 1-cm and 20-m (.4 in. and 66 ft., respectively) reach the Earth's surface (see graphic on p. 133). The examination of all other kinds of electromagnetic radiation at greater heights or from outer space became possible. An abundance of new astrophysical insights came about in the last decades with measurements in ultraviolet, X-ray, and infrared light (see p. 131).

Summary: Astronomy in the 20th century finds that outer space is in a constant process of transformation, in which radiation of different energy quality participate.

The Hubble Space Telescope (HST) is the first telescope in which, outside of the Earth's atmosphere, it researches the universe in a large wavelength range from ultraviolet to infrared. The HST was launched in April, 1990.

Detailed Events in the Sky

The following pages give a short overview of the visibility of large planets in mid-northern latitudes for the coming years. In addition are some special events in the sky—e.g., encounters of light planets or eclipses of planets by the moon. For the most important solar eclipses, you will find overview maps and observation information on the back jacket. Various astronomical year books have precise and comprehensive information about all events of a year.

Neptune: Opposition on July 23 in the constellation Sagittarius.

Pluto: Opposition on May 28 in the constellation Scorpius.

Special constellations: Conjunction of Venus and Jupiter on April 23 as well as of Venus and Saturn on May 29 in the morning sky. On March 26, the moon covers Jupiter during the day (see table on p. 166).

Eclipses: On February 26, a total solar eclipse that can be observed from Central America (see back jacket flap).

An encounter of planets on June 1991: Mars, Jupiter, and Venus with the crescent of the increasing moon in the evening sky.

1998

Mercury: It is visible three times in the evening sky and four times in the morning sky (see table on p. 167). The evening visibility is best in mid-March, and the morning visibility is best at the end of August.

Venus: From February through the summer, Venus is the morning star. On March 27, Venus reaches its maximum western elongation (46° 30′).

Mars: In the second half of the year, it appears in the morning sky in the constellations Pisces and Aries.

Jupiter: It is visible in the morning spring sky. It comes on September 16 into opposition and is visible the entire night in the constellation Aquarius.

Saturn: In late summer, it can be seen in the second half of the night. On October 23, it comes into opposition and is visible the entire night in the constellation Pisces.

Uranus: Opposition on August 3 in the constellation Capricornus.

1999

Mercury: The planet is visible three times in the evening and morning sky (see table on p. 166). Anyone south of the equator has a good chance to see it at the end of July in the evening sky.

Venus: From January through the summer, Venus is the evening star. On June 11, it reaches its maximum eastern elongation (45° 23′). In late summer, Venus appears in the morning sky. Its maximum western elongation (46° 29′) is on October 30.

Mars: It stands in opposition on April 24. Mars is visible throughout the entire year in the evening sky in the constellations Virgo, Libra, Scorpius, and Sagittarius.

Jupiter: In early summer, it appears in the morning sky. Jupiter comes into opposition and is observable throughout the night in the constellation Aries.

Saturn: On November 6, Saturn stands in opposition to the sun, and it is then visible in the constellation Aries throughout the night.

Uranus: Opposition on August 7 in the constellation Capricornus.

Neptune: Opposition on July 26 in the constellation Capricornus.

Pluto: Opposition on May 31 in the constellation Ophiuchus.

Special constellations: In February, Venus approaches the planet Jupiter. The two planets stand in conjunction on February 23. On March 20, Venus stands in conjunction with the planet Saturn. On November 15, Mercury passes by the sun, which can be observed from the South Sea and Pacific Ocean.

Eclipses: On February 16, there will be a partial solar eclipse that can only be seen in Australia. A total solar eclipse on August 11 can be observed from Central Europe (see back jacket flap).

2000

Mercury: It appears three times in the evening and morning sky (see table on p. 167). Evening visibility in mid-February is not very good. At the beginning of October, Mercury appears with the "evening star" Venus in the southwest sky.

Venus: It is morning star until August. Then it becomes evening star.

Mars: Only during the first months is it somewhat visible in the evening sky.

Jupiter: In summer, it appears in the morning sky. It reaches its opposition on November 28 in the constellation Taurus. Jupiter stands high in the sky and is very favorable for observation.

Saturn: On November 19, it stands in the constellation Aries in opposition and is observable under similarly favorable conditions as Jupiter.

Uranus: Opposition on August 11 in the constellation Capricornus.

Neptune: Opposition on July 27 in the constellation Capricornus.

Pluto: Opposition on June 1 in the constellation Ophiuchus.

Special constellations: In April, in the western sky, is the conjunction of the planets Mars, Jupiter, and Saturn. In the morning sky of May 31, there will be a conjunction between Jupiter and Saturn.

Eclipses: On January 21, a total lunar eclipse can be observed from Central Europe. On February 5, July 1, July 31, and December 25, there will be partial solar eclipses, which, though, cannot be observed in the United States or Europe.

Planets Covered by the Moon

Beginning and end calculated for Berlin in WT (by J. Meeus)
P = position angle (north point of the moon disk 0°, over east −90°, south −180°, and west −270°)
* = the covering takes place during the day
(a) = beginning 2° over the horizon, end under the horizon

Day	Year	Planet	Beginning			End		
			h	m	P	h	m	P
March 26	1998	Jupiter	11	34	81° *	12	39	233° *
July 29	2000	Mercury	17	37	82° *		(a)	
September 12	2001	Jupiter	13	07	139° *	13	44	232° *
November 3	2001	Saturn	21	09	62°	22	14	260°
December 1	2001	Saturn	2	37	77°	3	43	264°
February 23	2002	Jupiter	2	47	109°	3	34	256°
April 16	2002	Saturn	20	50	126°	21	31	221°
May 21	2004	Venus	11	24	61° *	12	41	277° *

2001

Mercury: There will be three visible eastern elongations in the evening sky and three visible western elongations in the morning sky (see table on p. 167). Favorable evening visibility in the last third of May, and morning visibility at the end of October.

Venus: On January 17, Venus reaches its maximum eastern elongation (47° 06′). Two months later, it disappears from the evening sky. In May, Venus becomes morning star. Maximum western elongation on June 7 (45° 50′).

Mars: On June 13, it is in opposition in the constellation Ophiuchus. It reaches a magnitude of −2.3 (see table on p. 167).

Jupiter: The first months, Jupiter is still visible in the evening sky. In the second half of the year, it is observable at first in the morning sky. Towards the end of the year, it is seen throughout the night when it moves into the constellation Gemini and stands high in the sky.

Saturn: On December 3, it stands in the constellation Taurus into opposition and dominates, with Jupiter, the night sky in the fall and winter.

Uranus: Opposition on August 15 in the constellation Capricornus.

Neptune: Opposition on July 30 in the constellation Capricornus.

Pluto: Opposition on June 4 in the constellation Ophiuchus.

Special constellations: On May 16, a conjunction between Mercury and Jupiter occurs that can be observed in the early evening sky; on July 15 a conjunction occurs between Venus and Saturn in the morning sky. Venus and Jupiter meet in the morning sky on August 6. On November 3 and December 1, the moon covers the planet Saturn (see table on p. 166).

Eclipses: On January 9, a total lunar eclipse occurs, which can be observed from Central Europe. A total solar eclipse is observable on June 21 from Africa and Madagascar (see back jacket flap).

2002

Mercury: It is visible four times in the evening sky in the beginning of May; visible three times in the morning sky in the middle of October (see table on p. 167).

Venus: In March, Venus becomes evening star. Its maximum eastern elongation is reached on August 22 (46° 00′). At year end, it appears again in the morning sky.

Mars: At the beginning of the year, it is visible in the evening sky. It appears towards the end of the year in the morning sky.

Jupiter: On January 1, it stands in opposition to the sun. It reaches its highest position in the constellation Gemini in the northern sky. Until the middle of the year, Jupiter is visible in the evening sky.

Saturn: For the first months, it is visible in the evening sky. In late summer, it reappears in the morning sky. The opposition takes place on December 17 in the constellation Taurus. Its visibility is very favorable.

Uranus: Opposition on August 20 in the constellation Capricornus.

Neptune: Opposition on August 2 in the constellation Capricornus.

Pluto: Opposition on June 7 in the constellation Ophiuchus.

Special constellations: On May 4, conjunctions take place between Mars and Saturn (evening sky); on May 7, between Venus and Saturn (evening sky); on May 10, between Venus and Mars (evening sky); on June 3, between Venus and Jupiter (evening sky); on July 3, between Mars and Jupiter (evening sky). The moon covers Jupiter on February 23 and Saturn on April 16 (see table on p. 166).

Eclipses: A total solar eclipse on December 4 is visible from South Africa and Australia (see back jacket flap).

2003

Mercury: Evening visibility three times in mid-April and morning visibility four times in mid-September (see table on p. 167).

Lunar Eclipses Seen from the U.S.

DayYear		Eclipse	EST
July 28	1999	partial	6:34
January 20	2000	total	23:45
July 16	2000	total	8:57
January 9	2001	total	15:22
May 15	2003	total	22:41
November 8	2003	total	20:20
May 4	2004	total	15:32
October 27	2004	total	22:05
October 17	2005	partial	7:04
September 7	2006	partial	13:52
March 3	2007	total	18:22
February 20	2008	total	22:27
August 16	2008	partial	16:11
December 31	2009	partial	14:24
June 26	2010	partial	6:40

Venus: At the beginning of the year, it is a magnificent phenomenon in the morning sky. Maximum western elongation is on January 11 (46° 58′). In the second half of the year, Venus becomes evening star.

Mars: In the first months, Mars appears in the second half of the night. On August 28, it comes into opposition and is visible throughout the entire night in the constellation Capricornus.

Jupiter: On February 2, the planet stands in opposition in the constellation Cancer and is visible in the evening sky into the summer.

Saturn: In the first months, Saturn is observable in the evening sky. In the fall, it is visible in the second half of the night. Opposition on December 21 in the constellation Gemini.

Uranus: Opposition on August 24 in the constellation Aquarius.

Neptune: Opposition on August 4 in the constellation Capricornus.

Pluto: Opposition on June 9 in the constellation Ophiuchus.

Special constellations: There will be a conjunction between Venus and Saturn in the morning sky on July 8. The moon covers the planet Venus on May 29 and October 26 (not visible in U.S.). Mercury passes the sun on May 7 (not visible in U.S.).

Eclipses: On May 16 and November 9, a total lunar eclipse can be observed from Central Europe. A total solar eclipse on November 23 is visible from the Indian Ocean and southern Pacific Ocean.

2004

Mercury: Evening visibility three times at the end of March and morning visibility four times at the beginning of September (see table on p. 167).

Venus: It begins as evening star. Its maximum eastern elongation is on March 29 (46° 00′). In the second half of the year, Venus is morning star. Its maximum western elongation is on August 17 (45° 40′).

Mars: At the beginning of the year, it is visible in the evening sky. It appears at the end of the year in the morning sky.

Jupiter: It is visible in the night sky at the beginning of the year up to the summer. Opposition on March 4 in the constellation Leo.

Saturn: At the beginning of the year, it is high in the night sky in the constellation Gemini. It is visible in the evening sky until June. In the fall, it is visible in the second half of the night.

Uranus: Oppositions on August 27 in the constellation Aquarius.

Neptune: Opposition on August 6 in the constellation Capricornus.

Pluto: Opposition on June 11 in the constellation Ophiuchus.

Special constellations: On May 24 in the evening there will be a conjunction between Mars and Saturn. In the morning sky, there will be conjunctions between Venus and Saturn on September 1, Venus and Jupiter on November 4, Venus and Mars on December 5. The moon covers the planet Venus on May 21 (day observation, see table p. 166). Venus passes the sun on June 8 (visible in U.S. and Europe).

Eclipses: A total lunar eclipse is observable from Central Europe on May 4 and October 28. On April 19 and October 14, two partial solar eclipses take place.

Mars Oppositions 1999–2010

Year	Opposition	Brightness	Diameter	Declination
1999	April 24	-1.5^m	16.2″	−11.6°
2001	June 13	-2.1^m	20.8″	−26.5°
2003	August 28	-2.7^m	25.1″	−15.8°
2005	November 7	-2.3^m	20.2″	+15.9°
2007	December 24	-1.7^m	15.9″	+26.8°
2010	January 29	-1.3^m	14.1″	+22.2°

Elongation of Mercury 1998–2005

(by J. Meeus)

E = eastern elongation (evening sky)
W = western elongation (morning sky)

	1998		**2002**
Jan. 6	23° 04′ W	Jan. 12	19° 01′ E
March 19	18° 32′ E	Feb. 21	26° 35′ W
May 4	26° 44′ W	May 4	20° 58′ E
July 17	26° 41′ E	June 21	22° 44′ W
Aug. 31	18° 11′ W	Sept. 1	27° 13′ E
Nov. 11	22° 57′ E	Oct. 13	18° 04′ W
Dec. 20	21° 38′ W	Dec. 26	19° 52′ E
	1999		**2003**
March 3	18° 11′ E	Feb. 4	25° 21′ W
Apr. 16	27° 35′ W	Apr. 16	19° 46′ W
June 28	25° 33′ E	June 3	24° 26′ W
Aug. 14	18° 48′ W	Aug. 14	27° 26′ E
Oct. 24	24° 17′ E	Sept. 26	17° 52′ W
Dec. 3	20° 23′ W	Dec. 9	20° 56′ E
	2000		**2004**
Feb. 14	18° 09′ E	Jan. 17	23° 55′ W
March 28	27° 50′ W	March 29	18° 53′ E
June 9	24° 03′ E	May 14	26° 00′ W
July 27	19° 48′ W	July 26	27° 07′ E
Oct. 6	25° 31′ E	Sept. 9	17° 58′ W
Nov. 15	19° 20′ W	Nov. 21	22° 11′ E
		Dec. 30	22° 27′ W
	2001		**2005**
Jan. 28	18° 26′ E	March 12	18° 20′ E
March 11	27° 28′ W	Apr. 26	27° 10′ W
May 22	22° 27′ E	July 9	26° 15′ E
July 10	21° 08′ W	Aug. 24	18° 24′ W
Sept. 18	26° 32′ E	Nov. 3	23° 31′ E
Oct. 29	18° 34′ W	Dec. 12	21° 05′ W

Favorable observation conditions:
For medium nothern latitudes in the spring evening, in the fall morning (for medium southern latitudes reversed) An observation location close to the equator is the most favorable.

2005

Mercury: It is visible three times in the evening and morning (see table p. 167). The evening visibility is best in mid-March.

Observing a total solar eclipse is one of the big events for every astronomer. By taking series of photos, you can capture the course of an eclipse. All you need is a small- or medium-sized camera with a telephoto lens.

Venus: At the beginning of the year, it is morning star. In the spring, Venus appears again as evening star. Its maximum eastern elongation is on November 3 (47° 06′).

Mars: The first half of the year, Mars dominates after midnight. On November 7, it is in opposition, and it reaches for the first time again a positive declination (in constellation Aries). For the rest of the year, it is a striking object in the evening sky.

Jupiter: At the beginning of the year, it dominates the second half of the night. It will be then, in its opposition, visible throughout the night in the constellation Virgo. It is visible in the evening sky into the late summer.

Saturn: On January 13, it stands in opposition to the sun and it is observable throughout the entire night in the constellation Gemini. In the evening sky, Saturn is visible up into the summer. It appears again in the morning sky in the fall.

Uranus: Opposition on September 1 in the constellation Aquarius.

Neptune: Opposition on August 8 in the constellation Capricornus.

Pluto: Opposition on June 14 in the constellation Ophiuchus.

Special constellations: There will be a conjunction in the evening sky on June 25 between Venus and Saturn, and between Venus and Jupiter on September 2.

Eclipses: On April 8, a partial solar eclipse takes place (not observable in Europe). A partial solar eclipse on October 3 is visible in Spain (partially in Germany).

2006

Mercury: It is visible three times in the evening and morning (see table on p. 167). The evening visibility at the end of February and the morning visibility at the end of November are the most favorable for observation.

Venus: It is morning star in the spring and summer. It reaches its maximum western elongation on March 25 (46° 32′). At the end of the year, Venus becomes evening star.

Mars: In the spring and summer, Mars is visible in the evening sky. An opposition does not take place this year.

Jupiter: At the beginning of the year, it appears in the morning sky. Its opposition is on May 4. It is then visible throughout the night above the constellation Libra. In the evening sky, Jupiter can be observed up into the fall.

Saturn: On January 27, Saturn comes into opposition to the sun; it is then visible throughout the entire night above the constellation Cancer. In the spring, it remains visible in the evening sky. In mid-summer, it becomes visible in the morning sky.

Uranus: Opposition on September 5 in the constellation Aquarius.

Neptune: Opposition on August 11 in the constellation Capricornus.

Pluto: Opposition on June 16 in the constellation Ophiuchus.

Eclipses: On March 29, a total solar eclipse can be observed in parts of West- and North Africa and Turkey (see map on back jacket flap). A partial solar eclipse can be seen on September 22 in Venezuela and above the Southern Atlantic Ocean near to the equator. On September 7, a partial lunar eclipse can be observed in the early evening in Central Europe.

2007

Mercury: It is visible three times in evening and morning (see table p. 167). The evening visibility is best at the beginning of February, and the morning visibility is best at the beginning of November.

Venus: In the evening sky, it reaches its maximum eastern elongation on June (45° 23′). In late summer, Venus is visible in the morning sky and reaches its maximum western elongation on October 28 (46° 28′).

Mars: At the beginning of the year, it appears in the morning sky. Visibility improves from month to month until Mars reaches its opposition position on December 24 and remains visible the entire night. At a high northern declination, Mars reaches only a modest apparent diameter (see table p. 167).

Jupiter: At the beginning of the year, it stands in the morning sky. In its opposition on June 5, Jupiter is visible the entire night above the constellation Scorpius. It is observable in the evening sky until late summer.

Saturn: On February 10, it reaches its opposition position and is visible the entire night in the constellation Leo. Saturn remains visible in the evening sky. In late summer, it becomes visible in the morning sky.

Uranus: Opposition on September 9 in the constellation Aquarius.

Neptune: Opposition on August 13 in the constellation Capricornus.

Pluto: Opposition on June 19 in the constellation Ophiuchus.

Eclipses: On March 4, a total lunar eclipse can be observed at midnight in Central Europe.

Glossary

Absolute magnitude Apparent magnitude of a star at 10 parsec (about 33 light years) distance. Symbol: superscript "m", e.g., $+4.8^{m}$ (absolute magnitude of our sun is 33 light-years distance).

Absolute zero Lowest temperature theoretically possible. At absolute zero (−273°C or 0 Kelvin), molecular motion almost ceases.

Absorption spectrum A spectrum that is produced when electromagnetic radiation has been absorbed by matter. Typically produced when radiation passes through cooler matter. In the spectrum, one observes dark lines, which are typical for the absorbing gases.

Accretion disk A disk of hot material spiraling into a compact massive object, such as a black hole or neutron star. Observable in X-ray light.

Achromatic lens Lens objective consisting of two different types of glass, which reduce chromatic aberration (same as apochromatic).

Albedo Measurement for reflection capability of the surfaces of planets and moons.

Alt-azimuth Coordinate system in which the observer becomes the point of reference. It is measured along the plane of the horizon: South direction Azimuth 0°, west direction 90°, north direction 180°, east direction 270°. See zenith distance.

Amplitude Difference between maximum and minimum magnitudes of a variable star.

Anastigmatic system Optical system that is free of astigmatism and spherical aberration.

Angular distance 1 degree (°) = 60 arc-minutes (′) = 3600 arc-seconds (″).

Anomalastic year The time interval of 365.25964 days between two sucessive passages of Earth through the perihelion of its orbit.

Apastron The largest distance between the stars of a double star system.

Aphelion The point in the orbit of a planet, comet, or artificial satellite which is farthest away from the sun.

Aplanatic An optical system that produces an image free from spherical aberration and coma.

Apochromatic Name for lens telescopes with almost complete correction of chromatic aberration (same as achromatic).

Apogee Largest distance of the moon to the Earth (about 405,500 km/251,410 miles).

Apparent magnitude The different brightnesses of the stars as the observer sees them (visually) or captures them (in photos). Divided into magnitude. In order to determine the absolute magnitude, the distance of a star must be known.

Apsides The two points in an orbit that lie closest to (periapsis), and farthest from (apapsis) the center of gravitational attraction.

AS Astronomical Society

Aspect The apparent position of any of the planets or the moon relative to the sun, as seen from Earth.
0° = conjunction
90° = quadrature
180° = opposition.

Astigmatism Defect of an optical system in which rays from a point do not meet in a focal point. Outside of the optical axis, light points are depicted, not as dots, but as small lines. Faulty optical systems show these lines also in the optical axis.

Astrometry Branch of astronomy that deals with measurements of celestial bodies. It includes the positions, movements and determination of the direction of incidence of the light of a celestial body.

Astronomical refraction Because of the ray refraction in the Earth's atmosphere, a star is raised above its actual location in the sky. In order to figure out the actual location, an adjustment must be calculated to the observed zenith-distance (astronomical refraction).

Astronomical triangle A star, which is not in the meridiam, forms, with the zenith and celestial North or South pole, a spherical astronomical, or nautical triangle. Navigation guide at sea.

Astronomical Unit A unit of length used for distances in solar system. AU = 149,597,887 km (approximately 93 million miles). Abbreviation: AU.

Big Bang The birth of the universe.

Blueshift Shift of the lines in a star spectrum towards the blue end. It indicates that the star is approaching the Earth.

Bolometrical magnitude Measurement of the apparent magnitude (see *magnitudes*) with a radiation receiver—e.g., bolometer and thermoelements—which measures the total of all types of radiation.

Calendar Recording and dividing time periods according to days, months, and years. The Gregorian calendar, the calendar we use, is based on the solar year.

Catadioptric Name for telescopic systems that consist of lenses and mirrors—e.g., Schmidt-Cassegrain telescope and Maksutov telescope.

Cataclysmic variables Binary stars whose distance corresponds to the radius of the larger star. Strong gravitational forces cause tidal forces and exchange of matter between the two stars. This results in large and unpredictable changes in light.

CCD (Charge-Coupled Device) A light-sensitive electronic detector.

Chromatic aberration The inability of a lens to bring all colors in white light to an exact point of focus.

Circumpolar star A group of stars that are permenently above the horizon. They never set. In northern latitudes, circumpolar stars are found in Cassiopeia, Ursa Major, and Ursa Minor.

Coelostat A flat mirror that tracks the light of a celestial body onto a firmly mounted telescope. The daily rotation of the Earth and the picture-view rotation are compensated.

Color-Brightness diagram A diagram with the apparent magnitudes of stars on the ordinate (y-axis) and their color indices on the abscissa (x-axis). Determines the age and distance, especially of star clusters.

Color index Measurement for the determination of the color of a star.

Coma An aberration of a lens or mirror in a telescope, when light rays fall on the objective or primary mirror at an oblique angle. The light is not seen as a point but as a fan-shaped area. The star starts to look like the tail of a comet, hence the name.

Concave mirror A reflecting hollow mirror with a shape rounded inward like the inside of a bowl.

Convection Form of the energy movement in stars—e.g., our sun—when the thermodynamic balance in certain layers is disturbed. Hotter matter rises, cooler matter sinks. Sun spots are an observable sign of convection.

Convex mirror Reflecting hollow mirror of global shape.

Coude telescope A telescope which, with the help of flat mirrors, projects the image onto a spot along the polar axis where it can be always observed at the same location, independent of the positioning of the telescope.

Culmination (or transit) During a total, 24-hour long, apparent rotation of the celestial vault, each star goes twice through the meridian: during the transition from the eastern onto the western celestial hemisphere and vice versa. The first meridian transition is the upper culmination; the second is the lower culmination.

Dawn/dusk Transition from day to night and/or night to day. In mid-latitudes (40–50°), dawn/dusk lasts, on average, half an hour (sun 6–7° below the horizon). "Astronomical" dawn/dusk lasts, on average, 2 hours (sun 17–18° below the horizon). From about an hour after sunset and/or before sunrise, stars of the 5th magnitude can be seen.

Dawn/dusk curves They are created through rays of the sun that are still below the horizon at the layers of the Earth's atmosphere. The dawn/dusk curve disappears at sunset and appears at sunrise.

Declination Coordinate in the equatorial system (see p. 20) that indicates the distance of a star from the celestial equator 0° up to the celestial pole ±90°. Declination axis of the equatorial mounting.

Density wave theory An explanation of spiral arms as compressions of the interstellar medium in the accretion disk of the galaxy.

Diffraction lattice Dispersing (light scattering) compositional element in a lattice spectrograph.

Eccentricity Characteristic of an orbit that describes, together with the inclination of an orbit, the size and form of the orbit of a celestial body—e.g., of a planet.

Eclipse year Time period between two sucessive passages of the sun through a certain node of the moon's orbit (see *node*).

Ecliptic system Astronomical coordinates system in which a basic circle is the ecliptic with the ecliptical North and South Pole.

Electron A subatomic particle that carries a negative charge. It moves around the nucleus of an atom.

Emission spectrum Strongly heated thin gases send out radiation in very specific wavelengths. In the spectrum, light lines appear.

Ephemeris A table showing the coordinates of a celestial body for a certain time period—e.g., daily 0 o'clock.

Ephemeris Time (ET) The uniform measure of time defined by the motion of the Earth about the sun. It is used as the time scale of celestial mechanics. An ephemeris second is defined as 31,556,925.9747th the Tropical Year for the epoch 1900 Jan. $0^{d}12^{h}$ ET. (See also *International Atomic Time*).

Equator coordinate system Astronomical coordinate system in which a basic circle is the celestial equator with celestial North and South Pole. See also moving equator system on p. 20.

Equinox Either of two intersection points of the celestial equator and the ecliptic, which occur twice a year when day and night are of equal length. Equinoxes occur in the Northern Hemisphere when the sun lies exactly above the equator. In the spring (the vernal equinox), it is March 21; in the autumn (the autumnal equinox), it is September 23.

Escape velocity Also escape speed. The minimum velocity required for the transportation of an object—e.g., for a space probe—to move away from the gravitational field of the Earth into outer space.

Evening star Name for the bright planet Venus during its maximum eastern elongation.

Extinction Weakening of the light of a star depending on its height above the horizon and wavelength as a result of absorption and scattering of the radiation by dust grains in space. Near the horizon, the stars shine and the sun and moon appear reddish, because the blue light is scattered more than the red.

First point of Aries Also known as zero point. Point of intersection of the celestial equator and the ecliptic. Starting point of right ascension.

Flash spectrum The emission spectrum of the solar chromosphere briefly before, during, or after a total solar eclipse.

Focal length The distance between the telescope objective (lens or mirror) and the focal point.

Focal point The point at which light rays converge (or from which they diverge).

Galactic coordinate system An astronomical coordinate system using the north and south latitude and longitude (right ascension) as coordinate points. The galactic latitude of an object is its angular distance north or south of the galactic equator. The galactic longitude is its angular distance from the galactic center measured eastward along the galactic equator to the intersection of the great circle passing through the body. The constellation

Sagittarius, at right ascension $17^h24.4^m$ and declination 28°55′ (equinox 1950.0), is the zero point of the galactic coordinate system.

Geocentric View of the Earth as the center of the solar system.

Geoid The form of Earth by taking the average sea level surface and extending it across continents.

Globules Dense compact dust clouds which are rich in mass, and between hot stars. Their increased density furthers the formation of new stars.

Gravitational center The largest and richest in mass of a solar system forms the gravitational center—e.g., our sun.

Gravitational collapse Collapse of a neutron star when its mass becomes larger than 3.2 sun masses. Their contraction causes a state in which the gravitational pull is so strong that not even light can escape—e.g., a Black Hole.

Heliocentric View of the sun as the center of the universe.

Herbig Haro objects Small emission nebulae, which are ten times larger than our solar system and which stream out of irregular, variable stars.

Hertzsprung-Russell diagram A two-dimensional graph that demonstrates the correlation between luminosity of stars and their spectral type. Shows, for example, development of stars. Abbreviation: HRD.

Horizonal system Astronomical coordinate system by which the location of an object is given in terms of its angle above the horizon (altitude) and the angle created relative to due north (azimuth).

Hour angle The angle measured westward along the celestial equator from an observer's meridian to the hour circle of a celestial body or point.

Inclination Measurement of the tilt of a planet's orbital plane in relation to the ecliptic (Earth's plane of orbit).

Inner planets Name for the planets with Earth's orbit: Mercury, Venus.

IAU International Astronomical Union.

International Atomic Time (TAI) The most precise time scale, based on the atomic second. The atomic time is measured with cesium atomic clocks. Atomic time is now common for the time indications of everyday life. Broadcasting and television send the coordinated universal time (UTC), which may not deviate more than 0.9 seconds from the UT1 (see *Universal Time*).

Julian date The number of days that have elapsed since noon GMT on Jan. 1, 4713 B.C. It was introduced in 1582 in connection with the Gregorian calendar to calculate the frequency of occurence or the periodicity of phenomen over long periods—e.g., for ephemeris calculations. Abbreviation: JD.

Libration The moon rotates irregularly due to minor distortions in its shape and inconstancy of its orbital velocity. Thus, one sees in the perigee (see *perigee*) more of the right side of the moon, in the apogee (see *apogee*) more of the left side. This is called libration in longitude. In addition, there is the libration in latitude: In the course of a lunation, the observer looks more beyond the North Pole or South Pole of the moon. The reason is the non-vertical moon-axis on the moon's orbital level. Finally, there is geometrical libration as a result of the different visual angles of the observer on the Earth's surface. Librations make it possible to see about 59% of the moon's surface at one time or another from the Earth.

Light spectrum Lines in the spectrum of light when it is emitted from a gas or when it goes through a gas.

Light-year Distance at which light travels in 1 solar year. That is equal to 9.46 trillion km (5.88 trillion miles) at a speed of 299,792 km (185,871 miles) per second. Abbreviation: l.y. 1 l.y. = 0.3066 parsec (see *parsec*).

Local group Accumulation of galaxy clusters in outer space. Between these accumulations, the galaxy density is low. Galaxy clusters are compared to "brick walls" or "walls" in the universe.

Magnitude (Abbreviation: ^{m}) The brightness of stars. The scale is a logarythmic measurement: a star of 2nd magnitude is 2.512 times less bright than a star of 1st magnitude, which, on the other hand, is 100 times brighter than a star of the 6th magnitude. Stars brighter than 1st magnitude are indicated with negative numbers.

Main sequence star Star within the main sequence in the Hertzsprung-Russell Diagram, which ranges from the blue, bright stars to the red, dimmer, stars. Most of the stars of the Milky Way have their place on this sequence.

Morning star Name for the bright planet Venus during its maximum western elongation.

Morning width The distance of the rising sun from the eastern point.

Moving equator system Astronomical coordinates system in which a basic circle is the celestial equator with the celestial North and South Pole. The first point of Aries, the zero point of the coordinates counting, takes part in the daily rotation of the sky (see p. 20).

Nadir The point on the celestial sphere that lies directly beneath an observer. Height: −90°. Stars below the horizon have negative height-angles. It is diametrically opposite the observer's zenith.

NASA National Aeronautics and Space Administration.

Nautical triangle See Astronomical triangle.

Nebula An interstellar cloud of gas or dust that can be seen in visible light as a luminous patch, dark hole, or band against a brighter background.

Neutron Elementary atomic particle with no electrical charge. Together with protons, neutrons form the atomic nuclei.

Neutron star A collapsed star with a radius between 5 and 10 km (3.1 and 6.2 miles, respectively).

Nodes Name for the two points of intersection of the orbital plane of a celestial body with the plane of the ecliptic. When the body, such as a planet, changes from the south side to the north side of the ecliptic, it is called an ascending node. The distance of the ascending node from the zero point in arc-measurement is used to describe an orbit.

Nutation A relatively small nodding motion of the Earth's axis, which is caused by gravitational forces of the moon, sun, and planets, with a period of 18.6 years. This motion is superimposed on precession (see *precession*).

Objective Light-gathering optical system of a telescope. The aperture (objective diameter) determines the efficiency of the telescope. It is measured in millimeters, centimeters, or inches (1 inch = 2.54 cm).

Oblateness Deviation of the spherical shape, e.g., the planet Jupiter. The view from a telescope is slightly elliptic.

Orbital elements Particular quantities that describe the size, shape, and orientation of the orbit of a planet, moon or comet. They are used to determine the location of the orbital body at any time (e.g., right ascension). Orbital elements include: the perihelion time, the numerical eccentricity, the orbital inclination, the distance of the rising node of the first point of Aries, the distance of the perihelion from the rising node.

Orientation When one looks through a refractor without a mirror or prism, north is on the bottom, south is on top, west is on the left, east on the right. With the naked eye, north is on top, south on the bottom, west on the right, east on the left.

Original pulp Composition of the particles at the point in time of the Big Bang: x particle, photons, neutrons, electrons, gluons, and quarks.

Outer planets Name for the planets outside of the Earth's orbit: Mars, Jupiter, Saturn, Uranus, Neptune, Pluto.

Parallax determination The angular displacement in the apparent position of a celestial body when observed from two widely separated points. If it can be measured, the distance of the object can be measured as well. Parallax is expressed as an angle. For objects of the solar system, parallax is measured from the surface of the Earth, which varies with the object's altitude and distance from Earth. For stars, it is measured from the Earth and the Sun. As basis line, the Earth-orbit radius = 1 AU (see *AU*) is used.

Parsec Unit of distance. Short for parallax second. Abbreviation: pc. Corresponds to the distance out of which an astronomical unit (see *AU*) occurs at an angle of 1 arc-second (1″). 206,265 AU or 30.8 billion km correspond to one parsec. A kiloparsec (1000 pc), abbreviated as kpc, and a megaparsec (1,000,000 pc), abbreviated as Mpc, are also common measurements in astronomy.

Peculiar stars Stars whose spectrums cannot be easily classified into the usual spectrum categories and/or magnitude categories (peculiar spectrum).

Periastron The shortest distance of the stars of a binary star system.

Perigee Point in the orbit of moon (or an artificial Earth satellites) that is nearest the Earth.

Perihel time Orbital element that describes the time at which a celestial body (e.g., planet or comet) runs through its perihel.

Perihelion The point in the orbit of a celestial body that is nearest the sun.

Phase angle The angle between the lines connecting the moon and sun, and the moon and Earth, or connecting and planet and the sun, and the planet and Earth.

Photons The elementary particles of electromagnetic radiation.

Precession Unequal gravitational forces of moon and sun cause a slow movement of Earth's axis with a period of 25,800 years around the ecliptic North Pole. A precession causes the movement of the celestial poles around the ecliptic poles and the retrograde movement of the equinox points. The nutation is a part of the precession.

Proper motion A "fixed star" moves with great speed in outer space. The idea of its immovability (fixed star) is wrong. The angle of the yearly proper motion is small (about 1/10 arc-second and less).

Proton Elementary particle with a positive electrical charge. Together with the neutrons, protons form the atomic nuclei.

Pulsar A rotating neutron star that emits pulsating radio waves, X-rays, or visible light at a high degree of regularity.

Quasar Quasi-stellar object (QSO). Extragalactic objects that resemble stars but are thought to be the most distant objects from Earth. They emit more energy than a hundred supergiant galaxies.

Radial velocity The movement of a star towards or away from the observer. Radial velocity is measured in km/s. Abbreviation: RG.

Redshift Shifting of the lines in a star spectrum towards the red end. It means that the star is receding. Interpreted as the withdrawal of the galaxies and expansion of the universe.

Reflector Telescope with mirror objective.

Refraction The change in direction of a light ray when entering a medium of a different density.

Refractor Telescope with lens objective.

Resolution capability Capability of a telescope to separate two objects that are standing close to each other. The smallest possible angle resolution in visible light is 11.7 arc-seconds divided by the aperture of the telescope in cm.

Right ascension The angle of an object around the celestial equator that is measured eastward from the vernal equinox in hours, minutes, and seconds: 24H =360°.

Saros cycle A time period of 18 years and 11.3 days in which solar and lunar eclipses recur.

Scintillation Visual disturbances caused by pockets of air turbulence. They come about, for example, as a result of temperature differences at the observation location. Objects in sky appear to blur or jump around.

Secondary mirror The smaller, second mirror in reflector telescopes (e.g., Cassegrain and Newtonian Telescope). It collects rays and brings them in an appropriate manner to the eyepiece.

Sidereal year The time period between two passings of the sun by a certain fixed star. Year length: 365 days, 6 hours, 8 minutes, and 8.55 seconds.

Solar constant The total energy radiated by the sun which comes onto the Earth's surface. Measuring value per surface: 1367 watt/m^2.

Solar day The time between two consecutive lower culminations of the sun through the meridian, or the time it takes for one complete rotation of the sun on its axis. The average value of all solar days of a year is called mean solar day. It is $3^m56.55^s$ longer than a stellar day.

Spatial interferometer An instrument or system in which a beam of light or other radiation is split and subsequently reunited after traveling a different path length so that an interference pattern is produced. The waves are collected by a group of separate, but linked, telescopes. The pattern can be used for the determination of star distances and diameters. The larger the distance of the receivers from each other, the higher is the resolution.

Speckle interferometry A procedure to determine the diameter of stars. Using short exposure-times (below $1/10^s$), several photos of a star are taken. They are overlaid and the "real" picture of the star is reconstructed.

Spectrum Any series of energies arranged according to their wavelength or frequency. The energy is subjected to dispersion with the help of a prism or a diffraction lattice. The examination of the spectrums (e.g., of star spectrums) allows important assertions about physics and chemistry in the universe (spectroscopy).

Spherical aberration Aberration of a spherical lens. Thereby, light rays converge not a single point but to a series of points whose distance from the lens or mirror decrease as the light rays fall nearer the edge of the optical element.

Spiral arms The curved, armlike structures that surround the nuclei of certain galaxies.

Star types With further research of magnitude deviations and of special spectral characteristics of stars, special star types have been discovered—e.g., white dwarfs, neutron stars, cataclysmical variables, peculiar stars, and others. See also pages 135 and 141.

Stellar day The time between two upper culminations of the zero point through the meridian.

Tides Different gravitational and centrifugal forces that the moon and sun exert on the Earth to cause low and high tides of the oceans. Tides can be observed in the atmosphere and Earth's body. Tides also cause tide friction (slowing down of the Earth's rotation).

Time zone A zone on the Earth's surface approximately 15° wide, within which the hour used is uniform.

Transit Passage of an inferior planet across the Sun's disc—e.g., Mercury passes by the sun.

Triple system Three stars forming a multiple star system. About 80% of all stars are part of a double and multiple star system, because the most elementary requirements of energy and rotation impulse preservation are met.

Tropical year Interval between two successive passages of the sun through the vernal equinox. Year length: 365.24219 days.

Umbra Shadow of the Earth that is observable during lunar eclipses.

Universal Time (UT) Also known as Greenwich time. This is the local time of the zero meridian that goes through the Greenwich Observatory.
UT1 = Universal Time corrected by polar movement.
UT2 = Universal Time corrected by seasonal deviations (changes of the rotation speed of the Earth).

Visual field Section of the sky that can be seen or photographed through a telescope.

Zenith A point on the celestial sphere directly over the head of the observer. Height 90°.

Zenith distance (coaltitude) First coordinate in the horizontal system. 90° − height of a star above the horizon = the zenith distance. Zenith distances that are greater than 90° indicate that the star stands below the horizon and is invisible.

Index

Page numbers in bold indicate the main reference

Photo Credits and Permissions

Astrofoto/Anglo-Australian Telescope Board: 130; Astrofoto/Kohlhauf: 8; Astrofoto/Shigemi Numazawa: 1, 16, 88; Photo Archive of the Baader Planetarium: 157 (top right); D. Bissiri 92 (bottom); California Institute of Technology and Carnegie Institution of Washington: 108 (top left), 109, 143, 149, 150, 151 (top); Deutsche Aerospace: 129 (bottom), 132 (top); Deutsche Museum: 162, 163; J. Dragesco: 95; European Southern Observatory (ESO)/R. West: 2/3, 4/5, 23, 114 (left series), 120 (bottom), 122, 132 (bottom left), 135 (bottom), 137 (left), 137 (right), 141, 148, 154 (right); B. Fogle: 127; Akira Fujii/Baader Planetarium: 32, 152; Tony Hallas & Daphne Mount/Baader Planetarium: 30; HST/NASA/M. Rosa: 104 (bottom right), HST/NASA/R. West: 111 (bottom); Intercon Spacetec: 26 (bottom), 159 (top); F. Kögel: 161; C. Leinert/Baader Planetarium: 126; W. Lille/Baader Planetarium: 6/7, 90 (bottom); N. Martin: 22 (bottom); Max-Planck-Institute for Extraterrestrial Physics (MPE)/K. Dennerl & W. Vogues: 28 (right), 29 (left), 29 (right); A. McEwen/Baader Planetarium: 115 (top); R.C. Mitchell/Baader Planetarium: 14 M, 15 M, 26 (top); MPG Pressebild: 132 (bottom right); NASA: 102 (bottom), 104 (bottom left) 113, 116 (bottom), 119 (bottom), 120 (top left); NASA/Archive T. Althaus: 164; NASA/Baader Planetarium: 96 (bottom), 99, 108 (bottom right), 116 M, 118 (left), 119 (top), 125 (bottom); NASA/ESO/M. Rosa: 117, G. Nemec: 92 (top), 101, 108 (top right), 114 (right), 116 (top); P. Parviainen/Baader Planetarium: 19 (top), 123 (bottom); R. Phildius: 27; L. Ream/Baader Planetarium: 125 (top); H. Rose: 28 (left); G.D. Roth: 91 (left), 91 (middle right), 93 (top right), 94 (top left), 94 (top right); 104 (middle left), 155, 156 (bottom right), 157 (top left), 157 (bottom), 158, 165; K. Rüpplein: 93 (top left), 139 (bottom); H.R. Salm/Baader Planetarium: 168; W. Schwartz/W. Lille: 96 (top left), 96 (top right); P. Stättmayer/Baader Planetarium: 142; Archive of Stars and Space/H.J. Staude: 102 (top), 110, 121, 123 (top), 134 (left), 134 (right), 135 (top), 154 (left); P. Stolzen: 90 (top); Fa. Treugesell: 156 (bottom left); U.S. Naval Observatory: 146, 147, 151 (bottom); K.-B. Veenhoff: 91 (top right); H. Vehrenberg: 18, 35, 37, 39, 41, 43, 45, 47, 49, 51, 53, 55, 57, 59, 61, 63, 65, 67, 69, 71, 73, 75, 77, 79, 81, 83, 85, 87, 128, 153;

Bode, Johann Elert, "Vorstellung der Gaestirne," 1782 (reproduced by Treugesell-Verlag, Düsseldorf): 10, 14 (top), 15 (top)
Moore/Hurt, Atlas des Sonnensystems, Herder Verlag: 103 (digital maps)

Drawings (diagrams & maps): Barbara von Damnitz with the exception of the following: D. Farnhammer: 19; Archive of Stars and Space/H.J. Staude: 136
Artwork front & back flaps: Viertaler & Braun, K. Löchel
Front cover photographs: Astrofoto/NOAO (background photo); Astrofoto/ROE/AAT Board
Back cover: Graphics: Barbara von Damnitz (middle map, upper left corner), Astrofoto/Shigemi Numazawa (upper photograph), Astrofoto/Anglo-Australian Telescope Board (bottom photograph)